BIOTECHNOLOGY

First MIT Press edition, 1986
© 1985 by Editions Hologramme

French edition published by Editions
Hologramme, Neuilly-sur-Seine, France,
under the title *Le Génie de la vie*.

World rights in English are held by
The MIT Press.

This book was set by SCG Paris.
Printed and bound by Dai Nippon in Japan.

Library of Congress Cataloging-in-
Publication Data
Antébi, Elizabeth.
    Biotechnology.
    Translation of: Le génie de la vie.
    Bibliography: p. 227.
    Includes indexes.
    1. Biotechnology. I. Fishlock, David. II. Title.
TP248.2.A57 1986   660'.62   85-30008
ISBN 0-262-01089-5

BIOTECHNOLOGY
Strategies for Life

Elizabeth Antébi and David Fishlock

Scientific advisors:

Saburo Fukui
Professor Emeritus at Kyoto University

Hiroshi Harada
*PhD, Institute of Biological Sciences,
University of Tsukuba*

Leroy Hood
*Chairman of the Biology Department,
California Institute of Technology*

François Jacob
*Professor at the Collège de France
and at the Institut Pasteur,
Nobel Prize Winner in Medicine 1965*

Severo Ochoa
*Roche Institute of Molecular Biology,
Nobel Prize Winner in Medicine 1959*

Mark Ptashne
*Department of Biochemistry and Molecular Biology,
Harvard University*

The MIT Press
Cambridge, Massachusetts
London, England

Contents

Introduction

1953: In Cambridge, England, two men discover the secret of heredity in the double-helix structure of the DNA molecule.

1973: In San Francisco, two other men succeed in creating the first "chimera" using genetic-engineering techniques.

Between these two dates something extraordinary occurred: men had learned to decipher, spell, write, and even correct the spelling mistakes of life itself. Today, in only a few weeks, a student can "learn to tinker in the laboratory with the very molecule of heredity as if it were a common auto engine," to use François Jacob's expression.

In industrial and university laboratories throughout the world, researchers are trying to get microorganisms to produce substances they would not secrete naturally, to nourish themselves on petroleum spills and other noxious wastes, and to concentrate lean mineral deposits. Biologists, microbiologists, biochemists, and geneticists are exploring the mechanisms that control and regulate every living being on earth, men and animals, plants and bacteria. All of this will have incalculable consequences for the diagnosis and treatment of such mysterious diseases as cancer and thrombosis, the diagnosis and correction of genetic deficiencies of children still in their mothers' wombs, and the improvement of the plant species used to feed mankind. Having explored matter and the infinitely small and the stars and the infinitely large, it is now time to explore "Man, the unkown."

A revolution or a renaissance? Have we not always produced fermented food and drink with the help of microorganisms? Have we not had vaccines since Pasteur? Or antibiotics since just after World War II? And have we not, from time immemorial, crossbred plants?

What has really changed is the possibility of tinkering with the genetic heritage of living matter, of recombining genes, modifying natural organisms, and domesticating microbes. And this has led to the birth of a bioindustry, which is not really a new industry but a complex web of enabling technologies giving new impetus to the conception, orientation, and strategy of traditional industries in such diverse fields as pharmaceuticals and medicines, food processing, agriculture, energy, and pollution control.

Bioindustry is in fact a series of paradoxes. The first paradox: Most major bioindustrial inventions are not the result of directed research but often answers to questions that have arisen in other fields. Thus the enormous vaccine industry stems from Pasteur's research on the refraction of polarized light in crystals. As François Jacob says, "If we had been specifically looking for a tool with which to cut DNA, we would never have found it as did Werner Arber in studying the phage, the restriction enzyme that has become the scissors of genetic engineering." Phages are viruses infecting bacteria, and only a handful of scientists were studying them during and just after World War II. "Purely an intellectual game, a form of folklore of interest to barely ten people on earth!" added Jacob, speaking of a friend who, passing him in the corridor, jokingly asked, "So, how are you getting along with your phage?" And it was exactly from this study of the phages that sprang molecular biology and the "phage" group founded by Max Delbrück, Salvador Luria, and Alfred Hershey. One of their students was a certain James Watson, who, along with Francis Crick, was awarded the Nobel Prize in Medicine for having determined the structure of DNA.

Another paradox: With the biotechnologies, bankers and investors are for the first time putting their money into products that not only do not exist but that no one can yet specify! For, although the electronics industry is the brainchild of research directed and financed by the government and the army, biotechnology owes its existence to private enterprise alone. Silicon Valley has now become Silly Clone Valley. Venture capitalists, investors with risk capital, avid for further "success stories," have in an instant turned from "chips" to "bugs." Wishing to avoid the error of the major manufacturers of vacuum tubes (General Electronic, RCA, Sylvania), which were ousted from the component market for not having negociated the switch to semiconductors in time, businessmen are now preparing for what promises to be a boom in the pharmaceutical and chemical industries, in energy and agroindustry. If, when J.J. Thomson discovered the electron in 1897, investors had proceeded to bet heavily on Jean Perrin, Robert Millikan, or Lee De Forest (inventor of the triode), we would have a good analogy with what has been happening in biotechnology since the late 1970s. And many university graduates, on hearing the sirens' song, have followed the example of Herbert Boyer, Walter Gilbert, or David Baltimore (and many others) in ferreting out capital, founding new biotechnology firms (NBFs), directing foundations, or sitting on boards of directors. Dramatic breakthroughs have been the rule since the very beginning, with Genentech stock being snapped up by Wall Street as soon as it was put on the market in 1980, then rocketing from $35 to $89 in a mere 20 minutes! Cetus beat still another record the following year when the value of its stock hit $115 million on first being offered. Yet perhaps it had all been exaggerated. Research proved long, difficult, and costly, many small companies were stillborn, and still others had to resign themselves to merging with large corporations and to becoming their own research laboratories, concerned above all with profitability. The initial analogy between electronics and biotechnology breaks down on closer analysis. With the electron microscope, memory systems, and data processing, electronics had provided molecular biology with the research tools it needed, but it also supplied a model, with the concept of a program and a code, which could be applied to living matter. And even back in 1943, had not the famed quantum physicist Erwin Schrödinger, in *What Is Life*, compared hereditary material to a coded message?

Contrary to the situation in electronics, biotechnology is not so much an industry in its own right as a set of enabling technologies for other industries to use. Even so, many technical and economic problems remain to be solved in the transfer from the laboratory stage to true mass production. Furthermore, the "raw material" of biotechnology is living matter, idiosyncratic, subject to instability and to the variability of biological laws, which are not always known or predictable. A microcomputer, on the other hand, is subject only to the laws of physics, which served as the basis for its design.

Living matter is subject to necessity, but chance or hazard also plays an important part. Hazard, undestood by scientists to be independent of cause or effect, has, through an erosion of meaning, become an ambiguous word. Over the past twenty years or so, a scientific literature has arisen that attempts to reconcile these two apparently irreconcilable views of the world around us. With varying degrees of success, physicists and historians of physics have produced an amalgam of science and Oriental philosophy bearing such titles as *The Tao of Physics, The Dance of the Elements, The Eye of Shiva*. To biologists, on the other hand, it is simply a matter of proving or disproving the existence of God or some other "higher intelligence." The most characteristic example of this was provided by Lysenko in the Soviet Union, who persisted in refusing to believe in DNA because it implied that men would no longer be able to determine their own destinies – as taught by historical determinism.

Those attending the many seminars held in Paris during the 1950s began to call themselves "Monod-theists," and to a journalist from *Omni* who came to interview him, James Watson declared, "We rewrote the Bible." Richard Axel, for his part, baptized his son Adam. These were jokes, of course, just as was Dali's contention that DNA proved the existence of God. But such whimsy cannot disguise the subtle change that occurred in our thinking and in our language, or mask our apprehension of the universe, which is so frequently commented upon by important scientists who also happen to be profound writers (François Jacob, Lewis Thomas, and James Watson, to mention but a few).

"In Japan we start from a quite different premise," says Kiyoshi Aoki, a professor at the Life Sciences Institute of Tokyo's Sophia University. "To Western biologists, life is matter; to us, who have been influenced by Oriental religions – in which everything, whether stones or bacteria or monkeys or men, has a soul – life is spirit. The Japanese mind shrinks away from the idea that there is a distinction between higher and lower forms of life or that there is a possibility of swaying evolution."

At the University of Tokyo's School of Medicine, a period is set aside each semester in remembrance of the animals that have been sacrificed, and the Hayashibara Company in Okayama has erected a monument to the hamsters that died in its laboratories. A Kyoto businessman has even raised a stele to the microorganisms (yeasts and molds) to which he owed his fortune. No one in Japan finds this surprising. But how would people in France or the United States react if the Institut Pasteur or Rockefeller University built a monument to their mice?

It is only by appreciating this fundamental difference between these two ways of thinking that it becomes possible to understand the odd phrase written by the famed micribologist (and translator of Jacques Monod's book *Chance and Necessity*) Itaru Watanabe, who concluded his remarks on evolution with the words "God is the future of mankind." He seemed to imply by this that transcendence might perhaps be a later stage in evolution, following energy and the Big Bang, matter and physics, life and biology, thought and the brain. In this he is opposed to other scientists, including Nobel Prize Winner in Physics H. Yukawa, who was brought up in China and to whom science is but one part of the infinitude of the universe, which is beyond the scope of science.

One of Watanabe's students and now director of social, natural, and environmental research at the Mitsubishi Kasei Institute, Keiko Nakamura, looks at things less rigidly. From her standpoint, Jacques Monod, if he were still alive, would have to take into account the recent discoveries in molecular biology that no longer make it possible to extend the knowledge we have of bacteria to mankind: "Western science and technology are direct offshoots of

Western philosophy. What we have had urgent need of since the Meiji era and the end of the last century is Western science and techniques, not Western philosophy. Today, young researchers are drawn to the new science and such books as Fritjof Capra's *The Tao of Physics.* Yet this reconciliation with Eastern ways of thinking seems familiar enough to us. The use of life as a raw material for technology naturally enough raises philosophical questions, but Japan is in the habit of assimilating everything – Shinto gods, Buddhism, Christianity – so why not this type of Western philosophy?"

The irruption of the engineer with his mechanisms and systems into a world thought to be immutable has roused mankind's ancestral fears. Though the public has become aware that the "intelligence" of a computer is artificial and consists of man-made programs and that its "memory" is defined by the amount of data it can store, this overly human language had already fostered confusion and imposed a logical sequence resulting in the present model of man: program-code-data. A film like *The Boys from Brazil,* in which Dr. Mengele creates clones – genetic twins – of Hitler, incipient child Führers resembling their "cellular progenitor" feature by feature, raised fantasy to the level of myth... a myth maintained by the heroes of molecular biology who had become legends in their own lifetimes. In *Double Helix,* a book published in 1968 that immediately became a bestseller, James Watson, alias "Lucky Jim" or "Honest Jim," recounted the remarkable history of the double helix "with a little of the innocence and absurdity of children telling a fairy-tale (J. Bronowski, *The Nation,* 18 March 1968, pp. 381-382). "The self-portrait of the scientist as a young man in a hurry," said J. Merton. And young Watson certainly was young at the time of his DNA discovery – only 24. Young, too, are the current superstars of molecular biology – David Baltimore, Mark Ptashne, Walter Gilbert, Leroy Hood, Richard Axel, and Robert Weinberg. This is even one of the outstanding traits of the attraction exerted by these remarkable people who, to parody G. Stent's remarks on Jacques Monod, combine the characteristics of Darwin with those of Prince Charming.

To this seductiveness has been added the golden legend of glory, the avalanche of Nobel Prizes that over fifteen years, has crowned the efforts of many dozens of scientists, and brought fame to research teams everywhere, from Cambridge, UK, to Cambride, US, from the East Coast to the West Coast of the United States and from Europe to Japan. Manna now showers down on these researchers, and over their cradles hover the fairy godfathers of industry and commerce, foundations of every kind, and the purveyors of venture capital. Genetic engineering, protein engineering, microbiological engineering are the subjects on everyone's lips. And as always at the start of any great new scientific adventure, a voyage into the unknown has begun. But with this difference: We ourselves are the unknown.

1

Fermentation: Mysteries Large and Small

Scotch, bourbon, sherry, port, beer and wine, sauerkraut, olives, pickles, yoghurt, cheese – all are everyday Western products of microbology. In the East, Japanese saki and soy-based natto, Chinese and Japanese soy sauce, and Indonesian tempeh are also commonplace foodstuffs. For thousands of years, microorganisms have been helping to make man's diet less monotonous, more spicy and stimulating.

Not until the middle of the nineteenth century and the work done by Louis Pasteur did we learn that beer and buttermilk were not result of a chemical reaction between compounds, but were due do the presence of a microscopically small living cell, yeast. Pasteur also showed that the origin of illnesses could be seen through a microscope as a swarm of living cells, bacteria. All of this unseen life of the cells that feed us and cure us operates through complex fermentation processes that harness minute chemical workers, enzymes.

The prehistory of molecular biology dates back to Pasteur's discoveries. Without the knowledge and mastery of the fermentation process, no product of biological engineering could have seen the light of day: neither antibiotics, nor human insulin, nor perhaps tomorrow the lymphokines, those chemical messengers of immunology. It all began thousands of years ago with the wine that inebriated Noah; the bread that mankind, having been banished from Paradise, had to earn by the sweat of its brow; and the poteen forged in Ireland by the divine smith.

Why does sherry have the aroma of hazelnuts? It is because of the spontaneous growth of certain microorganisms coming into contact with the air as the alcohol-enriched wine matures. The source of the strange flavors of Sauternes and Hungarian wines is the "noble rot" (mycoderma), discovered toward the end of the eighteenth century in the Weingau region. The bubbles in champagne are of carbon dioxide released by a process of secondary fermentation or "frothing" inside the bottle, and the natural autolysis, or self-destruction, of the yeast during the aging of champagne in the cellar. [1] Why does the flavor of beer vary from fruity to flowery, depending on the brand? The brewmaster carefully selects his yeast and its working conditions, such as temperature and acidity, to develop the flavor he wants. Similarly, the various aromas in leavened bread come from the fermentation process, and spontaneous flora leading to secondary changes in the dough after fermentation.

The process of alcoholic fermentation was understood for the first time in 1815 by Louis Gay-Lussac: when subjected to the action of yeast – and considered at the time a physical phenomenon, a sort of spark setting off a chemical reaction – sugars produce ethyl alcohol and carbon dioxide. In 1833, Anselme Payen and Jean-François Persoz isolated the first natural catalyst, an enzyme (*diastase* from malt) capable of liquefying starch and converting it into sugar. But it was not until the second half of the nineteenth century and the work of Louis Pasteur that the idea that fermentation was caused by living microorganisms, rather than being merely the decomposition of inert matter, finally became accepted.

Pasteur was considered "the white knight of science" by René Dubos, and a "soul on fire with an abstract passion no less romantic than carnal love" by Jean Rostand (*Hommes de vérité, C. Bernard, L. Pasteur, C.-J. Davaine,* Stock, 1966, p. 94). Pasteur was still a chemist at the time his master, Jean-Baptiste Biot, gave him the deflection of polarized light as a theme to study. Biot had shown that certain organic substances – sugars and tartaric acids – can, like crystals, deflect light, but also that, in contrast to crystals, they kept their optical qualities in solution. He also made use of the discovery of the German chemist Eilhard Mitscherlich that in the tartar formed during fermentation in wine one may distinguish, along with the large crystals of tartaric acid, the small, needle-shaped crystals of a neighboring acid, paratartaric, or racemic, acid (from the

Beer
Beer, the beverage of gods and heroes from Ancient Egypt to Ireland, from Africa to intertropical America, is undoubtedly the oldest fermented drink in the world. The very word "ferment" comes from the Latin verb to boil," and the general idea persists in some modern popular terminologies (as in the French bouilleur de crus). *But it was only in the seventeenth century that it became possible to attribute this phenomenon to carbon dioxide. Furthermore, the mysteries of fermentation have marked the religious beliefs of many peoples – the fermentation of bread with the leavening of the spirit, or wine as an initiation beverage from Dionysus to Ibn Arabi and in the legend of the Holy Grail. Today, however, fermentation is no longer a matter of myth but one of fierce industrial competition.*

[1] Lysis is the destruction of organic components as the result of physical, chemical, or biological activity.

11

Latin word for grape, *racemus*). These two groups of crystals show absolutely the same chemical composition, the same structure, the same specific gravity and the same refraction. Yet, mysteriously enough, a solution of tartaric acid always deflects polarized light, while a solution of paratartaric acid does not. Pasteur observed these solutions under a microscoscope and discovered that the facets of the polytartrates were asymmetrical: some deflected polarized light to the right, and others to the left. Pasteur was twenty-five years old when he first determined that there was a relationship between optical properties and molecular and crystalline structure – a discovery that was to have considerable as well as completely unpredictable consequences.

A change had begun, and Pasteur slipped from crystallography into biology. It was in this way that, for instance, instead of throwing out paratartrate solution that had been contaminated by mold, he explored its optical properties: the first isomer component (the one that deflects light to the right) is destroyed after a certain length of time, while the second (which deflects to the left) remains alone in solution and provokes intense optical activity.[2]

It was then that a new idea germinated in Pasteur's mind: only living things can produce such asymmetrical, "optically active" substances, and this is undoubtedly the basic distinction that can be made between the chemistry of organic and inorganic matter. "Life, as manifested to us, is a function of the asymmetry of the universe and of the consequences of this fact. The universe is asymmetrical for, if the whole of the bodies which compose the solar system were placed before a glass, the image in the glass could not be superposed upon the reality. Even the movement of solar light is asymmetrical.... Terrestrial magnetism, the opposition which exists between the north and south poles in a magnet and between positive and negative electricity, are but resultants of asymmetrical actions and movements.... Life is dominated by asymmetrical actions. I can even imagine that all living species are primordially, in their structure, in their external forms, functions of cosmic asymmetry" (René Dubos, *Pasteur and Modern Science,* Anchor Books, Doubleday, 1960, pp. 36, 38).

Biological activity thus depends on molecular structure: this idea was to be brought up again less than a century later by such physicists as Erwin Schrödinger, Niels Bohr, and Max Delbrück and gave birth to molecular biology in the 1950s. As for the asymmetry of the universe, this seems now to have been confirmed by experiments in nuclear physics.

In 1854, Pasteur became concerned with the problems raised by the fermentation of alcohol. At the University of Lille, to which he had recently been posted, he was obliged to devote his efforts to problems of interest to businessmen in that region. This permanent interaction between theory and practice fascinated him all his life, whether it concerned beer, silkworms, or vaccination. One of these businessmen complained of the contamination of beet alcohol during fermentation. Using a microscope, Pasteur detected optical activity, and to him this suggested the presence of living matter. Was yeast in fact something quite different from that complex chemical substance imagined by the leading chemists of the day, J.-J. Berzelius, F. Wöhler, and J. von

Liebig? Pasteur had observed that, during the transformation that occurs in the process of milk going sour, the lactic "ferment" that breaks down one sugar molecule into two molecules of lactic acide is nothing more than a swarm of microorganisms. He further showed that their numbers increased rapidly if they were given the appropriate food and that, depending on the ferment, their activity depended on such factors as the acidity (for ferments producing alcohol), neutrality (for lactic ferments), or alkalinity of the medium. He then summarized his ideas and observations in a 1857 "Memorandum on So-Called Lactic Fermentation" containing an early suggestion that "ferments" might also be the origin of infectious diseases. And in reply to a subsidiary question, he added that if the businessman's beet alcohol was contaminated, this was due to the presence along with the ferment of undesirable microorganisms – "infections."

1857 and 1858 were banner years, and it was during this period that Rudolf Virchow published his work on cellular pathology, whose importance we shall see in greater detail in chapter on the cell. It was also at this time that the concept of evolution formed in Darwin's mind. In a speech celebrating the twentieth anniversary of INSERM (Institut National de la Santé et de la Recherche Médicale), François Jacob stated, "A revolution in science is not simply an accumulation of data, a harvest of results, a change in the landscape. It is a change in the way people think, in the way they look at things.... It is a change in vision itself. After Darwin, one can no longer look at living beings the same way. After Pasteur one sees microbes and viruses in infectious diseases where before there were only morbid influences or properties."

In Pasteur's day, all the great chemists were fervent believers in spontaneous generation: that living matter could originate in inorganic matter. Francesco Redi had already fought successfully against this idea in the seventeenth century by demonstrating that worms were not generated from rotting meat but from eggs laid by flies in the meat. This demonstration was accepted for most vertebrates and invertebrates, but there remained one last bastion of spontaneous generation: the microorganisms themselves. Learning of Pasteur's work and his conclusions (that microscopically small beings gave birth to beings similar to themselves), Liebig cried out that one might as well try to explain the underlying currents in the Rhine by the movement of water mills on the Main! It is difficult today to imagine the violence of the intellectual debate Pasteur had to conduct for nearly twenty years to put an end to the myth of spontaneous generation. He finally did so by a series of innumerable systematic experiments and, an arduous task at that time, eliminating all contamination by airborne bacteria. He also showed that, while certain microorganisms need oxygen to grow and multiply, others can do so in the absence of oxygen: so-called anaerobic conditions.

Pasteur thus shed light on the fermentation processes that are at work in similar ways in the production of beer, wine, or saki, bread, cheese, or soy paste. It was this knowledge of fermentation that made it possible for industry to show its mastery in other fields, including the production of antibiotics

(2) Two isomer compounds are of the same chemical composition and the same molecular mass, but their atomic structures and properties differ.

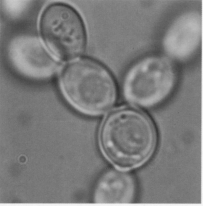
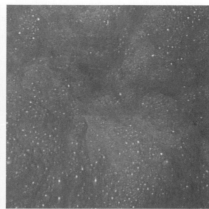

1   2   3

4

1, 2, 3. *Saki*
In the making of saki in Japan, rice is first washed, next seeded with the mold Aspergillus oryzae (1) to break down its starch, and then inoculated with yeast (2). The boiling activity of fermentation appears on the seventh day (3).
4. *Saki*
Early twentieth-century vats used in Japan for the fermentation of saki.
5. *Pasteur*
In this late nineteenth-century print, Pasteur is seen discovering the laws of fermentation.

5

The brewing of beer

Barley has been used to make beer since antiquity. But other cereals have also been used to make this beverage – sorghum in Africa, corn in America, rice and millet in China, and wheat in Germany.

After harvesting barley is allowed to germinate. This first stage is known as *malting*. The malt (germinated barley) contains starch and fermentable sugars, coloring matter, aromatic components (for smell and taste), tannins, which nourish the yeasts, the polypeptides, which give beer part of its foam, mineral salts, and above all the enzymes, which "digest" the starch. *Kilning* is a delicate stage whose biochemistry remains obscure and during which the maltster's talents come to the fore, for the goal is to balance color and flavor and partially to inactivate the enzymes present. These are reactivated in the next stage, *brewing*. During brewing the malt is ground and water added to form the "mash," a sort of slurry to which is added raw grain (corn, barley, rice, or wheat) and which is then heated in stages. The reactivated enzymes break down the chains of starch into soluble carbohydrates (glucose, maltose, dextrin, fructose, saccharose) and into proteins. The wort is filtered – the by-product, or draff, being used as animal feed – and it is then boiled with hops. The boiling stops the activity of the enzymes and brings out the aromatic components in the hops that give beer its bitter taste. The hops having been removed, the remainder, or the wort, is first cooled and exposed to the air and is then sprinkled with a yeast that continues its work after the oxygen in the air has been exhausted.

Five to seven days later, fermentation is stopped before the sugars completely disappear. The sugars will be reabsorbed – and this is essential to the stability and agreeable taste of the final product – during a second anaerobic (oxygen-free) fermentation under pressure and at low temperatures (the *Ruh*) lasting two to three weeks. The fizz and foam (due to carbon dioxide) appear at this point. The beer then need only be filtered, pasteurized, decanted, and bottled. One risk remains: a single smear of grease in a badly washed bottle can immediately destroy the foam upon pouring into the glass.

Pulque
The celebration of pulque (cactus beer) among the Aztecs (Codex Magliabechiano).

(3) A sort of bottom fermentation was already in use during the Middle Ages, but only locally in the German village of Einbeck (bock beer). It was revived in Bavaria in 1800, and the first light lager was brewed in Pilsen (Bohemia) in 1842.

and human insulin. To describe the various stages in this development let us take a few examples of traditional fermentation.

Fermented Beverages

Usually, a distinction is made between two major families of beers according to the way they are fermented. In the case of English, Belgian, and Dutch beers (bitters), which are more aromatic, as well as red beers like G. Killian's (the color comes from glucose sugar added) and the dark beers (ales and stouts) like Scotch Porter or Guinness, we speak of *top fermentation* as the yeasts rise to the surface. This method has been in use since the eighteenth century.

For the others we speak of *bottom fermentation*. In 1883, a Danish botanist, Emil Christian Hansen, working at the Carlsberg Laboratory in Copenhagen, isolated and cultivated a yeast that, in the course of fermentation at a temperature far lower than that used in top fermentation (6°-10°C as opposed to 15°-20°C), settled to the bottom of the vat. These flocculating yeasts avoid the need for expensive centrifuges and, if the temperature is raised to 15°C, make it possible to produce more fragrant beers. The medium to be fermented is thus artificially inoculated with a yeast prepared for this purpose. This is the transition from traditional to industrial fermentation. Bottom fermentation is employed for pale ales and beers, which account for 80-90% of the Western world's consumption.[3] There is also a method of *spontaneous fermentation*, though this has become quite rare (generally being found only in the Brussels region). It is very close to the original method of making beer, inasmuch as it is based on natural microorganisms found in the environment. Crick and Gueuse Lambic beers are made in this way.

As early as 1886, E.C. Hansen perfected an apparatus for propagating brewer's yeasts still used today in all breweries.

In 1909, and once again in the Carlsberg Laboratories, S.P.L. Sørensen discovered the method to measure acidity that made it possible to control the mashing process. His discovery was first applied in 1911/12.

Of the stages in fermenting, malting is the most costly, but today bacteria are extracted from amylases, which, when added, make it possible to malt more economically. An attempt is also being made to use proteases to eliminate residual proteins and to clarify beer. Among the latter enzymes, the most commonly employed is papain (extracted from papaya and also used to tenderize meat). In Japan, the Kirin Brewing Company has developed a mixture of papain and a protease obtained from a bacterium.

Fermentation methods similar to those used for beer are employed in making the Japanese wine, saki. Once again, but this time starting off with rice starch and using the *Aspergillus oryzae* mold, it is a question of obtaining fermentable sugars and then ethanol and carbon dioxide. The fungal spores are mixed into the steamed rice and the mixture incubated five to six days at a temperature of 30°C. The resulting product, koji, is in its turn mixed with steamed rice and a yeast is added – invariably a strain of *Saccharomyces cerevisiae* – which makes what is called moto. Starting with this culture, fermentation lasts three weeks.

Until only recently, wine too remained what it had been throughout antiquity and as far back as Noah's day. The principle is still the same, just as it is with bread : the production of ethanol and carbon dioxide resulting from sugars stemming from the conversion in the mass of grapes of starch and its derivatives produced by the carbon dioxide in the air as fixed by the leaves of the vine. The wild microflora present on each grape initiates alcoholic fermentation. Now that vineyards are being treated, these natural ferments are showing a tendency to disappear and be replaced by dried active yeasts sold in a plastic bag. They also make it possible to stimulate sluggish fermentation – particularly with *Saccharomyces bayanus,* whose high resistance to alcohol is made the most of (generally, an ordinary yeast is inhibited above 16°C) – to deacidify or to eliminate undesirable odors.

Alcoholic fermentation is frequently followed by malolactic fermentation (definitively proved in Bordeaux in 1953 with the introduction of chromatography on paper) This is produced by lactic bacteria giving malic acid (which is found in green apples, for instance) and lactic acid, making the wine smoother and less astringent. One of the dangers of this delicate operation is to see that the bacteria intervene before the end of the initial fermentation: there is then a risk of "lactic sting," the residual sugars being degraded by the lactic bacteria and producing, among other things, acetic acid (used for vinegar).

Certain enzymes make it possible to accentuate color and facilitate the clarification and filtration of wines; they are all produced by bacteria. Only the use of pectolytic enzymes is presently authorized, however. After decanting, the wine must be clarified and stabilized against the precipitation of tartar crystals (stabilization by cooling) and against cloudiness due to microbiological causes (by destroying the germs remaining in the liquid by adding sulfur dioxide or by eliminating by filtration through a membrane, or even by pasteurization at the time of bottling). Wines stored in vats are preserved by inert gases (carbon dioxide and nitrogen), which protect them from contact with the air, thus avoiding oxidation and sting.

Knowledge gained since the 1950s has made it possible to improve controls over temperatures and to produce drier white wines – for instance, by vinifying them at between 18° and 20°C, or making finer red wines (between 25° and 30°C). Mastering of fermentation temperatures has also made it possible to reduce the risk of certain accidents, such as lactic sting. Today, one of the causes of interruption

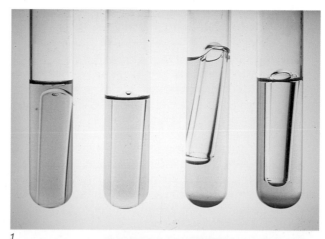

1

2

3

1. *The Hansen test*
To Pasteur, all of the yeasts at work in alcoholic fermentation were saccharomyces, without any great distinction between the various strains. Hansen, for his part, was a forerunner of modern microbiology in that he gave particular importance to cultural conditions and their effect on the physiology and morphology of yeasts. He also demonstrated that the method developed by Pasteur to eliminate the bacteria responsible for the diseases of beer and wine encouraged the development of pathogenic yeasts. It was as the result of his experiments that he was able to develop a strain that modified the characteristics of fermentation. To test the characteristics of the strains, glass tubes were inserted into the culture media and a yeast introduced. The carbon dioxide produced by the fermentation and accumulating in the tubes makes them rise. We can see here that the yeasts used for Carlsberg and Tuborg light lagers (on the right) can ferment a certain sugar (melibiose) introduced into each tube, while Saccharomyces cerevisiae (on the left) cannot.

2. *E.C. Hansen*
This portrait was painted in 1896, when Hansen was fifty-four years old.

3. *"Turning" champagne*
During the second fermentation, or "frothing," of champagne, the bottles are given a turn every twenty minutes (to eliminate the cloudiness caused by yeast deposits), and this process is continued anywhere from two weeks to two months. However, new techniques have ben invented to eliminate such tedious manipulation. These consist mainly of yeasts enclosed in porous balls, a process developed by ENSBANA of Dijon, a firm affiliated with Moët et Chandon and INRA (Institut National de la Recherche Agronomique).

in the fermentation process is attributed to the production of fatty acides formed by the yeast, which can be prevented by treatment with carbon (a fact known empirically but not understood scientifically). Fatty acids have, on the other hand, proved useful in stabilizing sweet wines (as an additive to sulfur dioxide), and fermentation activators have been perfected based on the skins of yeasts.

Parallel research is being pursued to improve strains and thus to eliminate abnormal scum at the beginning of fermentation (particularly in the case of Loire Valley wines) and to improve lactic fermentation (Bordeaux dry white wines, whites from Touraine and Alsace). But the assimilation of "foreign" strains still poses problems.

The genetic selection and improvement of vine stocks has been practiced over the past twenty-five years, but such specialists as Carole Meredith of the University of California at Davis have stressed the complexity of genetic engineering in the case of a plant like the grape vine. One particular obstacle lies in selecting one specific characteristic in relation to another: a gain in productivity at the expense of bouquet (Pinot Noir, Cabernet Sauvignon) or vice versa (Chenin, Muscat). One of the goals sought is the production "in vitro" of a wine capable of being produced without taking contraindications of soil and climate into account. Major universities (Cornell in the United States, Bologna in Italy, and Bordeaux in France) as well as world-famous viticultural centers, such as that in Geisenheim, West Germany, are continuing their research in this field.

There remains one unexplored sector: the relation between the biology of the soil and the properties of the raw materials derived from them.

Fermented Foods

Despite the fact that the fermentation process used for bread is similar to that used for wine, for the former we tend to speak of carbonic fermentation in that the alcohol is allowed to evaporate and only the carbon dioxide is left to leaven the dough. Generally speaking, a cultivated yeast is added to this dough, a mixture of flour, water, and salt. This yeast, more often than not *Saccharomyces cerevisiae*, is now produced industrially by such firms as Fould Springer and Lesaffre. Yet some bakers imitate the ancient Hebrew practice. Leaving dough intended for flat biscuits in the open air and subsequently baking it on hot stones (the oven was invented by the Greeks), they discovered that the cooked dough had not only risen but tasted even better. These modern bakers use leavening, a wild yeast closely similar to a mold. Considerable research has been undertaken in this field, particularly by Professor Spicher's team at the Cereal Research Center in Detmold, West Germany, as well as in San Francisco. One of the most famous French bakers of such products, Lionel Poilâne, who exports his bread to such distant countries as Saudi Arabia and Japan, has this to say on the subject: "The average Frenchman eats his own weight in bread in a year. In our language, nearly every expression refers to this form of food: as good as bread, as long as a day without bread, a little something to go with your bread.... We must preserve the ball-shaped loaf *(boule),* from which the term baker (*boulanger* in French) is derived."

Yet leavened bread is but a living fossil in our contemporary civilization, where the bread industry concentrates on new products for the "man in a hurry": sandwich breads, health breads, toasted breads, etc. The Pain Jacquet Company has even designed a special apparatus, the Little French Bakery, to produce bread on order instantaneously from frozen dough and yeasts. Biotechnological research, particularly in the field of yeasts, is being intensified with this end in view.

Just as it was by chance that the ancient Hebrews

Must
The juice has just been squeezed from the grapes. It is in the must of grapes (or the wort for beer) that fermentation originates.

16

discovered leavened bread, so it was also by chance that a young shepherd in the Causses region of southern France forgot his cheese sandwich in a cave one fine day. It was covered with mold when he eventually returned, but he hungrily ate it anyway – and found it delicious! Roquefort had been born. The fact is that in the "fleurines" or faults terminating in the caves where this famous ewe's milk cheese is now made, the mold called *Penicillium glaucum roqueforti* has been thriving for thousands of years. When the raw ewe's milk is treated with an enzyme, rennet, the work of the different bacteria produced cavities and holes in the cheese into which the *Penicillium* infiltrates. While modern techniques have made it possible to increase milk production through the careful selection of ewes and the use of milking machines, the milk tends to be less rich in useful bacteria. Once again, a selection must be made of appropriate strains. Today, a precise amount of spores is mixed into the milk along with the rennet and the seeded curd sent to Roquefort to allow the *Penicillium* to grow. Through its enzymatic metabolic action, the latter causes a profound change in the nature of the curd: proteolytic enzymes (those that break down proteins) give the cheese its creaminess; lipolytic enzymes (which metabolize fats) produce the aromas and aroma precursors. As to the blue veining in the cheese, this is simply the visible proof of the maturing of *Penicillium* spores.

Another famous cheese was born in the midst of the French Revolution, about 1790, when a Norman farm wife, in seeding some curd with a grayish-green mold, invented Camembert. The production of Camembert became industrialized toward the end of the nineteenth century, and *Penicillium candidum* was added to *Penicillium camemberti* Selected lactic and fungicidal yeasts are used with raw milk, and the duration of the aging is reduced, contributing to a smoother flavor for the cheese.

The principles of production remain unchanged today, but with a few variants. Milk contains casein (nitrogenous organic compounds); this is curdled with the help of an enzyme extracted from the fourth stomach of a calf and known as rennet, which metabolizes the casein, and lactic bacteria, which control draining, are added. Part of the whey (lactoserum) is exuded and forms the curdled milk, which is acidified by the lactic bacteria. The surface of the curd is dusted with a suspension of *Penicillium* spores and the curd is allowed to age for a period of about a month, corresponding to the duration of enzymatic activity. The components of the curd are broken down by the action of the enzymes produced by the microbian flora and the cheese develops its flavor and aroma. The type of milk (raw or pasteurized), yeast, brine (used to salt the cheese), and the local wild microflora all contribute to the formation of population of microbes that can attain a density of up to one billion per gram.

Cream cheeses (of the Fontainebleau or petit-suisse type) are produced by lactic acidification, are drained by centrifuging or filtration, and are not aged. Soft or semisoft cheeses (Camembert and Brie in the case of those with molded rinds), Roquefort and various blue cheeses for those with internal mold, and Munster or Pont Lévêque for those with washed rinds, whose curdling is accomplished with the aid of rennet, are drained slowly and allowed to ripen for varying lengths of time at different temperatures. In the case of processed cheeses (Saint-Nectaire and Reblochon), hard uncooked cheeses (Cantal), or hard, cooked cheeses (Emmenthal and Comté), rennet is also used, but draining is accelerated by various different methods and aging is carried out at speeds and temperatures that also vary widely.

In the Orient, too, as we already noted, foodstuffs have been based on fermentation since the earliest times. Not by pure chance is Japan the world's lead-

1. M. Harel
The monument erected to an unappreciated heroine of the French Revolution, Marie Harel, the inventor of Camembert.
2. Roquefort
The Penicillium roqueforti mold, details from a 2,250 X enlargement.

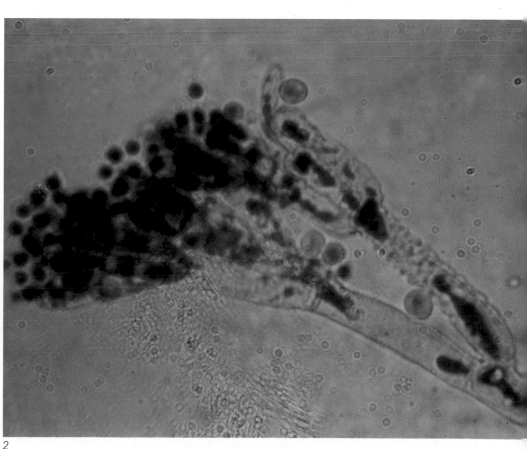

1

2

(4) It is in fact a mixture of soy beans and wheat fermented by means of the same fungus, *Aspergillus oryzae*, used in making saki. An equal quantity of brine is added to this preparation and the mixture allowed to ferment from eight to twelve months at a relatively low temperature, with yeasts and bacteria being added. The result is subsequently pressed and the soy sauce drawn off.

1

2
1. *C. Weizmann*
C. Weizmann, born in Byelorussia in 1874, worked as a chemist in Switzerland before engaging in scientific research work for the British Admiralty during World War I. Author of the Balfour Declaration (1917), he was Israel's first president from 1949 until his death in 1952.
2. *S. Kinoshita*
Dr. Kinoshita has become the president of Kyowa Hakko.

ing producer of amino acids and leads the field in enzyme technology. Its fermentation industry, with an annual turnover of $15 billion, accounts for 4% of its GNP (Gross National Product), a situation that Geoffroy Surbled, author of a June 1983 CCF (Crédit Commercial de France) report sumarizes in this surprising statement: "Pasteur's heirs are Japanese." Japanese culinary traditions (miso, shoyo, netto, saki, etc.) have also contributed to traditional industries: sodium glutamate for flavoring (Ajinomoto, a word meaning "the essence of taste"); dairy products (Yakult Honsha); whiskey (Suntory); beer (Kirin); saki (Kyowa Hakko); and soy sauce (Kikkoman), of which Japan is the world's leading producer, although it is of Chinese origin. (4)

Fermentation techniques have thus developed by a series of breakthroughs. With Pasteur, modern fermentation got a start and microbiology became a full-fledged discipline. The terme "microbe" was invented by Sédillot in 1878, but it was in 1855 that Thomas Escherich discovered a bacterium, which he called *Escherichia coli,* without in the least suspecting that he would be to biotechnology what J. von Sternberg was to be to the movie industry: the godfather of a great star.

It is odd in retrospect to think that Pasteur's work on beer and the initial attempts at industrial fermentation during World War I were due to the same cause: the enmity between France and Germany. If in 1857 the French scientist began to study the problems raised by fermentation, it was because French brewers were complaining about the poor flavor of their beer compared to those of their neighbors across the Rhine. France's honor was at stake.

During World War I, on the other hand, the Germans needed glycerol for their explosives and similarly the British needed acetone. The British blockade effectively prevented the Germans from importing the desired quantities of vegetable oils, the raw material of glycerol. Paradoxically enough, it was Pasteur who (posthumously) flew to the aid of the Germans: Carl Neuberg remembered one of the French scientist's observations – that alcoholic fermentation produced small quantities of glycerol. Neuberg discovered that if sodium bisulfite were added to the vat, glycerol would be obtained to the detriment of ethanol. This industrial fermentation (monthly production of 1,000 tons of glycerol) came to a halt just after the war.

Production of acetone, on the other hand, continued until the development of the petrochemical industry. It was due to Chaim Weizmann (who was to become Israel's first president in 1949). He developed acetanobutylic fermentation using an anaerobic bacterium. During the course of perfecting this fermentation process, a solution was found to the problem of contamination by undesirable bacteria. It was this that led to the development of pure cell techniques, which proved so valuable in the production of antibiotics.

Fermentation took a decisive turn during World War II and went on to another stage: the initial fermentation of penicillin and the cultivation of molds. The market was dominated by the United States and Europe immediately after the war, but the situation in Japan was particularly interesting. About seventy companies were competing there. Three years later only four of them were left, including Meiji Seika and Kyowa Hakko, still both at the top in

a country that has become the world's second leading producer of antibiotics. During the 1950s, microorganisms were forced to produce cortisone and sex hormones, and this gave researchers a slightly better understanding of the enzymatic mechanisms governing the metabolism of the cell.

And it was the Japanese who crossed the third threshold in 1956. Dr. Shukuo Kinoshita, working at Kyowa Hakko, had for many years believed that microorganisms might well be able to produce amino acids. He developed a process that made it possible to produce glutamic acid by fermentation and, by modifying the culture medium, to inhibit certain genes in order to curtail enzymatic operations. This is a new step inasmuch as it is an attempt to regulate microbian metabolism systematically in the light of our knowledge of biochemistry and microbian genetics. Olivier Fond, author of a report on biotechnology in Japan, has accurately written, "The whole trick in industrial fermentation is to change the direction of normal metabolism and pick up a molecule along the way." The fermentation of amino acids also makes it possible to obtain directly the L form of the acid, thus avoiding the stages required to separate it from the D form, which is found simultaneously in chemical synthesis. Furthermore, the production of amino acids stems from the discovery of a new strain, the *Cornybacterium* or *Brevibacterium* (depending on the patents involved). The discovery of new microorganisms has as a result become a top priority for businessmen and researchers alike – an indispensable line to be followed, according to Professor Teruhiko Beppu of the University of Tokyo, who works in close collaboration with industry: "We shall have to find new test and screening systems and new strains if we wish to produce new and competitive products. The transition from research to industry and to mass production depends on this. Genetic engineering is but a technique: it is up to us to find the target at which it should be aimed."

Escherichia coli has proved a disappointment to businessmen. Relatively unspecific, it can induce unfortunate secondary reactions, shows a slow growth rate, is hungry for nutrients (which are expensive), and raises problems in recovering the end product (with protein accumulating in the cell's interior). *Bacillus subtilis,* for its part, excretes proteins and exhibits active recombining mechanisms, but it raises other problems, such as those of vectors. Saccharomyces, streptomyces, and streptococci are still quite difficult to use. Microorganisms, "Man's finest conquest after the horse," stand a good chance of being the key to the industry of life. "Our most powerful weapon will be the combination of genetic engineering techniques and the mastery of strains and the fermentation process. Without one or the other the bioindustry of tomorrow would stand a strong chance of being crippled," concludes Hirotoshi Samejima, general manager of Kyowa Hakko and president of the management committee of BIDEC (Japan's development center for bioindustry).

1, 2. Fermenter
Interior of a fermentation vat and its closing mechanism.
3. Quality
Keizo Saji, president of Suntory, personally samples a new whiskey.
4. Separation
Purifying a substance or extracting it from a fermented "soup" requires increasingly more sophisticated separation systems. In this photograph, an automated system with variable controls (for temperature, etc.) is being developed in a pilot plant.

The Great Turning Point: Antibiotics and Secondary Metabolites

by Jean Florent, Agronomic Engineer, Assistant to the Head of the Biochemical Research Department at the Rhône-Poulenc Santé Research Center at Vitry/Seine, and Pierre-Etienne Bost, Chemical Engineer, Director of Research Programs and Projects at Rhône-Poulenc Santé.

The improvement in fermenters and the development of purification and separation techniques: with antibiotics, fermentation switched from simple "cooking" to industrial production. J. Florent and P.-E. Bost, coeditors of the book The Future of Antibiotherapy and Antibiotic Research (Academic Press, 1981), relate the story.

Behind this story lies that of a race against time between researchers and bacteria apt to acquire a resistance: in the few years following the initial use of penicillin in 1941, the majority of staphylococci were resistant to the antibiotic. Subsequent work, including that done since the early 1980s by two microbiologists of the Royal School of Medicine in London, Naomi Datta and Victoria M. Hughes, has shown that acquired bacterial resistance is due not to the selection of resistant strains but to a transmission of information (in the form of genes) from one strain to another and from one species to another. These two researchers have for the first time succeeded in making a comparative study of strains between "preantibiotic" and modern strains, for they have had access to a collection of a very special kind, that of the English doctor E.D.G. Murray, who, from 1917 to 1954, avidly collected bacteria of every kind from Europe, the Soviet Union, the United States, India, and the Middle East and preserved them in sealed glass tubes. Just how is resistance transmitted?

Penicillin acts by diversion: it becomes irreversibly attached to proteins, among which are several enzymes involved in the construction of the bacterial membrane, thereby making these enzymes incapable of fulfilling their role. Deprived of a proper membrane, the bacteria empty and die.

To stand guard and break down or destroy the penicillin molecules, the bacteria have learned to produce an enzyme, penicillinase, that protects the work of the enzymes building the membrane. But the question is immediately raised: How did they "learn" to produce this enzyme?

In the case of streptomycin, which works by inhibiting the operation of the ribosomes – the protein factories – the bacteria have "learned" to show the enemy genetically different ribosomes, almost as if one camouflaged a factory as a residential villa. They also know how to transform this antibiotic into derivatives that are inoperative, in that they are incapable of recognizing the ribosomes.

When another family of antibiotics was discovered, the cephalosporins (derived from a mold found in Sardinian sewer water in 1945 by G. Brotzu of the Institute of Hygiene of Cagliari), and an Oxford team isolated cephalosporin C and synthesized it during the 1960s, it was thought that the ultimate weapon had been found: a broad-spectrum antibiotic that the penicillinases could not recognize. And then it became evident that the bacteria were beginning to defend themselves.

The discovery of the resistance factor was a serious upset to traditional geneticists: not only could bacteria acquire genes "by contagion" when they were close to other bacteria that had them, but on top of this they could transmit this acquired characteristic in defiance of all established dogma.

It is in this way that an epidemic of resistance to a specific antibiotic may be propagated, to say nothing of an even more dangerous contagion phenomenon, in that the antibiotic sometimes immunizes the bacteria not just against their aggressor but against five or six others as well.

To go even further, "jumping" genes have been found among not only the procaryotes but the eucaryotes as well. In addition, we have seen that insects develop a resistance to insecticides and rats to rat poisons. Could we not imagine a similar process that would enable mankind to become resistant and adapt to new forms of chemical pollution, for instance? Without our being aware of it, could not genetic engineering be a very old natural mechanism? In some cases, as in cancer of viral origin, do not men acquire foreign genes, which become integrated with their own genetic program and modify it? This daring question challenges many an established dogma indeed, as we shall see in chapter 4.

Where antibiotics are concerned, the race between researchers and bacteria seems to have reversed, in that now it is a matter of finding the means to annihilate the new species stemming from this contest. Once again, mankind is becoming aware that the "industry of life" is not quite like any other industry.

"In the lower species, even more than in the higher animal and plant species, life prevents life."

(L. Pasteur and J.-F. Joubert, Académie des Sciences, 16 July 1877).

When Selman Waksman coined the word "antibiotic" in 1942, he was referring to "any chemical substance produced by a microorganism, which has the property in dilute solutions of inhibiting the growth of and even destroying other microorganisms." His definition has since been enlarged: products initially produced by microorganisms and later prepared by chemical synthesis, substances produced by the higher plant species or animals, and natural products that have been chemically modified may all be referred to as antibiotics as long as they meet the same criterion of activity. The terms antimicrobial and antiseptic are reserved for products whose action is antiinfectious (antibacterial, antifungal, antiviral) and that differ from antibiotics as such since they can only be prepared by chemical synthesis (sulfonamides, quinolones) or because they are only active at relatively high concentrations (ethanol).

The term "antibiotic" comes from "antibiosis," a word invented by Paul Vuillemin in 1889 to describe phenomena of antagonism between living organisms. This phenomenon had been known for many years; in particular, in 1877, Louis Pasteur and J.-F. Joubert had demonstrated that certain bacteria inhibited the growth of Bacillus anthracis, the vector of anthrax, and from this deduced the clinical potentiality of microorganisms as therapeutic agents.

In 1885, V. Babes demonstrated for the first time that microorganisms could produce substances capable of inhibiting growth of other microorganisms. The first antibiotic (mycophenolic acid) was isolated in 1896 by B. Gosio, and in 1913 C.L. Alsberg and O.F. Black isolated penicillic acid. Neither of these two substances was examined on the clinical level, although at the beginning of the century and up to 1914, pyocyanase, extracted from Pseudomonas aeruginosa, which contained a mixture of antibiotics, was used to treat human and animal diseases. For the next fifteen years, the phenomenon of antibiosis was treated as a laboratory curiosity with no great potential for practical application.

The great awakening occured in 1929 when Alexander Fleming, already known for his discovery in 1922 of lysozyme, a bactericidal enzyme found in tears and milk, revealed that a microscopic fungus, Penicillium notatum, produced an antibacterial agent, which Fleming called penicillin. He showed that culture media of Penicillium, active in vitro, were not toxic for animals; but for the next decade, further development of penicillin was inhibited by the failure to isolate this very unstable product. Fleming's discovery might even have been forgotten were it not for Howard Florey and Ernst Chain, who at Oxford University in 1940 inspired a group of eminent chemists to take up the challenge and isolate this remarkable substance. Clinical tests began in earnest in 1941.

During the preceding years, a number of other antibiotics had been isolated, in particular tyrothricin in 1939, extracted from *Bacillus brevis* by René Dubos; it was later shown to be a mixture of gramicidin and tyrocidin. Dubos was the first researcher to demonstrate the presence of microorganisms producing antibiotics in the soil. He thus made an important contribution to the "explosion" of antibiotics that began in the 1940s and that has continued right up until the present day.

In 1944, when they published their work

relating the discovery of streptomycin, S. Waksman and A. Schatz opened the way for the discovery of antibiotics produced by Actinomycetales (ramified bacteria), particularly the *Streptomyces*. Some fifty years earlier, in 1890, G. Gasperini had already noted phenomena of antibiosis between actinomycetes, on the one hand, and bacteria or fungi on the other. His observations had been confirmed starting in 1920 by a number of bacteriologists, and later by Waksman himself and H.B. Woodruff, who isolated actinomycin in 1941 and streptothricin in 1942.

This was the beginning of what we might call the golden age of antibiotics, which

Penicillium

was to last some fifteen years. Systematic programs were developed to select microorganisms, and led to the discovery of most of the major antibiotics in the *Streptomyces* group, most of which are still widely used today (chloramphenicol, chlortetracycline, erythromycin, kanamycin, leucomycin, spiramycin, neomycin, novobiocin, nystatin, oleandomycin, oxytetracycline, spiramycin, virginiamycin, etc.). By 1960, clinicians had at their disposal an arsenal of quality products, and had become increasingly dependent on them, although many of them were gradually losing their effect against certain pathogenic viruses. At this point the first chemically modified antibiotics (semisynthetic antibiotics), which were less affected by this phenomenon known as bacterial resistance, began to make their appearance.

The modern era of antibiotics began around 1970. It is characterized by the appearance of novel discoveries thanks to the application of new screening methods, the use of hypersensitive microorganisms, and the examination of new taxonomic families as potential sources of new antibiotics. It has also seen intensive development of semisynthesis (more than 70,000 derivatives were produced in twenty years, approximately 70 of which are being used in clinical medicine), and a widening of practical applications (oncology, parasitology, plant protection). In other words, over the last fifteen years, we have seen the development of research into naturally occurring bioactive metabolites that no longer fit the definition of antibiotics given by Waksman. In 1980, J. Berdy suggested enlarging the description to include any compound of microbial origin, capable, even in very dilute solutions, of inhibiting various growth processes in man, animals, plants, and microorganisms. Altogether, over the last fifty years, some 7,000 antibiotics fitting this new definition have been isolated, and every year some 300 new ones join their ranks. They occupy a very important place in the world of secondary metabolites (which will be defined later in this article).

The Major Families of Antibiotics
Antibiotics may be classified in various ways: in terms of their origin, effectiveness (activity spectra: antibacterial, antifungal, or antiviral), mechanism of action (molecular target), method of biosynthesis, and chemical structure. However, only the latter criterion allows us to distinguish satisfactorily between one antibiotic and another, and it alone gives a satisfactory classification, although some problems arise because of the complexity of certain structures. This is why the classification used should logically integrate both chemical and biological characteristics (antimicrobial activity). The part of the molecule that is primarily responsible for its activity constitutes its essential structural motif and allows researchers to define the family to which it belongs. J. Berdy has established a highly ramified system of classification.

Biological Activity of Antibiotics: Activity Spectrum and Mode of Action
All antibiotics inhibit growth of or destroy bacteria, fungi, and viruses. However, they do not have identical effects on all these life forms. Some antibacterial antibiotics, for example, which are known as "broad-spectrum antibiotics," act on several pathogenic species differing from each other in terms of structure and composition of the cell wall (for example, staphylococci, colibacilli, and mycobacteria). Others have a more limited spectrum and act on similar species.

For real effectiveness, antibiotic molecules must disturb a vital link in the microbe's metabolism, and this link is their target or point of impact. Targets are located on the bacterium's fundamental structures: wall, membrane, cytoplasm (ribosomes) and nucleus (chromosome).

Beta-lactams act primarily on the cell wall by inhibiting the final step in synthesis. Peptide antibiotics latch onto the phospholipids of the cytoplasmic membrane, disturbing the natural order and inducing abrupt osmosis, which lets the cytoplasmic components escape. Ionophoric polyethers modify the permeability of the membrane to certain cations, thereby creating ionic imbalance of the cytoplasm. Aminoglycosides, tetracyclines, macrolides, and chloramphenicol disturb protein synthesis in ribosomes. Among those antibiotics that act on nuclear DNA, we might note that novobiocin inhibits DNA gyrase, the enzyme implicated in DNA synthesis, while rifamycin inhibits RNA polymerase, the enzyme responsible for biosynthesis of messenger RNA, which transmits the genetic information contained in the chromosome's DNA.

A large number of bacterial strains are not as sensitive to antibiotics as in vitro experiments would seem to suggest. Since the introduction of antibiotics in therapy, the sensitivity of pathogenic bacteria has been modified, particularly in the hospital environment, and there are a large number of resistant strains in many species.

Resistance may be caused by a number of factors:
- reduction or elimination of the antibiotic's ability to penetrate bacteria through modification of the cell membrane, which means that the antibiotic is not sufficiently concentrated to act against its target;
- spontaneous mutation altering the antibiotic's target, so that the latter no longer recognizes its point of impact;
- modifications of the antibiotic by bacterial enzymes : thus acetylation, phosphorylation, or adenylation of certain sites of aminoglycoside molecules makes them inactive or inoperative, since they no longer recognize their normal target on the ribosome or are even ejected from the cell; hydrolysis of the beta-lactam nucleus by beta-lactamases results in penicillins and cephalosporins becoming totally inactive.

We might point out that apart from bacterial resistance, an antibiotic may be prevented from reaching its full theoretical potential, compared with in vitro results, for a number of reasons: defective absorption by the individual's tissues, modification by enzymes in the host's blood serum, bonding to seric proteins (serum effect), and so on.

Major Microorganisms That Produce Antibiotics
Although we are familiar with some antibiotics produced by the higher animal or plant species, the great majority of them are produced by microorganisms, especially those in the following three groups: Fungi, Actinomycetales (ramified bacteria), and Eubacteriales (simple bacteria). Respectively 20%, 70%, and 10% of isolated antibiotics belong to these three groups.

Antibiotics produced by Fungi (mainly *Penicillium, Aspergillus,* and *Cephalosporium),* although relatively few in number, are still the most widely used in current chemotherapy, in particular penicillins and cephalosporins.

Actinomycetales, which are easily the most prolific, are represented in 90% of cases by *Streptomyces.* Bringing up the rear we find *Micromonospora, Nocardia, Streptoverticillium, Actinoplanes,* and *Streptosporangium.*

Most of the Eubacteriales antibiotics come from the *Bacillus* (70% of cases) and *Pseudomonas* species (13% of cases).

Some of these species have a remarkable capacity for synthesizing antibiotics: *Streptomyces griseus* produces more than 40 different antibiotics, while *Bacillus subtilis* produce more than 60.

Birth and Development of a Family of Antibiotics
Although the discovery of penicillin was basically an accident felicitously exploited by Fleming, development of other antibiotics over the next fifty years was the fruit of painstaking and meticulous research.

The very long process that, in the best of cases, results in commercial distribution of an antibiotic, involves a whole series of steps.

1. *Screening of the microorganisms producing the substance:*
This consists of selecting the appropriate microorganisms from a heterogeneous population. To increase their chances of discovering novel molecules, researchers examine large sectors of microbial groups by collecting samples of soil, plants, or water in ecological microenvironments that vary greatly in terms of soil type, climate, and vegetation. The different populations found in a particular environment are developed, with researchers encouraging development of different families. Individual strains are selected and grown in a liquid

environment, and once fermentation is completed, the culture medium is screened by the most effective microbiological or biochemical processes available, depending on the desired activity. Experience has shown that among the strains selected, less than 1 in 10,000 produces a therapeutically useful antibiotic.

2. *Isolation of the active product:*
"Natural" strains found by screening produce only a small quantity of antibiotic, not much more than a few milligrams (mg) per liter of culture medium at the end of fermentation. However, at least 1 g of active, purified substance, crystallized if possible, is necessary to determine its structure and

physicochemical characteristics and to carry out an adequate examination of its biological properties. Fermenters with capacities of some tens or hundreds of liters are used to prepare this quantity of product. When placed in a suitable environment (i.e., optimal composition of culture medium, pH, air supply, agitation, temperature etc.), the microorganism first develops (trophophase), then secretes the active principle(s) sought (idiophase). Active substances are then separated from the culture medium and purified by conventional procedures, such as extraction by solvents; precipitation by change in pH, salinity, or polarity of solvents; exchange of ions; chromatography; ultrafiltration through

Streptomyces

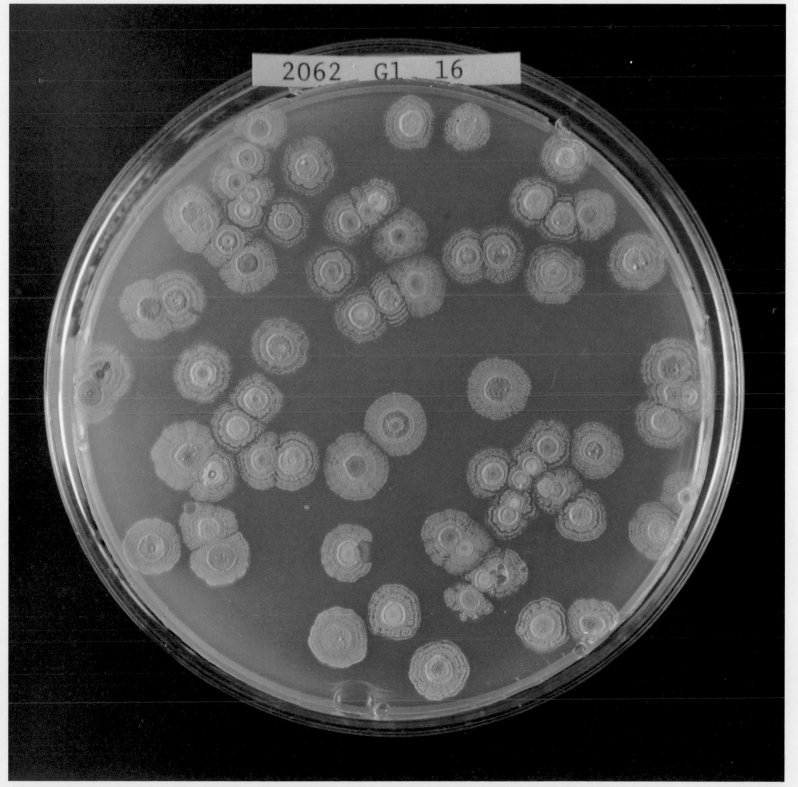

membranes; etc. If the substance produced remains inside the microorganism instead of being excreted into the culture medium, it must be liberated by breaking down the cell wall using suitable mechanical, thermal, or enzymatic processes. Once the active principle has been isolated and purified and its chemical structure elucidated, it is compared with the approximately 7,000 natural antibiotics in current use, to check whether it is novel (and patentable); then its biological properties (activity and toxicity for animals) are examined to ascertain its potential therapeutic interest.

3. *Development of the product:*
This stage follows confirmation of the product's biological qualities in preliminary in vivo analyses. These analyses cover both the biological aspects (in-depth studies on animals: metabolism, long-term toxicity, local tolerance, pharmacokinetic and galenical tests, and later clinical tests) and the technical aspects (feasibility studies and work concerned with improving the performance of the processes involved). Technical analyses cover three different areas: improving the strain producing the antibiotic (genetic engineering), improving the culture medium (microbiological engineering), and finding novel techniques and equipment for both fermentation and extraction (biochemical engineering).

There are two major ways of improving strains:
● by conventional mutation techniques: mutation-inducing agents (ultraviolet light, chemical agents, and so on) are used to modify the microorganism's genes to improve biosynthesis of the active product;
● modern genetic recombinant techniques: researchers attempt to fuse two strains (or clones) producing the antibiotic in question so as to combine their genetic instructions (fusion of protoplasts, hybridization, or in vivo genetic recombination); genetic engineering can also be used to incorporate fragments of DNA taken from one clone into the genetic code of another clone (this is the strict definition of the recombinant DNA technique); to achieve the desired results using this type of technique, the researcher must have an excellent understanding of the biosynthesis of the antibiotic.

Whichever method is used, the final objective is to obtain strains with improved properties: better growth, faster biosynthesis of the antibiotic concerned, higher yield, lower consumption of raw materials and the possibility of using less costly raw materials, and adaptability to simple and more economic fermentation conditions (particularly in terms of air supply and agitation).

Improvement of culture media covers the nature of the raw materials, their concentration, optimal balance, interactions, preparation, and use, as well as the pH factor, air supply, etc. Microorganisms must be provided with the best possible working condi-

tions, which means supplying the nutrients they need at the right moment.

Biochemical engineering is concerned with improving the performances of conventional fermenters, defining and designing new types of fermenters, designing techniques for large-scale extrapolation of production, and developing foolproof techniques to combat contamination and provide automated systems for continuous fermentation and biosynthesis.

4. *Large-scale preparation:*
Industrial fermentation is carried out in very large containers, with capacities exceeding 200 cubic meters (m^3). The system for extracting and purifying the active product must be adapted to this scale, which implies imposing and very expensive installations.

At this stage, the productivity of the strain is of course much greater than when it was discovered. Systematic study of strains, culture media, and equipment has gradually increased penicillin production from 5 mg/liter to approximately 30 g/liter (almost 6,000 times more); streptomycin production has increased from 100 mg/liter to more than 10 g/liter; leucomycin from 300 mg/liter to 15 g/liter. Around 5,000 kg of pure antibiotic can be obtained from fermentation of penicillin G in a 225-m^3 fermenter.

5. *The search for more powerful derivatives or analogues:*
Quite commonly, an antibiotic with extraordinary therapeutic and commercial potential must be withdrawn because of the appearance of long-term toxic phenomena or because of bacterial resistance. Rather than purely and simply abandoning the product, researchers may try to correct its negative aspects while still retaining its basic characteristics. Alternatively, research may be undertaken to increase the potential of an already valuable product with the aim of investing it such properties as broad spectrum, bactericidal effectiveness if it is merely bacteriostatic, greater in vivo activity, absence of bacterial resistance, respect for the nonpathogenic flora normally found in humans or animals (saprophytic flora), solubility in water, stability in water and tissues, administration by several different routes (oral, intraperitoneal, intramuscular, intravenous), higher levels in blood and tissues, low toxicity, good local tolerance, absence of dangerous interactions with blood-serum proteins (serum effect, enzymatic deterioration), absence of allergic reaction, and so on.

In some cases, derivatives developed by relatively simple chemical reactions (esterification of certain functions, for example) may considerably increase the antibiotic's potential for use (by increasing its solubility or tissue absorption, etc.).

The search for analogues of a given anti-

biotic to the same family with the aim of finding products with enhanced performance is a dynamic area of investigation and a number of different techniques are employed – microbiological, biochemical, and chemical:
● exploration of novel ecosystems and use of special culture media to promote development of populations outside conventional taxonomic groups;
● genetic techniques: mutation of the strain, fusion of protoplasts within a single species or from distinct species, in vitro recombinant DNA techniques;
● the use of mutants, blocked at a certain stage of biosynthesis (idiotrophic mutants), which are assisted during the productive phase of fermentation by supplying a key element of the desired molecule (this technique, known as mutational biosynthesis or mutasynthesis, was used extensively for aminoglycosides);
● bioconversion, which consists of modifying the chemical structure of a molecule by subjecting it to the action of carefully selected microorganism;
● controlled biosynthesis by addition of precursors, i.e., fragments of the molecule not synthesized by the microorganism involved; by changing the type of precursor, new antibiotics can be formed (this is the technique used in the production of natural penicillins);
● semisynthesis, or modification of the antibiotic by chemical means.

Of all these techniques, semisynthesis has given the most spectacular results, particularly for penicillins and cephalosporins, and to a lesser extent for aminoglycosides and tetracyclines. This involved very extensive research: some 40,000 semisynthetic penicillins and cephalosporins were prepared, and produced "generations" of products with improved biological performances (insensitivity to bacterial inactivation enzymes, effectiveness with respect to the more formidable hospital germs, such as *Pseudomonas aeruginosa*, increased lenght of life, i.e., increased persistence in the individual being treated, etc.).

Secondary Metabolites
To obtain the matter and energy it needs for growth and reproduction (in other words for synthesis of its components – proteins, polysaccharides, lipids, nucleic acids, and sterols), as well as for indispensable vitamins and coenzymes, a microorganism uses products of relatively low molecular weight, usually under 1,500. These building blocks are found in the microorganism's immediate environment if it is grown in a synthetic medium; however, more often they are metabolic intermediates synthesized inside the cell from elements making up the complex culture medium surrounding it. These metabolic intermediates (nucleotides, amino acids, simple sugars, fatty acids, terpenes) are known as primary metabolites, and are indispensable to growth. In contrast, the so-called secondary metabolites are not indispensable to growth of micro-

organisms developed in vitro in a pure culture, or at least are not recognized as being indispensable as far as current scientific knowledge is concerned, although this view may eventually be modified. However, in the natural state they do play an important role in the organisms that produce them, since they function as sexual hormones and transporters of cations; defend against other bacteria, fungi, amoebae, insects, or plants; or act as agents to induce symbiosis or cell differentiation. Secondary metabolites, including antibiotics, are produced more or less directly from primary metabolites, which often exert a positive or negative effect on their biosynthesis. They are generally produced during the stationary phase (idiophase), which, in liquid culture, follows the trophophase or exponential-growth phase of the microorganism. For this reason they are often called "idiolites." Their molecular weight ranges from 100 to 1,500, and their structures may vary considerably.

In the decade between 1950 and 1960, many industrial microbiologists working in pharmaceutical research believed that antibiotics represented only a small proportion of secondary metabolites. Convinced that these metabolites possessed as yet undiscovered potential, they attempted to use them to treat illnesses other than those caused by bacteria, fungi, or viruses.

Cephalosporium

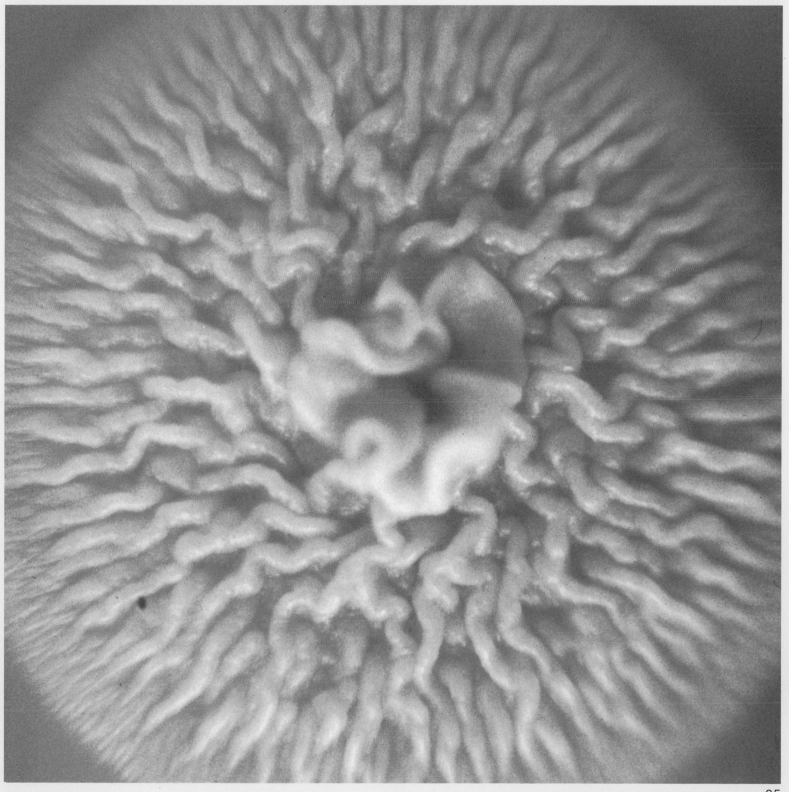

This initiative did bear fruit, since at the present time we can identify about 2,000 nonantibiotic secondary metabolites that possess properties of interest in various sectors of human and veterinary medicine or plant protection.

Many of these secondary metabolites possess antibiotic properties that are of no practical therapeutic use but that do allow them to be detected in the fermentation liquid and isolated; according to the normal procedure for products of synthesis, these are then subjected to a wide range of tests on animals in various fields (oncology, immunology, pharmacology, parasitology, etc.). In the same way chemical screening, which uses tests not correlated to any particular biological activity, allows identification and isolation of products possibly worth of interest. The difficulties involved in identifying and purifying secondary metabolites that cannot be detected by either microbiological or chemical means remained a handicap to development for many years, since researchers could not reasonably envisage large-scale administration of culture media or extremely complex unrefined extracts to animals; the weak concentration of active products in the solutions examined and the danger of secondary effects caused by other substances (the result of which is falsification of results) proved insuperable, and in any case this type of testing remained extremely expensive.

A solution to this dilemma was found by H. Umezawa (of the Institute of Microbial Chemistry, Tokyo) in 1972. He suggested using enzymatic in vitro tests to detect inhibitors of key enzymes in the human metabolism in culture media of a microorganism. Since functional disorders of the organism can be associated with excessive or irregular enzyme activity, it seemed quite logical that enzyme inhibitors produced by microbes might manifest pharmacologically interesting properties. This supposition proved to be well-founded, so much so that current research into antimicrobial, anticancer, and antiparasitical therapies tends increasingly to use this technique. Its major advantage is that it makes for very fine screening by focusing research on products with a specific mode of action, and acting on a precise enzyme target. For example, researchers would look for fungicides or insecticides that inhibit chitin synthetase, or antiviral agents inhibiting viral neuraminidase. Tests based on linkage to specific cell receptors use the same approach.

At the present time, and not counting antibiotics, a large number of secondary metabolites are either already on the market (some have been for many years) or are about to be marketed. In the field of human medicine, we might mention antitumor agents (daunorubicin, doxorubicin, bleomycin), immunosuppressants used particularly in organ transplantations (cyclosporin), antiobesity agents (acarbose), and utero-

contractants (alkaloids of ergot derived from lysergic acid); in the field of veterinary medicine there are anticoccidians (monensin, lasalocid, salinomycin), anthelminthics (avermectin), growth inducers (monensin for grazing animals), and anabolizing agents for animal husbandry (zearalenone); and in the field of plant protection, fungicides (kasugamycin, polyoxins), insecticides (toxin from the Bacillus thuringiensis spore), and growth regulator (gibberellins).

A large number of products discovered through screening for enzyme inhibitors are currently being tested. Among them are amicoumacin A, forphenicin, and esterastin (antiinflammatories), mevinolin (hypocholesterolemiant), ascofuranone (hypolipemiant), dopastin (hypotensor), bestatin (immunostimulant used in the treatment of cancer), antipain (inhibitor of carcinogenesis), pepstatin (antiulcer), muraceins (inhibitors of the enzyme converting angiotensin 1, potential antihypertensors), etc. In the field of agriculture, the herbicidal properties of herbimycin are currently being examined.

There is no need to discuss the remarkable commercial success of certain antibiotics. However, results obtained with some other secondary metabolites should be mentioned, since the general public is less familiar with them. In particular, we should mention daunorubicin and doxorubicin in oncology; cyclosporin in immunology; and in the field of veterinary medicine, monensin and avermectin.

Daunorubicin, extracted from Streptomyces coeruleorubidus by Rhône-Poulenc researchers in 1962, was the first anthracycline used in clinical medicine to fight leukemia. It was also the first drug to allow prolonged, if not complete, remission in acute leukemia (which, up until the discovery of daunorubicin, was always fatal). Doxorubicin, derived by hydroxylation from daunorubicin, was discovered some years later by Farmitalia, and is active against a wider range of tumors, including solid tumors. The major cause of activity in both these products is their positioning between the pairs of bases of the original DNA. Use of both is unfortunately limited by chronic or acute cardiotoxicity, and by spontaneous or acquired resistance of certain tumors. These problems have stimulated extensive research into the relationship between structure and activity of molecules. Hundreds of structural analogues have been prepared by semisynthesis and examined in the hopes of finding products with improved therapeutic ratios (higher activity/ toxicity ratios and wider spectra of activity). 4'-epi-doxorubicin (epirubicin) may be a suitable candidate. Among the most recently discovered natural anthracyclines, aclarubicin would appear to have an interesting future.

Cyclosporin, fruit of Sandoz's research, was isolated in 1970 from a fungus, Tolypocladium inflatum. It is a cyclic polypeptide made up of eleven amino acids. This product is characterized by powerful immuno-suppressant activity and specificity of action against lymphocytes: it prevents their transformation into T lymphocytes prior to multiplication. In other words, it suppresses the mechanism of cell immunity, which causes tissue and organ transplants to be rejected. Cyclosporin has been applied with spectacular success in transplants of bone marrow and kidney and heart transplants.

Monensin, discovered by Eli Lilly in 1964, is an ionophoric polyether produced by *Streptomyces cinnamonensis*. Because of their specific structure, products in this family have novel chemical and biological properties based on their chelating action with metals. It has no applications in human therapy, but has acquired an enviable position in the market because of its ability to control fowl coccidiosis, which until then had regularly decimated industrially reared poultry. It is also used as an additive in cattle feed, where it considerably increases growth and improves the ratio of food absorption by acting on rumen bacteria. The commercial success of monensin has attracted the attention of competitors, who are attempting to cash in on its success with two other major polyethers, lasalocid and salinomycin.

Avermectin (Merck) is a macrocyclic lactone produced by *Streptomyces avermitilis*. It was discovered in 1977 after intensive research for natural products with anthelminthic properties. This substance is inactive in vitro but was discovered to be active through in vivo tests on a mouse infected with threadworm. Ivermectin, a semisynthetic derivative of avermectin, which is considerably more active, is the market version of this product. Its action in very dilute solution against a large variety of threadworms and parasitical arthropoda is due to its action on the mediation of neurotransmission by gamma-aminobutyric acid. It is used in the treatment of parasitosis in cows, sheep, and horses and can also be used for plants. Researches are now looking at its potential for treating filariosis in man.

Economic Importance of Secondary Metabolites

World production of antibiotics and other secondary metabolites is around 30,000 tons a year. Approximately 60% is used in human medicine and 40% in veterinary medicine and animal feed.

By 1983, antibiotics for use in human medicine represented a market of around $8.5 billion (industrial price), or around 15% of the world pharmaceutical market ($55 billion), with the major share of this revenue going to cephalosporins ($3 billion) and penicillins ($2.4 billion), followed by macrolides ($0.8 billion), aminoglycosides ($0.7 billion) and tetracyclines ($0.7 billion). Among anticancer drugs, which represent only 2% of the pharmaceutical market, the most important product is doxorubicin ($0.1 billion).

In the field of veterinary medicine, antibiotics represent a market of $0.5-$1 billion, with $2-$2.5 billion for animal feed additives, distributed between the tetracyclines (chlortetracycline, oxytetracycline), the polyethers (monensin, lasolocid, salinomycin), the macrolides (tylosin, spiramycin, etc.) and others (avoparcin, bacitracin, bambermycin, virginiamycin, etc.).

Among the major world producers of antibiotics are (in alphabetical order): in the United States – Abbott, American Cyanamid (Lederle), Bristol-Myers, Eli Lilly, Merck, Parke-Davis, Pfizer, Schering Plough, Squibb, and Upjohn; in Europe – Antibioticos, Bayer, Beecham, Ciba, Farmitalia, Gist-Brocades, Glaxo, Hoechst-Roussel, and Rhône-Poulenc; in Japan – Fujisawa, Kyowa, Meiji, and Takeda.

The Future of Secondary Metabolites

The long list of bioactive secondary metabolites of microbial origin speaks for itself; furthermore, it is far from exhaustive. Many substances active against microorganisms, parasites, insects, and weeds, and others capable of encouraging plant or animal growth or manifesting various pharmalogical effects remain to be discovered. As early as 1963, J.W. Foster advised his peers, "Never underestimate the power of the microbe." In 1979, when laying down the laws of applied microbiology, D. Perlman formulated the same concept, and demonstrated the same humor, when he said, "Microorganisms can and will do everything; they are smarter, wiser, more energetic than chemists, engineers and others." In fact, as A. Demain recently remarked, the discovery of new and interesting secondary metabolites will in the future depend on the ingenuity of researchers and their ability to develop simple in vitro screening tests well suited to the activities sought.

2

Enzymes:
The Little Chemists in the Cells

In 1906, in the little German village of Esslingen, a thirty-year-old chemist, Otto Röhm, was trying to find a way of eliminating the foul smells emanating from tanning factories, which ware distressing the entire neighborhood. At that time tanners were in fact using dog turds to soften skins! Röhm told himself that the active ingredient in this concoction must be the enzymes that remained in the feces, and consequently decided to extract these directly from animal pancreas, thus discovering that it was trypsin that made it possible to soften skins. But his original assumption proved false in that it appeared that the active proteases were not animal enzymes but enzymes from bacteria lining their intestinal tracts.[1] But this in no way detracts from the importance of the work done by Röhm, who, along with his friend Otto Haas, founded a company on 10 September 1907 to exploit this tanning enzyme under the name of Oropon. Two years later, they left their little shop in Esslingen for a real factory in Darmstadt. There, between 1914 and 1934, Röhm produced an entire range of enzyme products for tanning and textile concerns, those dealing in detergents, pharmaceutical companies, and producers of fruit juice. His detergent, Burnus, the enzymes of which were still too volatile and ineffective, remained on sale until the early 1960s, when the first of a host of biological detergents appeared that broke down the amino acid chains of proteins and thus made it possible to eliminate stains of biological origin (such as blood). When he died on the eve of World War II in 1939, little could Röhm have guessed that he was the first important businessman to stake his money on products of basic importance to contemporary biotechnology, the enzymes.

But just what are enzymes? They are the little workers within the cells that we saw at work throughout our chapter on fermentation. Natural catalysts, they are far faster than any others now known (capable of increasing reaction speeds by a factor of 10^{12}-10^{15}). Like all other catalysts, they initiate or accelerate a chemical reaction and are again intact at the end of the reaction. Furthermore, while they are also extremely selective, nontoxic, and biodegradable, they do have their disadvantages: sensitivity to temperature and acidity and a potential for becoming inhibited by chemical reagents. They may consist entirely of amino acids, or of amino acids and nonprotein cofactors (metallic ions or an organic molecule). Most of the cofactors indispensable for the work of certain enzymes – such as ATP (adenosine triphosphate) – are themselves consumed int he course of the reaction. One of the solutions to this would be to produce them by genetic engineering, and Genentech has recently announced that it has successfully obtained ATP in this way.

As we have already seen, the first enzyme, the diastase of malt, was identified in 1933 by Payen and Persoz. In 1874, the Danish chemist Christian Hansen provided the first relatively pure enzyme preparation for industrial use by extracting rennet from a calf's stomach. Since antiquity, in fact, cheese had been made using the stomach of sheep and goats (which contained this enzyme), as well as the sap of fig trees (containing another enzyme, ficin), but it was only in the nineteenth century that the functions of enzymes were recognized as such. For these ferments separated from the bodies in which they were formed, Wilhelm Kühne found a name – "enzyme" – meaning "in the yeast." And the famous argument between Pasteur and Liebig on spontaneous generation was only to find a definitive answer in the work of Eduard Büchner, who demonstrated that, in fermentation, the active principle of change was not the microorganism itself but within each cell, the enzyme.

In 1884, the chemist J.C. Van Marken founded one of the pioneer laboratories in this field as part of the Royal Dutch Yeast and Spirits Factory, which he had established in 1870 on his return from a fascinating trip to Vienna (where, under the auspices of Professor Félix Hoppe-Seyler, remarkable work

Renin
This enzyme is synthesized by the kidneys and plays a part in controlling blood pressure by producing angiotensins (factors determining blood pressure). To counter hypertension, laboratories throughout the world are trying to produce renin inhibitors. Various research teams in France, the United States, Australia, and Japan have already cloned the gene coding for human renin. In January 1985, California Biotechnology announced the in vitro production of human renin by genetic engineering, and this should make it possible to have sufficient quantities of this enzyme available for work on inhibitors.

(1) Today, however, the enzymes used in the tanning industry are derived from animal pancreases.

Enzymes: The Little Chemists in the Cells

1

2

3

1. Protease
A model, designed by Professor Martin Ottesen of the Carlsberg Laboratories, representing Novo's proteolytic enzyme "subtilisin."

2. Amylase
A chemical solution poured over petri dishes coated with gelose makes it possible to select the microorganisms that secrete alpha-amylase: a light halo is formed around them, though it does not appear around those that do not secrete the enzyme.

3. Glucose isomerase
Glucose isomerase in solution (left) has a half-life of thirty days – which, in simple terms means that its activity decreases every thirty days. When glucose isomerase is immobilized (right), that is to say, when it is trapped in a certain substance, its half-life doubles to sixty days, and the enzyme is far easier to manipulate.

was being done on what was not yet known as DNA). In his company, Van Marken introduced the ne plus ultra of contemporary techniques: pasteurization, microbiology, genetics. This was to be the nucleus of the future Gist-Brocades Company. In Japan, Takamine invented the first industrial technique for preparing enzymes and produced amylases from a strain of *Aspergillus oryzae* by means of a "surface culture" on moldy rice. This process was never totally abandoned, even after the appearance of "submersion" techniques developed during the war for penicillin, and were to be of benefit to the enzyme industry. Another essential contribution was that made by the German biochemist Richard Willstätter, a specialist on alkaloids and vegetable and animal pigments, who succeeded in achieving a high degree of enzymatic purification (between 1920 and 1928). The first biochemist to prove that an enzyme was a protein and that it could be obtained in the form of a crystal was an American, James Sumner, with urease (which is today used in artificial kidneys). After this, John H. Nothrop obtained pepsin and trypsin in crystalline form. Crystallization has been achieved for nearly 200 enzymes, and this has made it possible to determine their three-dimensional structure (responsible for their specific activity). Enzymology became a full-fledged discipline during the 1930s, above all due to the work of the German physiologist Otto Warburg, who received the Nobel Prize in Medicine for this work on the enzymes responsible for cellular oxidation during breathing. The first synthesis of an enzyme (ribonuclease from bovine pancreas) was achieved in 1969 by two teams, one directed by Stanford Moore and William Stein at Rockefeller University, the other at Merck.

Of the 2,200 enzymes that have been identified until now, only 1-1.5% are used in industry, and of these five have assumed major importance. First come the proteases, often found in detergents, the market for which reaches $100-$120 million. Then come the amylases, which attack starches and are used for beer, bread, and textiles, and glucoamylase and glucoisomerase, which along with amylase transform corn into fructose syrup; the market for theses three enzymes together comes to $100 million. Finally there are rennets for cheese making, for which the market is now $30-$40 million. Increasing interest is being concentrated on the lactases, which break down the lactose in milk and buttermilk; it appears that 80-90% of the world's population suffers from an intolerance to milk because of the absence of lactase in the human digestive system.

The role of enzymatic engineering is thus to effect biochemical changes in substances. The major disadvantage of enzymes being their instability, an attempt was made to use them not in a "free solution" but by "attaching" them, making it possible to use them continuously and to reuse them. Following the work of Ephraim Katchalski at the Weizmann Institute, immobilization techniques were perfected by Dr. Chibata at Tanabe in Japan.

Continuous flux bioreactors provide a solution to the stability and the recuperation of the enzymes employed. But they still raise numerous problems. Covalent attachment and intermolecular reticulation are undoubtedly the surest methods, but they are also the most complex. Adsorption and immobi-

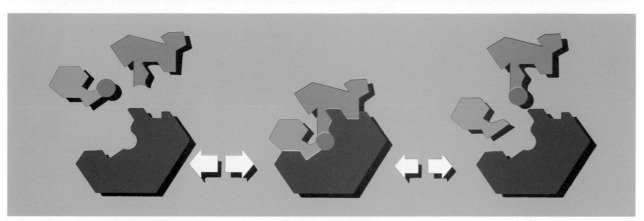

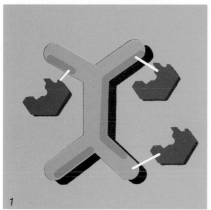

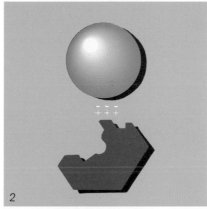

The enzyme (in red) is shown here as the piece of a puzzle with a built-in "hollow" like the bolt of a lock. Two molecules of different natures make up the substratum and hook into the enzyme like the key in the lock. The enzyme acts as a catalyst on the substratum, and the products of the reaction are released. As can be seen, one part (in orange) of the two molecules has been transferred to the other molecule. It is also worth noting that the enzyme remains intact at the end of the reaction and is ready to act as a catalyst once again.

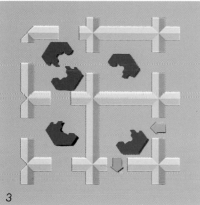

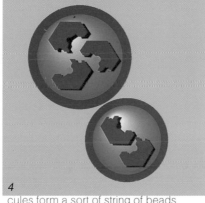

Various methods are employed today: (1) chemical methods, by direct covalent attachment of the enzyme and its matrix (generally a polymer), or by immobilizing the enzymes in polymeric gels, or even by reticulation of the enzyme with a chemical reagent (the system used to immobilize glucoisomerase) – in this case, the enzymatic molecules form a sort of string of beads. There are also physical methods: (2) using electrostatic attraction to maintain the enzyme in the neighborhood of an inert support (adsorption); (3) introducing the enzyme into porous materials the cavities of which are surrounded by chemical barriers that prevent extrusion while allowing contact with the substratum (inclusion); (4) microencapsulation, the enzymes being enveloped in membranes that are impermeable to enzymes and macromolecules but permeable to chemical products of low molecular weight – this method is used above all for urease in artificial kidneys.

lization in gels are simpler, but there is the risk that the enzyme may be more easily detached from its support. Henceforth, however, the key to the enzymatic system will be the bioreactor, where biochemical reactions (bioconversions) take place. On the basis of these technologies, laboratories have been able to develop measuring apparatus (enzymatic captors) for industrial continuous fermentation processes: they use the enzyme's property of reacting with a single substratum among several. This method was commercialized for the first time by Yellow Springs Instruments of Ohio.

The first great product of enzymatic engineering, and one made possible by enzyme immobilization techniques, was the corn syrup with a high fructose content used to sweeten such products as Coca-Cola. The market for sugar experienced a serious crisis in November 1974 and prices soared. A major importer of cane sugar, the United States, reacted vigorously: Why not get sugar from America's number one crop, corn? It was as a result of this that, thanks to enzymes, a syrup far sweeter than table sugar was produced. The stakes were considerable, as this made it possible to get rid of surplus corn in exchange for a product with a very high sugar content. The 1976 slump in the price of sugar came too late: fructose syrup has already captured 30% of the market. Starting in 1979, a galactose/glucose separation process sold by Mitsubishi Chemical Industries that made it possible to produce fructose continuously became responsible for 40% of America's production as well as 90% of Japan's production of HFCS (High-Fructose Corn Syrup). Major American producers are Cargill, E.A. Staley Manufacturing, American Maize Products, and Archer Danields Midlands.

In addition, fixed enzymes have also proved valuable in the production of amino acids, which are used in food additives, animal feeds, and medicine. In this instance use is made of the property of certain enzymes that act only on one of two types of isomers, isomers being compounds having the same atoms in the same order but with different structural arrangements. When chemists synthesize an amino acid, they obtain a mixture of the D form (inactive) and the L form (active). Only the L amino acid is of value to industry (except for methionine). To obtain these L amino acids by fermentation, use is made of the property of certain enzymes to synthesize asymmetrically, that is to say, to produce but one of the two isomers, the L form. It was thus that Dr. Chibata developed a chemical synthesis process that made it possible to obtain only L amino acids, leading to a gain in production of 60%. In 1973, Dr. Chibata also industrialized the synthesis of L aspartic acid, starting with immobilized aspartase, and in 1974 he perfected the synthesis of L malic acid.

Immobilization makes it possible to cope with some of the limitations placed on the use of enzymes such as temperature restrictions. While in nature enzymes may support temperatures up to a threshold of about 40°-100°C, to increase productivity and avoid contamination industry may require even higher temperatures, as in the starch industry, where thermostable alpha-amylase is used at temperatures above 100°C. Another solution to this could be genetic engineering.

In the relatively near future, genetic engineering should be capable of providing enzymes more stable, more active, and less expensive, and also less dependent on the vagaries of the climate or political factors: did not recent troubles in Zaire cut imports of papaya and as a result the production of papain? Even now, the gene of chymosin, along with pepsin a component in rennet (derived from the fourth stomach of the calf), has been secreted by a bacterium (Genencor, Collaborative Research – under contract to Dow Chemical), and Novo has announced obtaining a thermostable malogenic alpha-amylase. This formula may lead to shorter fermentation periods with inexpensive culture media and simple selection methods.

Biomimics

Still another hope is the chemists' dream of creating artificial enzymes, which would not be true enzymes but molecules of a totally different chemical structure that mime the enzyme's behavior as a catalyst. Actually, the enzyme exerts its catalytic effect on a molecule, the substratum, which becomes attached to a precise point on the enzymatic protein, the active area, like one piece of a puzzle fitting into another or a key fitting into a lock. There is then a transfer of electrons at the interface between the two molecules, the product becomes detached from the active area, and the enzyme returns to its original structure. As we have already

Restriction enzymes
Enzymes have proved to be
fantastic "tools" for the
genetic engineer.
Restriction enzymes above
all have provided them with
highly precise molecular
scalpels.

seen, the enzyme's activity is determined by its sequence of amino acids and by its three-dimensional structure, like any other protein. But the entire reaction takes place solely in one part of the enzyme, the active area: it is therefore the activities in this active area that the chemists are trying to reproduce.

But what are the characteristics of this active area? Two things essentially: the nature of the chemical groups implicated in the attachment of the substratum, but also their geometric arrangement in space. Nevertheless, it soon became apparent that copying this structure was extremely difficult. Consequently, an attempt was made to synthesize a molecule that would be comprised simply of those parts of the active area implicated in the chemical transformation.

The simplest technique consisted of starting with a compound existing in nature and exhibiting no enzymatic activity and, by synthesis, subjecting it to the structural changes that would confer on it the catalytic abilities of an enzyme. This was true in the case of the cyclodextrins, which are produced by the breakdown of starch and look a little like the toric (tire-shaped) structures of glucose, with a hole in the middle where the substratum can become attached. These were studied during the 1950s by Friedrich Cramer in Heidelberg and then mainly by Myron Bender at the University of Illinois at Evanston, Ronald Breslow at Columbia University, and Iwao Tabushi at the University of Kyoto.

Another method consists of entirely synthesizing a substance capable of carrying out a specific reaction. The initial research into such substances was carried out by D.J. Cram at the University of California at Los Angeles and by J.M. Lehn at the University of Strasbourg on the basis of the macrocyclic compounds called "crown ethers" synthesized in 1967 by C.J. Pedersen at Du Pont de Nemours and the cryptants obtained in Strasbourg. But many problems arose: reaction times were far too slow, and above all there were difficulties in obtaining the two characteristics of enzymatic catalysis: finding the catalyst intact at the end of the reaction, and liberating the product in the medium. It was as if, in a vaudeville show, a magician causes to appear in an empty cage a bird that is to fly out over the audience and be caught at a desired spot: the bird is there (the guest molecules do enter the cavity), the door of the cage does open, but the bird cannot leave the cage.

The competition between chemical synthesis and enzymatic engineering will basically depend on cost levels: in some cases, as in the production of glutamic acid, fermentation has proved far less costly than the hydrolysis of the protein. In other cases, as for riboflavin, both methods compete. Genetic engineering may upset this balance. For many questions are still raised in enzymatic engineering, some of which will undoubtedly be solved thanks to recombinant DNA techniques. What, for instance, takes place whithin the cell during an enzymatic reaction? Into how many phases does a reaction break down? And, of course, how can we create new enzymes for new products (a question of vital interest to the enzyme industry).

The field of application of industrial enzymes is broad indeed. The tanning industry no longer stinks, for instance. Textiles now avoid damage or discoloration due to bathing in acid for several days by encouraging the activity of wild microbian flora in breaking down the starch used during weaving (to prevent the threads from breaking). The necessity of providing massive quantities of standardized products has modified the traditional fermentation industries. Enzymes (alpha- and beta-amylases) are added to beers in which raw grains replace some of the malt, while beta-glucanases are added to "pure malt" beers to prevent the formation of macromolecules that could plug up filters during clarification of the wort or fungic alpha-amylases in bread to free the fermentary sugars necessary for the yeast's activity. Amylases and pectinases are used to clarify fruit juices, lipases to enrich inexpensive oils. But above all, fixed-enzyme technology opens the way to a wide variety of industrial applications: the synthesis or modification of molecule (antibiotics, vitamins, hormones, ATP), steroid chemistry, flavors and aromas, perfumes and cosmetics, water purification, the fixation of nitrogen, quantitative analysis, etc.

In the early 1970s, what the "greedy" enzymes (as they have been called in France) were really eating were the profits of the companies that made them; they had become the victims of a consumers' movement led by Ralph Nader, who accused these same enzymes of causing allergies. It was true, at that time, that the free proteases caused allergies, not among the housewives who used them, but among the workers who labored in their clouds of dust, frequently ignoring all safety precautions (not wearing masks and gloves). In France, where enzymes had a bad name, businessmen quickly changed the names of their detergents, which became "biological." But this legerdemain came too late to save smaller firms from going under, including one of the pioneers, Rapidase, which was bought up by Gist-Brocades. Far surer processes have been developed (granulation), and sales of enzymes are again rising. But the storm cleared the decks. Henceforth, only two companies will be able to dominate the market: Novo and Gist-Brocades.

At present, 80% of the market for enzymes is held by only three companies: Novo in Denmark (which entered the business in 1940 with trypsin extracted from bovine pancreas, from which insulin had already been derived), Gist-Brocades in Holland, and Miles Laboratories (an American firm taken over by the German Bayer group). In certain specific sectors, 10-20% is held by Denmark's Christian Hansen (rennet) and by the Japanese pool of Amano, Nagase, Sankyo, and Oriental Yeast. We might also mention Boehringer Ingelheim in Germany and Sturge Enzyme in Great Britain. To say nothing of the research companies, Genex and Genencor. Both in the number of patents applied for (a systematic national strategy) and in industrial applications, Japan seems to dominate the enzymatic-engineering field (closely followed by the United States) – except that Novo can claim 60% of the Japanese market and has the only glucoisomerase system of real importance on a world scale.

1. *Miles Laboratories*
About 1912, Dr. Miles (third from the left), founder of Miles Laboratories – since bought up by Bayer – posed for this souvenir photograph. One of his greatest successes was the production of Alka-Seltzer. Between 1902 and 1942, 14 million of Dr. Mile's calendars were sent to drugstores everywhere.
2. *Immobilization*
In Japan, where the technique of immobilized enzymes was invented, they are used in many production units, like this one at Tokyo Jozo.
3. *Production*
Novo, the world leader in the production of enzymes, began to produce enzymes in about 1940 and achieved its top ranking in the mid-1960s. These enzymes are used in the detergent, leather, textile, starch, and food-processing industries.

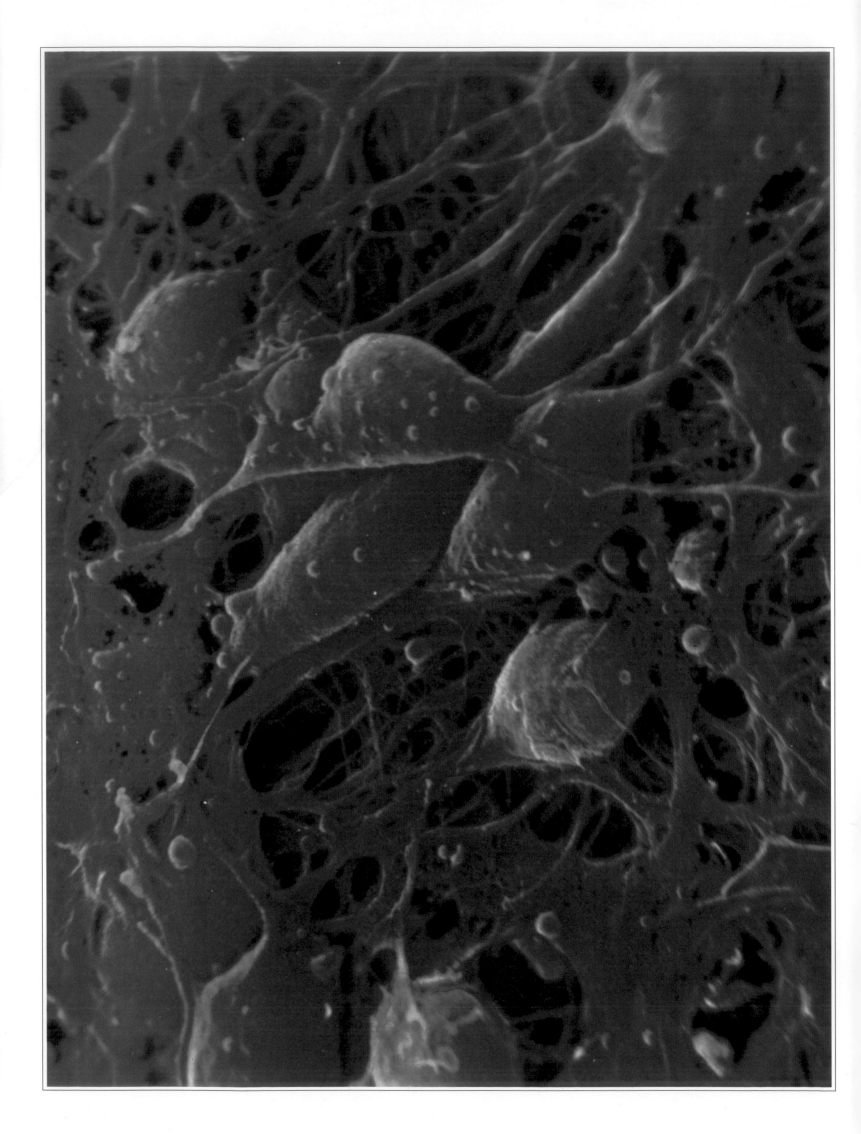

3

The Cell:
A Beehive of Workers

"Creation must be considered in its entirety, nature as an accessory fact," wrote Theodor Schwann on 5 April 1881, a few months before his death. According to his biographer, Marcel Florkin, this was an indication that "the scientist had surrendered." With the botanist Matthias Jacob Schleiden, the twenty-nine-year-old Schwann had promulgated the theory of cells and then for the next forty years maintained a complete silence . Why?

Theodor Schwann was the offspring of a family with a strong religious tradition. Destined for the priesthood, he was educated at the Tricoronatum of Cologne, one of the most brilliant of Jesuit schools and a bastion of the Counter-Reformation. He chose to concentrate on science, and later at the University of Bonn attended the classes of Johannes Müller, one of the most prestigious physiologists of his day. One evening in 1836 he dined with Schleiden, also a student of Müller, who told him his ideas concerning the role of the nucleus in the development of the cell.[1] This led Schwann to enunciate his cellular theory in 1839: "Cellular origin is common to all living things.... The individual life of the cells has its source in the forces inherent in each molecule," he wrote in a letter to Du Bois-Raymond. As only rarely happens in the world of science, the idea was immediately accepted, despite some controversy (such as the conflict between Schleiden and the German physician and psychologist G.T. Fechner, who wanted to demonstrate scientifically the identity of spirit and matter). In 1856 Fechner published a humorous tract, "Professor Schleiden and the Moon," to illustrate his viewpoint on this argument over "the soul of plants."

Schwann left for Louvain the same year he announced his cellular theory, and nine years later left for Liège, where he taught until his death. This was undoubtedly the starting date of his silence and his prolonged religious crisis. For this theory once and for all put an end to the vitalistic doctrine (which held that the origin of life was some mysterious "breath of life"), and Schwann was thoroughly upset by this: if life could be explained mechanically, where did that put God, or free will, or what Schwann called "the irrational"? Prey to the same conflicting feelings that made Albert Einstein, faced with the consequences of his own theories, cry out, "God doesn't play dice with the universe!" Schwann was torn between his love of science and his love of Christ. In an unpublished autobiography brought to light by M. Florkin, Schwann offers a curious explanation of himself and his scientific career according to which the latter was an ordeal imposed by God: "From birth I have had a tendency to self-examination.... It is the way my mind works at any given moment: I reflect on what I am doing at the very moment I am doing it.... I consider this tendency to be the result of a congenital weakness in the upper portion of my spinal cord whose effect on the brain is such as to control the blood vessels of the brain. This thus produces a continuous cephalic congestion and hyperexcitation.... Proof that the illness is congenital I find in the fact that, even as a child, I often sighed.... I think that in my case, this is the form assumed by original sin, while for others it may be something else.... One wants to be virtuous according to one's own strength. This tendency of mine... clearly shows the way I thought at that time: my virtue was a pagan virtue very close to stoicism. To cure me of this, God abandoned me to my own devices" (M. Florkin, "Theodor Schwann and the Beginnings of Scientific Medicine," Palais de la Découverte, 4, February 1956, pp. 7-9). As we have seen, during the last year of his life Schwann bowed down to the mystery of creation.

The story of Schwann is typical of something that has occurred repeatedly during the long history of biology: an attempt to reconcile two irreducibly opposed languages, that of science and that of philosophical or religious thought. When Schwann wrote, "The individual life of the cells has its sources in the forces inherent in each molecule," he felt himself, because he put the two languages on the same level, in total contradiction with his beliefs,

Neurons
Cells can assume different shapes (round, rectangular, compact, elongated, etc.). The nerve cell (neuron) has a long thin extension at one end (the axon) and other ramified extensions at the other (dendrites). A human's brain contains tens of billions of neurons.

(1) The existence of a nucleus in the cell, noted as a curiosity by Leeuwenhoek in 1702 in the red blood corpuscles of fish, was confirmed in 1832 by Robert Brown's observations on plants.

A. van Leeuwenhoek
This Dutch naturalist, born in Delft in 1632, was one of the first explorers of "The Arcana of Nature," the title he gave a book he wrote between 1715 and 1722 and from which this portrait has been taken.

(2) It would seem that this term was employed for the first time some ten years earlier by the Czech physiologist J.E. Purkinje.

which he expressed in the following terms: "If, all of a sudden, God endowed bees with the principle of free thought, they might quite rationally think: my ancestor built hexagonal cells, but I can mathematically prove that a cylindrical cell is better and has a greater capacity for a smaller surface. If the bee did in fact follow this line of reasoning, if it placed its limited intelligence above its instincts, we can readily see that it would be end of the species" (ibid., p. 17). It was from this same viewpoint, but reversed, that James Watson stated that he had rewritten the Bible.

Our knowledge of cells and their various components has grown enormously since Schwann's day. Others means of observation, the interchange of ideas between scientists in many disciplines – biopaleontologists, biochemists, geneticists, embryologists, etc. – as well as incredible gains in technology have made it possible to explore ever more thoroughly the infinite complexities of this universe. And first of all, how and when did the cell make its appearance on this earth?

"Seeing" a Cell

It was well before the beginning of the Cambrian era, in a period known as the Phanerozoic from the Greek for "when life appeared," that the very first cells made their appearance, small, spherical, and anaerobic, as there was not as yet the least trace of oxygen. They were heterotrophic bacteria, in that they fed on organic substances. They were followed by photosynthetic bacteria, still anaerobic, but endowed with a new process for fixing nitrogen (an element present because the ultraviolet radiations, which were not yet filtered by a layer of ozone, broke down the ammonia). With the advent of blue algae over two billion years ago, oxygen began to be produced by photosynthesis, but it took several hundred million years for it to accumulate in the atmosphere and for cells to be able to use it to breathe. And it was nearly 1,450 million years ago that the first cells without nuclei developed. Multicellular creatures first appeared at the beginning of the Cambrian era. Thus bacteria and blue algae were "prenuclear" organisms (procaryotes), with but a single chromosome, while vegetables and higher animals, as well as fungi, molds, and yeast have a "good nucleus" (eucaryotes) and a minimum of two chromosomes. The presence of DNA in the components of the cell, such as the mitochondria or the chloroplasts in the eucaryotes, seem to support the hypothesis that these are former procaryotes, once free, absorbed by the cell and living in symbiosis with it.

Cells draw their energy from foods based on glucose. The metabolic system of the eucaryotes is based on respiration – combustion of the glucose by oxygen – part of the freed energy being captured by phosphate links in the form of adenosine triphosphate (ATP) molecules, the fuel of the organism, with the production of carbon dioxide and water. It is thus, in the case of plants, that light energy must be converted into chemical energy for the plants to make use of it: the ATP stores solar energy. Animal cells, which have no chlorophyll, manufacture ATP from the glucids in the plants they consume. These glucids, burned by the oxygen in the "combustion chamber" of the mitochondria, produce ATP. It is the ATP that provides the cell with not only the energy to build the links of the molecular chains, but also chemical transfers between cells, muscular activity, etc.

While certain procaryotes also breathe, many of them derive their energy from fermentation.

We know much about the prehistory of cells from the recent work of paleobiologists working with fossils. But the study of live microscopic creatures was only made possible with the invention of the optical microscope in 1590 by a theatrical producer, Zacharias Hansen, and above all by the work of the most celebrated microbiologists of the seventeenth century, the Englishman Robert Hooke, the Dutchmen Anton van Leeuwenhoek and Jan Swammerdam, and the Italian Marcello Malpighi. In 1663, the British astronomer and mathematician Robert Hooke became the first to describe what he called cells ("little rooms") found with his microscope in the stems of plants and of cork. Twenty years later, van Leeuwenhoek in his turn described living cells (bacteria, red blood corpuscles, spermatazoa, protozoa, etc.) before the Royal Society. Thanks to his "microscope" (enlarging lenses set between silver strips), he succeeded in getting an enlarging power of 300, the minimum required to observe individual cells. Along with Malpighi, the great burgher from Delft is considered one of the founders of histology (the study of tissues). Over a century was to elapse before the appearance of cytology (the study of cells) and the statement of the first cellular theory (1838-1839). Meanwhile, there had been great improvements in the microscope, particularly with the introduction of achromatic lenses and as a result of the hypothesis put forward by F.V. Raspail and H. Dutrochet. With the theory advanced by Theodor Schwann and Jacob Schleiden, the cell became the elementary unit of life. But at that time it was still only considered a bladder filled with a liquid, the nucleus being but one component among many others. It was not until the 1930s and the development of the electron microscope that viruses and proteins could be distinguished.

In 1854, the German biologist Max Schultze for the first time defined the cell as "a small mass of protoplasm (2) enclosing a nucleus" and established structural and functional analogies between the cells of plants and animals. A year later, in his book *Die Cellularpathologie,* Rudolf Virchow, the greatest of all nineteenth-century German physicians, wrote the celebrated phrase *"Omnis cellula e cellula"* (All cells come from cells). Virchow was still another of Müller's students. To him, life was the sum of physical and chemical activities and essentially the expression of the cells' activity. In 1875, cellular division was demonstrated experimentally by the German botanist E. Strasburger, who observed mitosis and the fragmentation of the nucleus of plant cells under the microscope, and subsequently by the German zoologist W. Flemming, who has observed the same thing in amphibians.

Let us imagine a country consisting of billions and billions of cities. Each of these cities has its own communications network, a control center connected to a data bank, power stations using clean fuels, assembly units, storage areas, and a multitude of control stations. Science fiction from H.G. Wells or George Orwell? Not at all. This country is a

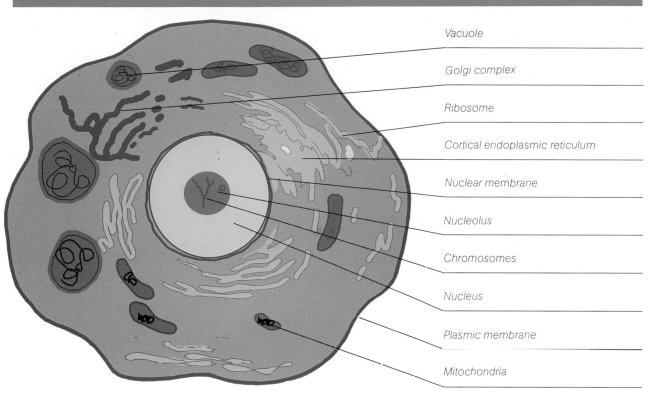

Vacuole

Golgi complex

Ribosome

Cortical endoplasmic reticulum

Nuclear membrane

Nucleolus

Chromosomes

Nucleus

Plasmic membrane

Mitochondria

The cells of men and the higher animals are isolated from each other by a *membrane*, which plays an active role in their nourishment and in their communications between themselves and with the outside world. Each cell consists of a command center connected with the data bank of chromosomes (the *nucleus*); an internal communications system (the *endoplasmic reticulum*); power stations (*mitochondria*, and in the case of plants *chloroplasts*), where ATP is made; storage areas (*vacuoles*); control stations (*centrioles*); chemical processing centers (*lysosomes* and *Golgi bodies*); protein assembly units (*ribosomes*); and *centrosomes*, which play a part in division or mitosis, making sure there is the same number of chromosomes in the cells. (Another method of cellular multiplications is meiosis, or the fusion of parental gametes.)

This whole molecular society communicates at a distance through the nervous system and the hormonal system: signals are transmitted and trigger a specific response in an organ whose cell has the appropriate receptor.

In contrast to the eucaryote cell, shown previously, the cell of a bacterium does not contain a nucleus: it is a procaryote organism. Other differences: the fine cellular membrane is enclosed in a far stronger shell, which protects the cytoplasm from exterior aggression. In addition to extrachromosomic DNA, the bacterial cell also contains the *plasmid*, which has proved to be a choice vector for genetic engineers. In many bacteria, a mobile filament, the *flagellum*, serves as a an organ of locomotion; that is to say, it helps the bacterium move in a liquid environment.

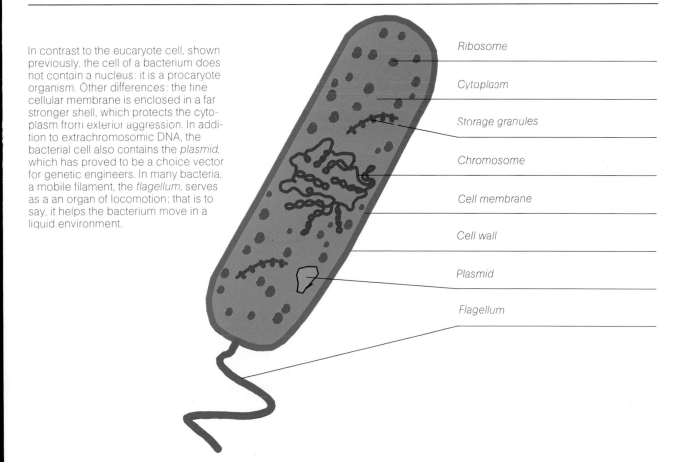

Ribosome

Cytoplasm

Storage granules

Chromosome

Cell membrane

Cell wall

Plasmid

Flagellum

(3) Amino acids are designated by the first three letters of their names, a list of which follows: phenylalanine, leucine, isoleucine, methionine, valine, serine, proline, threonine, alanine, tyrosine, histidine, glutamine, asparagine, lysine, aspartic acid, glutamic acid, cysteine, tryptophane, arginine, and glycine.

(4) At the turn of the century, A. Garrod demonstrated that a urinary infection was caused by the absence of an enzyme.

man with his 60,000 billions cells. The cells of which he is made are like all living creatures – with one exception: the virus, which acts as a parasite on a cell, transforming it into a zombie and forcing it to obey its instructions.

The cell is made up of nucleic acids, sugar (polysaccharides), fatty acids (lipids), water, salts, and metals, but above all of proteins, which account for over 50% of its dry weight. The Dutch chemist Gerardus Johannis Muller discovered proteins in the 1830s and gave them that name, thinking they were the basic (*protos,* first) elements of live metabolism, and for many years they were considered the chemical basis of heredity. They are in fact macromolecules with a high molecular weight of 100,000 (the unit being the weight of the hydrogen atom), and are made up of twenty amino acids.(3) As each chain of amino acids forms a peptide, the protein is a long, complex chain of polypeptides extending three-dimensionally in space. The first synthesis of the dipeptide was achieved by Emil Fischer and Ernest Fourneau toward the end of the nineteenth century. But it was only between World War I and World War II that any research was devoted to the structure of proteins, when Thomas Astbury established in Great Britain a department that he wished to call "molecular biology," a discipline that, according to him, was more an approach than a technique.

In 1934 at Cambridge University, J.D. Bernal and Dorothy Hodgkin began the x-ray examination of crystalline proteins. In 1936, Max Perutz worked with Bernal on the structure of crystalline hemoglobin and then with Sir John Kendrew on myoglobin. The 1958 Nobel Prize in Chemistry was awarded to Frederick Sanger, the Cambridge University biochemist who had succeeded in establishing the insulin sequence, and in 1962 it was given to Perutz and Kendrew for having deciphered the three-dimensional structure of hemoglobin.

The Medical Research Council's White Building in Cambridge, England, represents the same thing that the Cavendish Laboratory did for electronic and nuclear physics at the end of the nineteenth century and the beginning of the twentieth century: an exceptional mulch in which the genius of a Bernal, Perutz, Kendrew, Watson, or Crick could flourish. Eight Nobel Prizes have been awarded for work done in this Laboratory. On the other side of the Atlantic, still another center of intellectual ferment, and one already made famous by such physicists as R. Millikan, was to become the rival of the charming little university town in the English countryside: the California Institute of Technology, or Caltech, in Pasadena. It was there that in 1950 Linus Pauling and Robert Corey determined the three-dimensional structure of the alpha helix, the essential structural components in synthetic polypeptides and proteins: all proteins have a spiral structure due to the formation of hydrogen bridges between the various groups of peptide bonds, with precise characteristics. This discovery, for which Pauling was awarded the Nobel Prize in 1954, paradoxically led him into error and made him lose his lead in the race to elucidate the DNA structure to Crick and Watson.

While chemical research made it possible to determine the primary structure of proteins (the peptide sequences) and research in physics the alpha helix, it was the crystallographic research of Perutz and Kendrew that showed that, in addition to its helical structure, each protein has its own spatial configuration due to the chains shrinking in on themselves as the result of the attraction between the various residues of amino acids, with hydrogen, electrostatic, hydrophobic, or covalent links between these residues. Furthermore, in many proteins, a still more elaborate level of organization associates several chains of polypeptides. It is the links between the chains in the latter two cases (shrinkage in space and the association of several chains of polypeptides) that make proteins unstable, for they can be relatively easily broken down by physical or chemical agents.

All of these observations are important where genetic manipulation is concerned. As we shall see, it is important to obtain the stablest proteins possible. Furthermore, over 100 known hereditary diseases are due to a protein deficiency:(4) Parkinson's disease and certain cases of mental retardation are due to a lack of enzymes.

Anemia, for its part, may be caused by the accidental substitution of one amino acid by another in hemoglobin, the chains of proteins charged with carrying oxygen in the blood. This first molecular explanation in pathology was made in 1949 by L. Pauling, who observed the difference in charge between mutant hemoglobin and normal hemoglobin, and W. Ingram subsequently (1956-1958) demonstrated that the mutation had led to the change of a single amino acid. It thus became evident that the examination and mastery of proteins had unlimited potential in therapy and preventive medicine.

The cultivation of mammal cells has become one of the interesting techniques of biotechnology. The recombination of genes does not make it possible to produce such a large number of interesting proteins, which are frequently giant molecules that the microorganism finds it impossible to secrete or excrete. As for the glycoproteins (which are long chains of sugars), bacteria do not have the enzymes necessary for the construction of such chains, in contrast to yeasts, which, for their part, have the disadvantage of being difficult to discipline and make regrettable errors in construction.

While the cultivation of bacteria, yeasts, and molds – small independent units sheltered by strong cell walls – do not raise any special problems, the same is not true for the cultivation of mammalian cells: their cell walls are fragile, their nutritional requirements are far more stringent (carbon, vitamins, mineral salts, small quantities of antibiotics to prevent contamination and blood plasma), their vulnerability to temperature variations enormous (above 37°C they stop dividing and die), and above all they are not used to reproducing in an isolated medium. Despite all of these difficulties, Wellcome in Great Britain has since the 1960s used kidney cells from baby hamsters to produce a vaccine against hoof and mouth disease. Even ten years prior to this, in the course of the research that led to the development by Dr. Jonas Salk of a polio vaccine in 1954, it became apparent that the polio virus could develop in the cells of men or monkeys and that such cell cultures would be of considerable industrial importance.

Wellcome, Genentech, Damon, Searle, Invitro (of the Monsanto group), Celltech, and Hayashi-

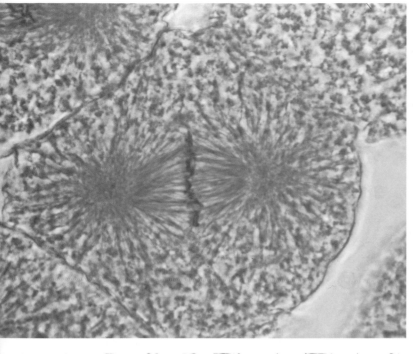

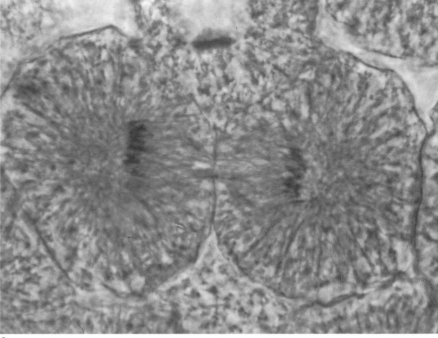

2

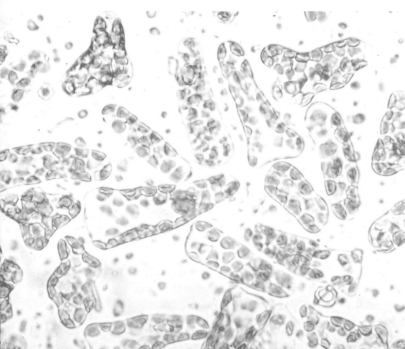

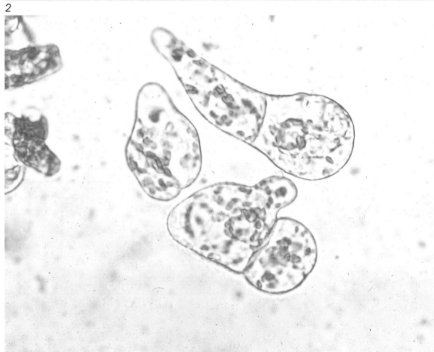

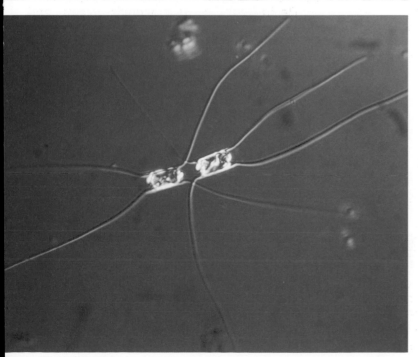

4

1, 2. *Mitosis*
An animal cell dividing:
second phase (1), then third
phase (2), known as
"metaphase" and
"anaphase."
3, 4. *Plant cells*
Cells of Ipomoea
heredipolia *separated (3) by*
grinding up the leaves,
during the initial stage of
division, or prophase (4).
5. *Microalgae*
A marine diatom, a
unicellular brown alga that
forms a part of the vegetable
plankton. Its cell membrane
is enclosed in a silicon shell.

39

bara are working today on mastering this highly promising technique. With the development of bioreactors, researchers were able to obtain extremely dense cell cultures (with impressive results using fixed cells).

The Enigmas

The cell still poses one great question: that of cellular differentiation. How are the many cell families (we customarily distinguish 200 of them), which are highly specialized, formed? Why are muscle cells the only ones to produce the myosin that causes the muscle to contract? Why do the cells of red blood corpuscles alone produce the hemoglobin that makes it possible to carry oxygen to every part of the body? Why do the cells of the skin alone manufacture the keratin that gives it its strength? Does not the DNA in each cell contain coding genes for myosin, hemoglobin, keratin, and all the other proteins? Yes. Then why are they not expressed everywhere?

In the process of division of, say, the fertilized egg in a mouse's uterus, cells divide into two, then four, eight, sixteen, etc. All of these cells are visibly separate from one another, and they show no differentiation. After a certain period of division, however, they suddenly start to adhere strongly to one another. If, as can now be done, one succeeds in preventing this adhesion, no differentiation, no embryo, is formed. If, on the other hand, after forcing them to separate, one rejoins them in a compact mass and reimplants this cellular mass in a mouse's uterus, a baby mouse will be born. The phenomenon is thus reversible. But what we do not know right now is what happens in each cell. We only know that ten days after fecundation, the human embryo is a hollow ball with three kinds of cells distributed in layers of inner, middle, and outer tissues. At this stage the little human being could just a easily become a starfish (if it is not the embryonic part of the placenta being viewed in the microscope) or, a few days later, a chicken. The genes then manufacture the proteins appropriate to each of the three tissues and the cells begin to specialize: in the outer layer, for example, into nerve cells, epithelial cells, etc. Then, for example, the nerve cells begin to specialize into brain, spinal cord, eye, etc., building what will subsequently become a human being. Why do they stop when an organ is complete? A mystery. Why do bone cells, for instance, make longer or shorter bones depending on whether they belong to the finger or the leg? Again a mystery. How do cells "know" that they must start dividing again to reconstruct a damaged organ or tissue (the liver, for instance) or an amputated limb (the lizard's tail or the salamander's foot)? A mystery still once again. "For the moment," says François Jacob, "we are trying to create the tools that will enable us, say, in 180 years or so, to take on the problem, and then to solve it in 200 or 300 years. The three great problems facing experimental biology are heredity, the embryo, and the brain. In respect to genetics, we have a few ideas; for the others, virtually nothing. Understanding how molecules work does not tell us the algorithm of the brain; understanding how cells interact at a certain stage does not explain the development of the embryo."

In addition to solving problems concerning pregnangy, birth control, and prevention in utero, research in embryogenesis could shed light on the mechanics of cancer. The body of the human adult is made up of three types of tissues: those in which the cells never divide (like those in the brain), those in which the cells may divide in exceptional circumstances (regeneration), and those in which the cells are constantly renewed (external – skin – or internal – lungs, intestines – envelopes, the blood, and the mammary glands). And 95% of cancers are found in the last tissue type and in multiplying cells that have not yet differentiated. An anomaly occurs. The cell no longer responds to certain regulatory mechanisms. It does not complete its differentiation and subdivides endlessly. It has become cancerous.

In the case of bacteria, the regulatory mechanism of the gene was brought to light at the end of the 1950s through the famous Pajamo (nicknamed "pyjama") experiment, so named after the three researchers involved: Arthur Pardee, François Jacob, and Jacques Monod. The experiment's results were as follows: If cells are placed in suspension in a mixture of glucose and galactose, the cells divide and their numbers increase until the glucose is exhausted. Growth stops, then resumes again in twenty minutes until the galactose is exhausted. It was inferred that these twenty minutes have thus allowed the cells to adapt to galactose, and that the lack of glucose has thus removed an inhibition that repressed the gene coding for enzymes capable of metabolizing galactose. In 1960, the transfer of a gene of galactosidase from a male to a female bacterium established this inference by proving the existence of regulating and repressing genes that act as switches on other genes.

The phenomenon is unfortunately far more complex in the case of eucaryote organisms, and we do not have any general theory that, for instance, would make it possible to induce a cancerous cell to become normal again by differentiating and ceasing to divide.

We scarcely know much more about the aging and death of cells; their declining ability to reproduce, regenerate, or grow; the slowdown in the synthesis of proteins; the fragmentation of the Golgi apparatus; or the lengthening and breaking of mitrochondria with age. But henceforth these questions will be approached on the molecular level.

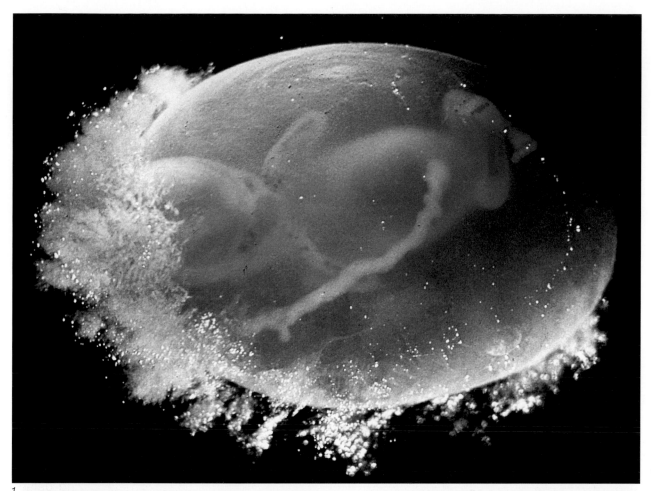

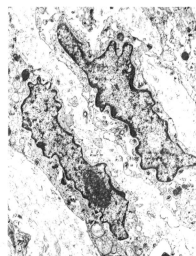

1. *Human embryo*
An eight-week-old embryo.
2. *Nucleus of a cancerous cell*
In a normal eucaryote cell, the nucleus is round or oval. In a cell from a tumor, it is broken into two parts and appears abnormally lengthened.
3. *Cell culture*
Human cells kept "alive" in a nutrient solution. Large-scale cell cultivation like that being carried out at Hoffmann-La Roche could lead to the production of interferon. The water around the jar is kept at body temperature, and the balls prevent evaporation.

41

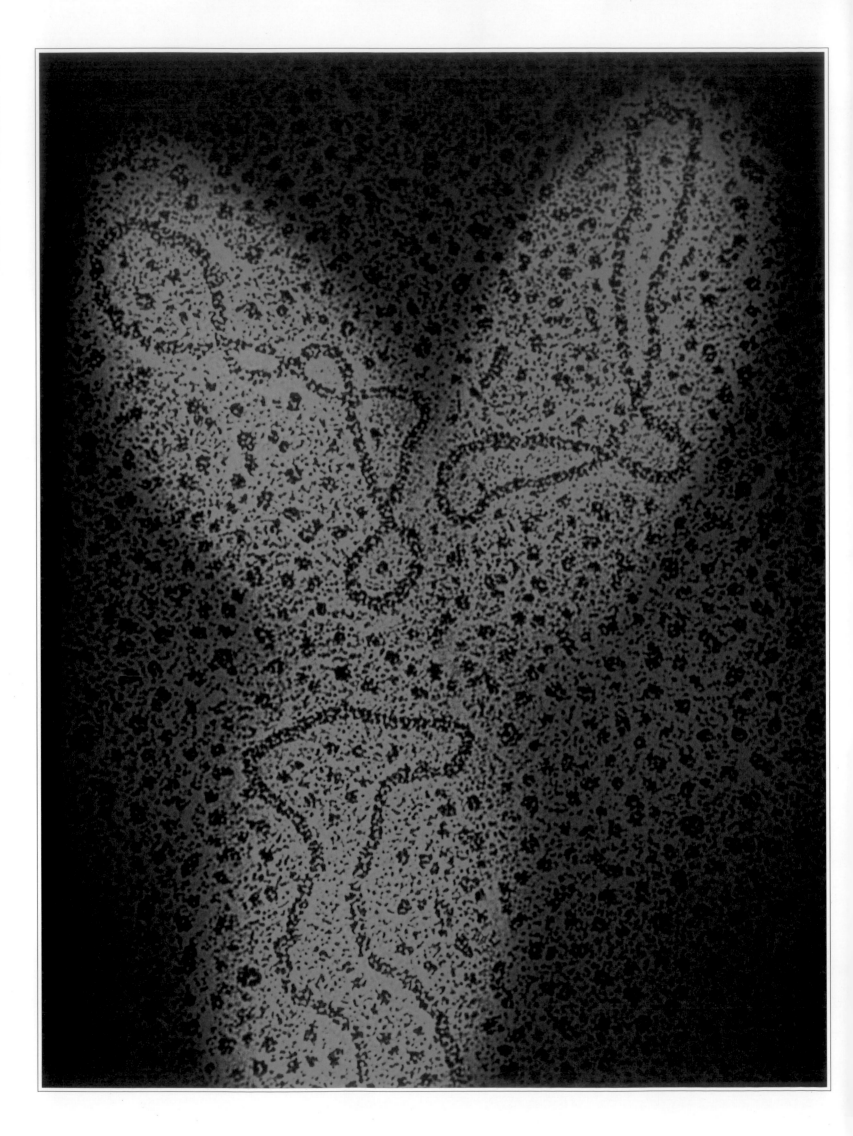

4

Four Letters to Write Life

1953 marked a turning point in molecular biology: Francis Crick, James Watson, and Maurice Wilkins established the structure of DNA (deoxyribonucleic acid) and the four-letter alphabet of our genetic heritage (A-T, C-G; see the caption for the figure on p. 48).

Almost a century had elapsed since the discovery in 1869 by the German Friedrich Miescher of a substance produced by the nucleus that did not correspond to any other known category of protein. He called it "nuclein." Miescher quickly concluded that he was dealing with a reservoir of phosphorus for the cell, but did not question the concept generally accepted at that time that proteins were responsible for the transmission of hereditary traits. Yet three years earlier the German naturalist Ernst Haeckel had suggested that the nucleus might well contain the factors governing heredity. But Haeckel was a staunch believer in spontaneous generation.

At the same time, in 1866 and 1869, the monk of Brno, Gregor Mendel, published his two treatises setting forth the transmission of hereditary characteristics and established the first laws of heredity – the basis of modern genetics, though they may have been partially invalidated by subsequent discoveries. Working with generations of garden peas smooth or wrinkled, green or yellow, and crossbreeding them, Mendel determined the uniformity of the first generation of hybrids, or the law of the purity of the gametes (for example, the "children" of yellow and green peas are all yellow); the heterogeneity of second-generation hybrids in which the parental characteristics reappear – that it to say, the law of segregation of characteristics, transmitted indirectly to descendants; and the principle of the dominance of one of two characteristics. The discovery of the recombining of chromosomes in part invalidated the second law. In 1887, the Belgian Edouard van Beneden discovered that the reproductive cells (gametes) contained half as many chromosomes as the eggs resulting from their fusion, which was a start in explaining Mendel's laws. But Mendel's laws were totally ignored at that time, as they did not coincide with contemporary ideas; they had to be rediscovered, about 1900, by three botanists, the Dutchman Hugo de Vries, the German Carl Correns, and the Austrian Erich Tschernak.

Another German, Albrecht Kossel, continued Miescher's work on the chemical composition of nuclein. Kossel, who was studying the cells of yeasts, in 1880 isolated xanthine (a derivative of uric acid discovered four years earlier), about which it was already known (from Adolf Strecker's work in 1858) that it is a product derived from *guanine* after treatment with an acid. He had previously discovered traces of hypoxanthine in the nucleus (derived from *adenine* after treatment with acid) and, in 1887, *adenine*. In 1893, with A. Neumann he isolated *thymine*. And finally he discovered *cytosine* and *uracil*.

Kossel subsequently proposed the tetranucleotide hypothesis (that there were four bases present in nucleic acids, in equal molecular proportions). But he still thought that proteins were the most important constituents of the nucleus. This considerably delayed matters for it encouraged scientists to concentrate on the products of the breakdown of DNA, rather than to focus on its intact structure. Phoebus Levene, a student of Kossel, established that DNA had a high molecular weight.

The Dutch botanist Hugo de Vries, who had discovered the importance of mutations, or sudden hereditary changes, put forth the idea of communication between the nucleus and the cytoplasm (confirmed seventy years later by the discovery of messenger RNA) and postulated that only part of the cell's information was used at one time – which was definitively proved fifty years later.

In 1905, the English biologist William Bateson introduced the term "genetics" to describe the study of heredity. But at that time, the word "gene," which had ben introduced in 1860 by Wilhelm Johannsen, referred to an abstract concept without

Plasmids

Plasmids (ring-shaped strands) are among the favorite vectors used by genetic engineers in recombining operations. It is one of these rings that is opened, using a restriction enzyme, so that a foreign gene can be inserted into the genetic heritage of a microorganism and make it produce, for instance, a substance that it would not secrete naturally.

43

any close connection with the chromosomes (which had been so named by Waldeyer in 1888). In 1920, Thomas Hunt Morgan showed that the genes were carried by the chromosomes located in the nucleus of the cell. Ten years later, George Beadle and Edward Tatum, continuing the work done in 1909 by Archibald Garrod on congenital errors of metabolism, advanced the hypothesis that the genes regulated and controlled the enzymes and made them specific. This was the famous formula: One gene, one enzyme – which was to become: One gene, one polypeptide. Beadle and Tatum's experiments on mold had been made possible by the artificial mutations induced in DNA in 1927 by the American biologist Hermann Muller by exposing living organisms to x rays.

Over a half-century, the sizes (as well as the time needed for observation) of the organism being studied were greatly reduced: from Mendel's garden peas, to the drosophila flies used by Morgan and his students C.B. Bridges, H.J. Muller, and A.H. Sturtevant, to the genes distributed on chromosomes, to fungi (neurospora), to bacteria. But the same question always arose: What chemical molecule is responsible for the organism's behavior?

On the Track of DNA

The beginnings of an answer were provided by O.T. Avery at the Rockefeller Institute of New York. A serious illness was prevalent in the United States at that time: pneumonia. It was already known that there were several types of pneumonococci, some of them virulent and others not, and that dead virulent pneumonococci could communicate their virulence to inoffensive strains, which remained pathogenic and transmitted this characteristic to their descendants. O.T. Avery thought that the virulence lay in the cell itself. He isolated the molecules liberated by the dead bacteria and added to them an enzyme that destroys deoxyribonucleic acid (DNA). The mixture can then no longer transmit the virulence. In 1944, along with C. MacLeod and M. McCarty, he published an article in the *Journal of Experimental Medicine,* in which using chemical, enzymological, and serological analyses in conjunction with electrophoresis, ultracentrifugation, and ultraviolet spectroscopy, it was concluded that "the active fraction consists essentially – if not exclusively – of a viscous, highly polymerized form of deoxyribonucleic acid." For the first time, scientists had suggested that there existed a medium for heredity that consisted not of proteins but of DNA

Human chromosomes
Human beings have twenty-three pairs of chromosomes containing all of the genetic information coded by the DNA. Etymologically, their name means "colored bodies," for they absorb coloring agents selectively thanks to the presence of chromatin. They have been made visible here by fluorescence.

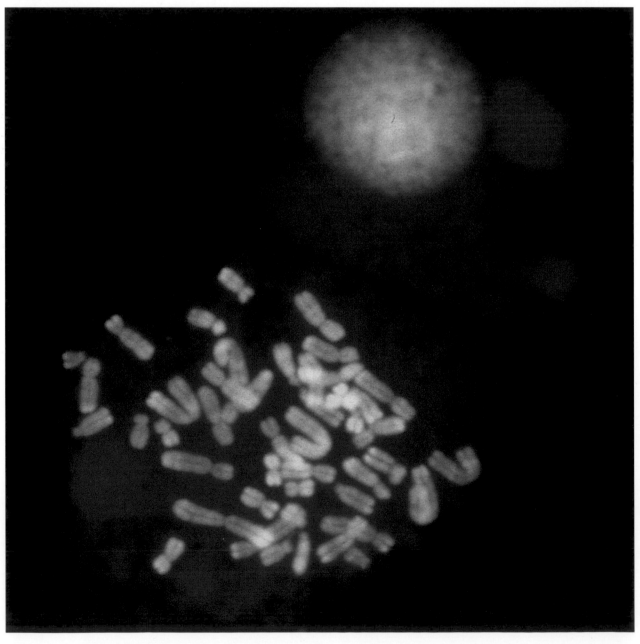

and that this lay behind genetic phenomena. Adherents of the protein hypothesis protested for a long time after this that a fraction of active proteins resistant to the proteases remained in the mixture being studied or that the experiments were limited to bacteria. Despite the immediate enthusiasm of such scientists as Sir MacFarland Burnet, G.W. Beadle, and André Lwoff, Avery was condemned to what René Dubos called "scientific apartheid." And Dubos added, "Several schools of biologists, inspired by physicists who had moved into biology, made it fashionable to think about biological problems in terms of theoretical constructs, rather than of anatomical structures, physiological processes, and behavioral patterns; some biologists talked as if they were more concerned with cosmic riddles than with living organisms. In contrast, Avery questioned the validity of biological generalizations and was even reluctant to use the word gene. He was virtually ignored by the theoreticians of genetics, precisely because he made no effort to communicate with them or, more exactly, to communicate to them what he had discovered by working at the bench instead of speculating about the secret of life" (René Dubos, *The Professor, the Institute and DNA,* The Rockefeller University Press, 1976, p. 155).

The biologists "inspired by physicists" that Dubos was talking about were essentially the phage group that had grown up in the United States in 1940 around Max Delbrück, Salvador Luria, and Alfred Hershey, who shared the Nobel Prize in Medicine in 1969 for "discoveries concerning the mechanisms of reproduction and the genetic structure of viruses." These phages – or bacteriophages – which gave their name to the group, are viruses that infect bacteria and were discovered in 1915. In 1917, F. Twort and F. d'Herolle showed that they would be capable of fighting the bacteria responsible for typhoid and dysentery. But above all they were to prove a fantastic research instrument and a vector in splicing genes, by the 1970s one of the tools of the genetic engineers.

Max Delbrück, a physicist and a student of Niels Bohr, was born in Berlin in 1906. In 1935, in a article he published with two biologists, Timoleef-Rossowsky and K.G. Zimer, on the mutation and the structure of the gene, he used the quantitative mechanical model of the physicists. It was one of the first interdisciplinary studies of its kind.[1] Delbrück emigrated to the United States in 1937, where he was joined in 1940 by the medical doctor Salvador Luria, who had been born in Turin in 1912. Between 1940 and 1945, Delbrück and Luria determined the general way in which phages multiply (duration of the infection process, number of bacteriophage descendants produced by the infected bacteria, the phases of the infection, etc.). They began to explore the interaction that occurred when viral particles of different species infect the same bacteria. Their work drew the attention of the American biologist Alfred Hershey, who had for many years studied phages as part of his research on immunology. So the phase group was informally established at the American molecular biology laboratory at Cold Spring Harbor. By that time, work on viruses was relatively well advanced.

To try to understand the multiplication of viral particles in the cell of a bacterium, an attempt was

1

2

3

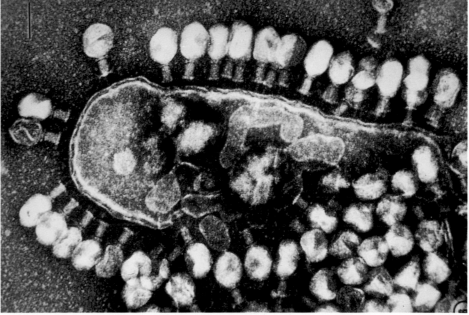

4

1. *O. Avery*
O. Avery's team in about 1932. Left to right: (seated) T. Francis, O. Avery, W. Goebel; (standing) E. Terrell, K. Goodner, R. Dubos, F. Babers.
2. *Founding Fathers of the Brotherhood of the Phage*
Max Delbrück (left) and Salvador Luria, at Cold Spring Harbor.
3. *The Brotherhood of the Phage*
Max Delbrück (standing) and the happy Brotherhood of the Phage.
4. *The phages*
Hue and cry of an E. coli bacterium attacked by a commando unit of bacteriophage T2 viruses; the viruses have just glued themselves to the bacterium and injected their own genetic material into it. The bacterium bursts, releasing a whole clan of viruses that have multiplied within it.

(1) We should mention Mendel's use of statistical observations based on a theoretical mathematical model, traceable to his training as a physicist; he was a physics teacher at the University of Vienna and had as professor Christian Doppler, discoverer of the acoustical effect bearing his name, now employed in, among other things, radar technology.

(2) It was toward the end of the 1970s
that D. Olins (United States), R. Kornberg (Great Britain), and Pierre Cam-
bon (France) observed this winding of
the DNA around globules formed of
special proteins, the histones.

made to identify the genetic material of the virus, and in 1947 Seymour Cohen isolated the DNA of phages T2 and T4. In 1952, the famous Hershey-Chase experiment convinced the group that the genetic component of the virus was DNA and nothing else. By using radioactive tracers in the protein or in the nucleic acid, they showed that only the nucleic acid of the virus penetrates the bacterium and that this penetration is enough to achieve complete reproduction of the phage. The experiment demonstrated that the genetic material of the virus was nucleic acid.

The tetranucleotide hypothesis took some time to die. It was disproved by the work of Erwin Chargaff at Columbia University betwen 1948 and 1952 and by that of his student Ernst Fischer and R. Hotchkiss at the Rockefeller Institute: the relationship of purines and of pyrimidines varies, but the total of purines is equal to the total of the pyrimidines; the quantity of adenine is equal to that of thymine, and the quantity of guanine is equal to that of cytosine. It was this that led Watson and Crick to proclaim their principle of "complementarity" and the pairing off of the bases. Thus while the fact that the bases were grouped in pairs was announced for the first time in 1950, no one could explain this phenomenon, nor did they have the least notion of the spatial structure of DNA. At Caltech, T. Astbury, working with Florence Bell, was able to take the first photograph of the diffraction of DNA by x rays; he obtained a molecular weight of from 500,000 to 1 million (roughly the same as that obtained by Signer, Caspersson, and Hammarsten a few years previously).

The Discovery

That was the situation when two teams entered the race to determine the structure of DNA: one led by Linus Pauling at Caltech in the United States and another consisting of Watson, Crick, Wilkins, and Rosalind Franklin in England. Linus Pauling was then at the summit of his glory. Furthermore, he had just discovered the alpha helix. In Cambridge, England, on the other hand, the adventure of the discovery of the "double helix" was recounted by James Dewey Watson in a brilliant and controversial book, *Double Helix*, published in 1968: "When I wrote the first version of my text, I thought: 'I'm in the process of rewriting the Bible – I'm going back to the origins and finding out what it's all about'" (from an interview in the may 1984 issue of the magazine *Omni*). And he mentions Crick rushing into a pub crying, "We've discovered the secret of life!" In 1962 the Nobel Prize Committee awarded Watson, Crick, and Wilkins (R. Franklin had died the same year) "for their discoveries concerning the molecular structure of nucleic acids and their significance to the transfer of information within living matter."

Just what is DNA? A double helix whose A-T and C-G strands are connected by a sugar molecule (deoxyribose, hence deoxyribonucleic acid) and a molecule of phosphoric acid. It is a huge molecule (which could have been predicted by its molecular weight) consisting of four nucleotides connected in a very precise order by extremely strong valent bonds to form chains. Each nucleotide consists of a sugar and a phosphate – identical for all of the nucleotides in the DNA – and a third component, the

base, which gives each of the four types its individuality: adenine, thymine, cytosine, and guanine.

The bonds between the bases (weak hydrogen links) always work in the same way: adenine with thymine (two possible links), cytosine with guanine (three possible links). The DNA is wound into small rods, the chromosomes; [2] in each human chromosome, the DNA chain may contain 3 billion nucleotides, and the unwound DNA in all of our cells, if put end to end, would form a filament that would stretch 8,000 times the round-trip distance from the earth to the moon (according do David Weatherall in the *New Scientist* of 5 April 1984).

Genes are, in fact, pieces of DNA containing the instructions necessary to make proteins. This is a major discovery, in that, once the relationships between genes, DNA, and proteins are understood, an attempt can be made to modify the instructions in recombining DNA so as to produce new proteins. A gene is formed of about 1,000 letters (a letter being a base pair). A bacterium contains about 3,000 genes, or roughly the contents of an encyclopedia. As for a human cell, it contains nearly 3 billion letters, corresponding to an encyclopedia of 1,000 volumes!

The invariable combination of the sequences of the bases of the two strands of DNA (A-T, C-G) is used by the cell partly to produce exact copies of the DNA, as Meselson and Stahl have shown, and partly to assemble the proteins.

Why an exact copy of the DNA? So that the entire program appears again in the new cell and runs no risk of being divided at each mitosis. How is this copy made? If we consider one of the two strands of DNA as being comparable to the negative of a photograph and the other strand to a positive, we can then "print" a "positive" strand from the "negative" and a "negative" strand from the "positive." We thus find ourselves again with two new perfectly identical DNA molecules connected together by an enzyme, like a zipper.

The Genetic Code

The question of the genetic code quickly arose. How does the DNA go about making a protein? Or, more precisely, how can a 4-letter alphabet be transcribed into a 20-letter alphabet (the 20 amino acids making up the protein)? Combining the 4 letters two by two, we only get 16 combinations. But three by three we get 64 combinations of 3 letters (the codons): to each codon there corresponds an amino acid or a punctuation mark, and as we have 64 codons, several can simultaneously designate the same amino acid (as in the case of valine, coded by GTT, GTC, GTA and GTC, etc.). In the late 1950s George Gamow and Francis Crick advanced the hypothesis that a recombination of bases could determine the order in which the amino acids were linked together, and then Crick conceived of the triplet code.

In 1952, an article by Alexander Dounce – preceded by the work of Torbjörn Caspersson and Jean Brachet – suggested that another nucleic acid played a part in the synthesis of proteins: RNA, or ribonucleic acid, in which the sugar is ribose (and not deoxyribose) and the base uracil (U) replaces the thymine but is connected in the same manner to the adenine. The experiments carried out between

1955 and 1960 by H. Fraenkel-Conrad and G.S. Schramm on the tobacco mosaic virus showed that RNA also carries genetic information. In 1956, in Cambridge, England, Mahlon Hoagland and Paul Zamenick discovered transfer RNA (see below).

But what was still not understood was how the message from the DNA that led to the synthesis of proteins was transmitted. At the time, as the only RNA known was the RNA contained by the ribosomes, it was thought that it acted as a template in which the sequence of proteins was in some way predetermined, somewhat like the grooves on a photograph record. As this concept unfortunately was proved false in a serie of experiments, the Pasteur team with Jacques Monod and François Jacob offered the hypothesis of a metabolically unstable RNA, initially called "the magnetic tape" but later to be known as messenger RNA. They then proposed the model of the operon, a unit of transcription constituted by the genes by the action of an operator that is itself subject to the expression of a repressing agent produced by another regulating gene. Regulation of the synthesis of proteins is thus governed by two genes: a regulating gene, which produces a repressing agent, and an operating gene, which transmits the signals and enables the structure genes of the operon to be expressed.

During Easter, 1960, François Jacob was in Cambridge, England, discussing the hypothesis of an intermediary RNA with Sydney Brenner and Francis Crick. Their response was immediate: had not a rapidly regenerated RNA, distinct from ribosomal RNA, been brought to light by Hershey, and then by Volkin and Astrachan in bacteria infected by phages? A few months later, Sydney Brenner and François Jacob were in Max Delbrücks's laboratory at Caltech on the West Coast of the United States, where they undertook a series of experiments that was to prove the existence of this messenger RNA. It was found in *E. coli* at roughly the same time at the Harvard laboratory of James Watson on the East Coast, where one of the members of the Pasteur group, François Gros, had just encountered Walter Gilbert: "An amazing young man with a Martian's face.... He appeared to be highly intelligent from what we could judge, for he didn't say a word throughout the entire day" (François Gros, "L'Histoire du messager," in *Hommage à Jacques Monod, les origines de la biologie moléculaire,* Editions Vivantes, 1980, p. 127). There remained the repressing agent, a protein that was to be brought to light several years later at Harvard by this same Walter Gilbert and Benno Müller-Hill. Messenger RNA was the missing link in establishing the chemical nature of the code, which was accomplished in 1961 by Severo Ochoa and Marshall Nirenberg. In his testimonial to Jacques Monod, Francis Crick wrote, "From the time we realized that the ribosome was fundamentally a playing head, the world looked different" (ibid., p. 106), and Jacques Monod told François Gros, "This time, François, we've got a really good story, and we'll be able to have some fun" (ibid., p. 128).

The Jumping Genes

The study of genes provided many surprises for the scientists involved. The first was the discovery that, contrary to common belief, genes could jump

1. **Watson and Crick**
The "big Erector set" built by J. Watson (left) and F. Crick (right). This photo was published for the first time in 1968 by Watson in his book Double Helix.
2. **A, G, C, T**
The four letters of life.
3. **DNA**
An x-ray photograph of crystalline DNA (A form), which enabled Watson and Crick to guess the double-helix structure of DNA. It was taken by Rosalind Franklin in the laboratory of Maurice Wilkins, a third Nobel Prize Winner.

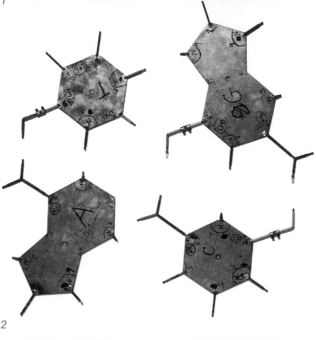

4. **F. Jacob**
Nobel Prize Winner in Medicine for 1965 along with J. Monod and A. Lwoff for their discoveries concerning genetic transmission and above all that of messenger RNA.
5. **B. McClintock**
In a field with Stephen Delaporta, "the old corn lady," who discovered the existence of jumping genes.
6. **Reading DNA**
Fragments of DNA have been separated by electrophoresis: they can be "read" thanks to coloring agents that, bound to the DNA, glow under the effect of ultraviolet light.

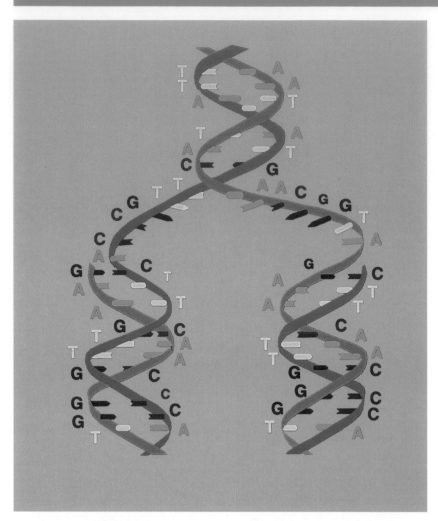

The DNA double helix consists of two chains curled around each other. The adenine (A) always pairs with thymine (T), and cytosine (C) with guanine (G). The connection between these bases being weak, the two initial chains may separate and form two "daughter" double helices. Chain A (in blue) "fishes" for new nucleotides in the molecular "soup" provided by the cell and pairs off with a brand-new B strand. It is thus that DNA recopies itself through the intervention of enzymes that separate the two chains and then reunite the bases of the two newly formed chains (always with the pairings A-T, C-G).

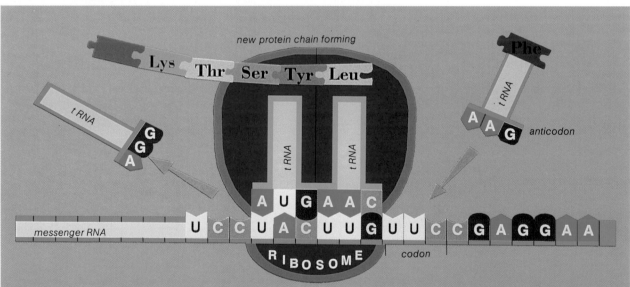

How is the code transmitted? A certain type of RNA, polymerase RNA, recopies DNA as messenger RNA. The starting signal is given by special sequences of nucleotides that trap the enzymes. This signal sets things off. The series of genes whose transcription is regulated by a "repressing agent" constitutes an operon. Once the copy is made, the messenger can leave and moves to the interior of the cell to the spot where its message will be decoded. The ribosomes take charge of the messenger:

they are made up of different kinds of proteins and a new form or RNA, ribosomic RNA, which acts as a translator for the message carried by the messenger RNA. It is then that a fourth kind of RNA, transfer RNA, comes into play, for it no longer translates but transcribes the language of the genes into the language of the proteins, more or less as one might transcribe Morse code (from a 2-sign into a 26-letter alphabet): each type of transfer RNA carries along a specific amino acid; at one end it must

recognize the codon to which to become attached (initiatory codon, generally ATG), and on the other it must have the corresponding amino acid. The whole process takes place sequentially, with each codon causing each transfer RNA to lay down its load of amino acid as if stringing a pearl on a necklace. The completed necklace would then be the protein, the sequence of nucleotides making up the gene that determines the sequence of amino acids constituting the protein.

from one spot to another and were not fixed, as was previously believed. This discovery dated back some forty years in fact, but did not cause a stir until the discovery of the double helix and more thorough research in molecular biology. It was only given the offical seal of approval in 1983 when the Nobel Prize in Medicine was awarded to Barbara McClintock, by then eighty-one years old!

Barbara McClintock began her work on the genetics of corn (maize) at Cornell University at the very beginning of the 1940s and continued it at the University of Missouri before moving in 1942 to Cold Spring Harbor on Long Island in New York, a center for the study of molecular biology, directed today by one of the Founding Fathers, James Watson.

McClintock quickly came up against a rule laid down by geneticists at the beginning of this century – Genes are nicely strung in a neat row on a chromosomic rod – a rule as firmly rooted as Leibniz's prequantum dictum that "Nature does not make jumps" (that is, Nature is never discontinuous). It was a revolution of the same kind as that made by the discoveries of quantum physics that McClintock initiated when she studied the spontaneous changes that appeared on corn in the variations in the colored spots transmitted from generation to generation, from which she concluded that the genes have a certain degree of freedom. She noted that these differences are repeated in regular sequences, with unstable mutations accompanied by chromosomic discontinuities in specific areas, and inferred that these areas must thus correspond to new genetic elements, to a new category of genes, which are mobile and which "jump" during the course of successive generations. All of which was confirmed toward the end of 1982 by P. Starlinger and his team at the University of Cologne. The presence of jumping genes was subsequently demonstrated in many species, including higher forms of life: transposons in bacteria, but also retroviruses in mammals.

"A large proportion of the mutations in a living being," says Werner Arber, "are not due to errors in replication but to transfers of fragments of DNA which have a specific biological function and which fuse with other fragments of DNA which have an equally well-known biological structure, this fusion being the origin or a new function of the species receiving the fragment of DNA. A good example of this mechanism is provided by bacteria which have acquired a resistance to antibiotics: most of the time, resistance genes are found on viruses or plasmids which can transfer from one organism to another. They also frequently form transposons which can jump from a chromosome to a transferable plasmid and from a transferable plasmid to a chromosome. Bacteria which forty years ago had no resistance have since acquired the genes for this. Where have the genes come from? Undoubtedly from other microorganisms carrying them, as, in their ecological niche, they must defend themselves against antibiotic substances produced by one of their neighbors. The natural vectors of the transposons carrying the genes have always existed in nature. From time to time there occurs a horizontal exchange of these genes from bacteria which resist antibiotics to those which do not. If no specific attempt is made to select the resistant forms, this goes unnoticed. But, over the past forty

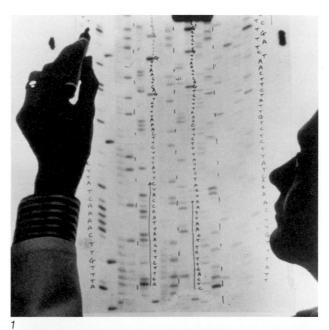

1

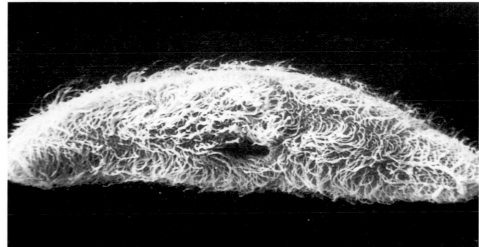

2

3

1. **Code**
A researcher at Searle in High Wycombe, in England, runs through the sequence of letters that will enable him to determine, thanks to the genetic code, a gene that will code for a specific protein.
2. **Paramecium**
The ciliate protozoa that has cast doubts on the universality of the genetic code.
3. **Victor McKusik**
The man who created the first DNA atlas.

49

years, as mankind has spread massive amounts of antibiotics into the intestinal tracts of men and animals, a massive selection of resistant microbes has taken place. This is a highly interesting indication in the case of microorganisms of an evolution by an influx of genes from all other microorganisms which makes me look on biological evolution as a gigantic interactive workshop. This transfer of genes also occurs in higher organisms by means of retroviruses, for instance. The transfer of genes which we carry out systematically today is thus nothing more than something which occurs naturally, perhaps in a slower and more complex manner – but there is nothing unnatural or against nature in this: it is a process which we can observe in nature."

Jumping genes may thus prove essential in understanding the mechanisms governing cellular differentiation, immunity defenses, and parasitism. And perhaps also in shedding light on an enigma that arose toward the end of the 1970s: that of the nature of introns, those silent genes discovered within the genetic heritage.

The Silent Genes

Behind this discovery lay an observation made at MIT (Massachusetts Institute of Technology) by Philip Sharp, who one day noticed to his amazement that, in an adenovirus, the chain of nucleotides of a specific gene was longer than that of its messenger RNA. Soon after, he noted the same odd situation in the SV40 virus. Simultaneously, in 1977, teams led by Pierre Chambon in Strasbourg, Susumu Tonegawa in Basel, Philip Leder in Bethesda, Maryland, and R. Flavell in Amsterdam showed the existence in higher organisms of "a mosaic of genes" in which, alongside the genes that express themselves and code for a protein (the exons), there are other genes that seem to have no purpose at all, the introns. Just what are these introns? Dormant agents, capable of being awakened during a mutation? Do they have other functions, such as regulating the expression of the genes?

The study of the genome of eucaryote cells initially inclined researchers to conclude that, contrary to the situation with the procaryotes, the DNA is first transcribed in the nucleus as a precursor of messenger RNA and undergoes changes before winding up as mature messenger RNA, which then travels in the cytoplasm. What changes? An excision and a splicing: the introns are excised from the copy of the DNA as precursor RNA: then the coding sequences of extrons are spliced in, that is to say, connected end to end by enzymes that have not yet been identified.

With the discovery of gene mosaic, what is true for a bacterium is no longer true for an elephant, as Jacques Monod was fond of saying. Furthermore, it was discovered that there were fragments of DNA that coded for nothing whatsoever, and that a large part of the DNA in higher organisms contained fragments that repeated themselves, virtually identically, up to hundreds of thousands of times, creating an enormous potential for sudden genetic variations. In short, the genome presents a considerable degree of mobility and the genes can not only jump but express themselves, keep silent, or even yawn! It appears that 80-90% of human DNA seems to serve no purpose. Another troubling fact is that the DNA of a salamander is thirty times as long as human DNA. What is the purpose of all this DNA in any case? With transposons and mosaics of genes, the breech through which the chance of genetic variability can enter becomes still broader.

The Double Helix
Can Turn to the Left

At about this time it became clear that the double helix could turn to the left! The elegant structure of DNA is in fact more complex than it appeared. It is subject to interaction with other biological substances, to incessant movement, the fastest detected to date being in the neighborhood of a picosecond (that is to say, one-millionth of one-millionth of a second). In response to the stresses to which it is subjected, it twists itself into a figure 8. If the torsion is too great, the two strands may even separate locally, the bases then becoming exposed to interaction with the enzymes or proteins, or to changes in structure, the right-hand helix changing structure to a left-hand helix, as if a corkscrew were suddenly to twist in the opposite direction. Furthermore, the phosphate-sugar links are no longer regular but start to zigzag: this is Z DNA, detected by A. Rich at MIT and R. Dickerson at Caltech. The replication of DNA thus poses far greater problems than those expected: polymerase DNA, which plays a part in this process, cannot alone unravel the double helix. The stresses created when one attempts to open the chains lead to increasingly great torsion as the two strands are separated. The elimination of the stresses at the dividing point is carried out by a batallion of enzymes, including the enzyme gyrase, discovered by M. Gellert in 1975, which induces torsion in the opposite direction.

The Code Is No Longer Universal

An international congress on the genetics of ciliates (unicellular microorganisms of the protozoa family) was held in Cold Spring Harbor in May 1984, and to the astonishment of those assembled, a Frenchman from the CNRS (Centre National de la Recherche Scientifique), François Caron, and an American from Bloomington, Illinois, John Preer, presented very similar reports with a stupefying conclusion: that the genetic code, which had always been assumed to be constant, could in fact be variable. The scandal originated with the paramecium, in that this little infusorian with vibrating cilia uses but a single arresting codon, TGA. But, as we have already seen, three codons signal a stop in the genetic code: TGA, but also TAA and TAG. The paramecium does use the latter two codons, but to make an amino acid, glutamine. Preer and Caron arrived at this conclusion independently after each had sequenced the ends of the very long gene (8,000 bases). Fraternally, they decided to share the columns of the scientific journal *Nature* side by side. Reading the code by three-letter codons remains universal, but for the first time the rule of correspondence was invalidated for the entire genome of an organism.

Thus the paramecium cannot exchange its genes with another organism, for each would be incapable of correctly reading the other's message.

	U		C		A		G		
U	UUU Phe	UUC Phe	UCU Ser	UCC Ser	UAU Tyr	UAC Tyr	UGU Cys	UGC Cys	U / C
	UUA Leu	UUG Leu	UCA Ser	UCG Ser	UAA UAG ponctuation terminaison		UGA	UGG Trp	A / G
C	CUU Leu	CUC Leu	CCU Pro	CCC Pro	CAU His	CAC His	CGU Arg	CGC Arg	U / C
	CUA Leu	CUG Leu	CCA Pro	CCG Pro	CAA Gln	CAG Gln	CGA Arg	CGG Arg	A / G
A	AUU Ile	AUC Ile	ACU Thr	ACC Thr	AAU Asn	AAC Asn	AGU Ser	AGC Ser	U / C
	AUA Ile	AUG Met Ponct	ACA Thr	ACG Thr	AAA Lys	AAG Lys	AGA Arg	AGG Arg	A / G
G	GUU Val	GUC Val	GCU Ala	GCC Ala	GAU Asp	GAC Asp	GGU Gly	GGC Gly	U / C
	GUA Val	GUG Val Ponct	GCA Ala	GCG Ala	GAA Glu	GAG Glu	GGA Gly	GGG Gly	A / G

The genetic code

In this table of the code of correspondences, known as the genetic code, the nucleotides are represented by their nitrogenous bases A, U (for uracil which corresponds on messenger RNA to thymine, T, of DNA), G, and C. A group of three of these letters (CAU, for instance) is known as a codon. The amino acids for their part are represented by a three-letter symbol – generally the first three letters of their names – Ala for alanine, Lys for lysine, etc. Looking at the table, one can see that there are "synonymous" codons.

"It is almost as if," states François Caron, "we tried to initiate a dialogue between two computers programmed in different languages: the one speaking PASCAL would be unable to process information in ADA."

This discovery raises questions as to at least part of the history of the evolution of species, inasmuch as two organisms having different genetic codes must be separated by a greater evolutionary distance than are two organisms that use the same code. And other ciliate protozoas undoubtedly share the same code as the paramecium. Thus it would seem that the ciliates are situated on a branch of the evolutionary tree that emerged far earlier in the history of life, far earlier in fact than anyone might have supposed until now. Would other species be capable of obeying genetic codes different from the universal code or that of the ciliates? "That seems, "says Caron" to be the conclusion from the observations of Japanese researchers who have found a mycoplasm using the TGA codon to code for another amino acid, tryptophase. All of these recent discoveries, even if they have not been entirely explained, will thus call into question the validity of many assumptions concerning the origin and evolution of species."

"These examples," says Werner Arber, "show that in biology most theories are of short duration. The theory of DNA is obviously not invalidated by the discovery of Z DNA. But in biology, in contrast to what occurs in physics (where one starts with a theory which is then validated or invalidated by experiments), nature always surprises you. Mosaic genes are a striking example of this: no one had even thought of this complication until it was actually observed. In dealing with living matter, it is particularly difficult to elaborate long-lived theories."

5

The Engineers of Life and Their Chimeras: Recombinant DNA

"When I come to the laboratory of my father, I usually see some plates lying on the table. These plates contain colonies of bacteria. These colonies remind me of a city with many inhabitants. In each bacterium there is a king. He is very long but skinny. The king has many servants. These are thick and short, almost like balls. My father calls the king DNA, and the servants enzymes. The king is like a book, in which everything is noted on the work to be done by the servants. For us human beings these instructions of the king are a mystery.

My father has discovered a servant who serves as a pair of scissors. If a foreign king invades a bacterium, this servant can cut him in small fragments, but he does not do any harm to his own king.

Clever people use the servant with the scissors to find out the secrets of the kings. To do so, they collect many servants with scissors and put them onto a king, so that the king is cut into pieces. With the resulting little pieces it is much easier to investigate the secrets. For this reason my father received the Nobel Prize for the discovery of the servant with the scissors" (from *Les Prix Nobel 1978*, Almquist & Wiksell International, pp. 175-180).

Such is "The Tale of the King and His Servants" told by Silvia Arber, ten years old, when her father Werner Arber was awarded the Nobel Prize along with Hamilton Smith and Daniel Nathans for their discovery in 1970 of restriction enzymes (or endonuclease enzymes), the scissors of genetic engineering without which any recombining would be impossible.

In the course of his studies of the phage in Zurich, Werner Arber had identified an enzyme that recognized a specific site, that is to say, a given succession of letters, and attached itself to the DNA there. The helix then started to twist, up to a point where the chain broke. We know today that the enzyme in fact recognized whether or not the site of attachment was methylated – and, if it was not, that this indicated the invasion of a foreign DNA, which had to be immediately destroyed. The problem was that

the chain was not cut at the precise point where the enzyme had become attached, but on a spot on the molecule that was never the same.

"The way I look at it," says Werner Arber, "this is a pretty good answer to the old question of predestination: If one day we knew all of the genes of a living individual, could we predict his entire life? No, not even at the molecular level of an enzyme with its substratum. As to the activities of certain restriction enzymes in vivo, for instance, though the specific site where the enzyme is attached is completely predictable, the action of the enzyme is, for its part, completely unpredictable."

At Johns Hopkins University in Maryland, Hamilton Smith, following up on Werner's results, became the first to obtain restriction enzymes in vitro that would cut DNA at the same spot every time, and Daniel Nathans began to use the enzymes systematically for the molecular study of viral genes.

But once the pieces of DNA had been cut, how could they be reconnected? "Servants with glue," enzymes capable of repairing the "nicks" in DNA, had been discovered in laboratories throughout the world three years previously. These were the ligase enzymes.

The Pioneers

Now all the tools required to cut, glue, and recombine genes were available. The work of Paul Berg at Stanford initiated the idea that genetic information might be transferred from one organism to another. In the course of his research, Berg studied above all the nature of mutations in certain transfer RNAs, the "suppressors," whose task is to detect false or absurd codons – typographical errors – that might introduce an error into the chain of proteins. An entire section in the School of Medicine where he worked devoted itself to studying phages carrying bacterial genes (the transductants), so Berg, along with D. Jackson and R. Symons, decided to

Probe DNA
Fragment of DNA that hybridizes with the RNA of an adenovirus. Gen-Probe in the United States has begun to develop ribosomic RNA probes that may prove considerably more sensitive than probe DNA and avoid the necessity of separating DNA's two strands. Health care (and above all the prenatal diagnosis of hereditary diseases) will be the principal beneficiaries of these new analytical methods.

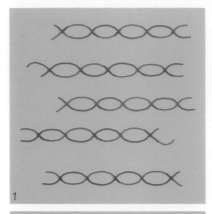

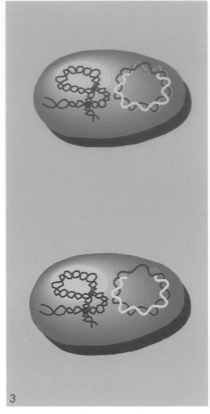

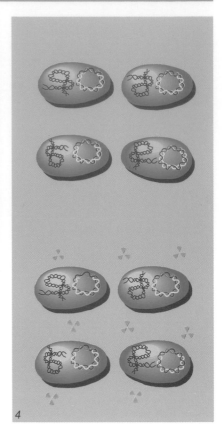

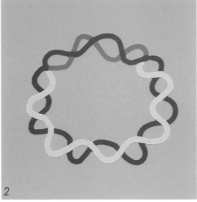

Using restriction enzymes, researchers can extract from cells containing a gene they wish to study a molecule of DNA, which they then break down into fragments the size of a gene (1). Subsequently, using ligase enzymes, they bind this foreign gene to a vector, a bacterian plasmid, for instance, which they open up to accommodate the gene (2). The recombinant plasmid is then reintegrated into the host strain of bacteria (3), which multiplies in a culture, producing clones (4). For the researchers, it is then a matter of detecting the clones containing the DNA they wish to study. They can do this, for instance, by using a molecular probe that has been made radioactive (in orange). The clone containing this foreign gene is then multiplied on a large scale, or amplified, separated from the plasmid by restriction enzymes, and purified.

produce virus hybrids of the SV40 bacterial gene in vitro: this was the first method of cloning the DNA sequences. A year later, in 1973, Stanley Cohen and Annie Chang at Stanford University and Herbert Helling in San Francisco succeeded in the in vitro synthesis of genetic molecules from two different sources, and called these composite molecules "chimeras": for indeed they were reminiscent of the creature in Greek mythology with a lion's head, a goat's body, and a serpent's tail.

It then became possible to clone, that is, to reproduce, a specific gene so as to obtain many exact copies, and in sufficient quantities to provide enough material for in-depth molecular study of the gene.

As it is extremely complicated to extract and purify the DNA of eucaryote cells, an indirect method was adopted for the latter, which consisted of synthesizing it in vitro by choosing to recopy the RNA (simpler in structure and found in smaller quantities) using an enzyme discovered in 1970 by H. Temin, Mizutani, and David Baltimore: reverse transcriptase. The desired messenger RNA must first be isolated from among thousands of others: a (soon radioactive) probe isolates and "fishes" it out. Next the RNA is recopied into DNA, and then eliminated with soda. To manufacture the double helix from a single strand of DNA, polymerase DNA is used; then the end of the helix is cut with a restriction enzyme, giving a copy of the gene (cDNA).

This bit of synthetic gene must then be glued to a "vector" capable of introducing it and making it express itself in a microorganism. A bit of this vector, whose nature we shall consider later, is then cut off and replaced by a piece of the synthetic gene, which is glued on with the aid of ligase enzymes, more or less as one would cut off a piece of magnetic tape and replace it with another in editing a musical tape recording or adding another piece to the completed tape.

To determine the sequences of bases that constitute the genetic information, Fred Sanger, A.R. Coulson, and S. Niclen, working at the Medical Research Council in Cambridge, England, were the first to develop an enzymatic sequence method in 1975; this resulted in the elucidation two years later of the sequence of a phage, the lambda bacteriophage, which consisted of no less than 48,502 pairs of bases! Since then, Sanger has determined the sequence of a virus of the herpes family, the Epstein-Barr virus, with over 140,000 pairs of bases.

1977 could be considered as Year One for the biotechnologies. It was during that year that, first of all, Walter Gilbert and A. Maxam at Harvard developed a system different from Sanger's, in which chemical reactions made it possible to interrupt the DNA chain at specific places. The Maxam-Gilbert

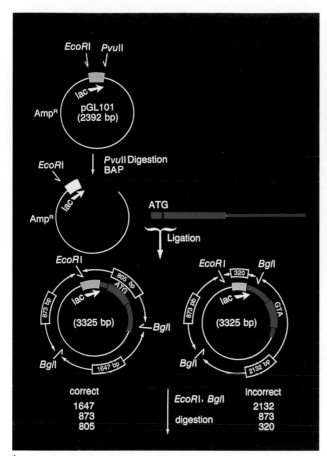

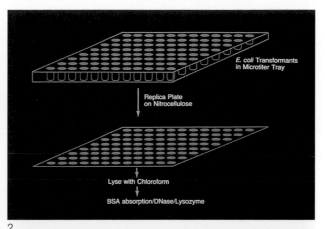

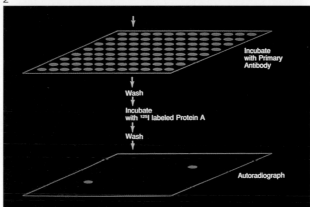

1

method makes it possible to read directly the DNA sequence by electrophoresis of the products of the breakdown of nucleic acids on Sanger's polyacrylamid gels, and then by autoradiography. In 1980, the Nobel Prize in Chemistry was simultaneously awarded to Paul Berg, Walter Gilbert, and Fred Sanger. (It was Sanger's second Nobel Prize, as he received his first 1958 for determining the structure of a protein, insulin.) 1977 was also the year when Baxter isolated and analyzed the gene with the code for the human growth hormone, whose importance we shall subsequently see. 1977 also saw the first practical application of genetic-engineering techniques with the expression by a microorganism of the gene of somatostatin (a human hormone isolated by R. Guillemin in 1972). Herbert Boyer, Itakura, and Riggs synthesized the gene producing the hormone and introduced it into a vector (a plasmid), joining it with the aid of an amino acid – methionine – to the gene of a protein – betagalactosidase – to prevent the breakdown by the bacteria of this very small peptide, which consists of but fourteen amino acids. On 2 December 1980, a patent for this process of inserting a gene in a bacterial plasmid was granted Stanford University after a prolonged argument over whether or not it was possible to patent living matter.

Vectors

The problem of the vector was critical in genetic engineering and remains so today, for once a chimera has been produced, the essential thing is to put it in a microorganism and to liberate the protein then expressed. The most commonly employed vector is a plasmid, that is, the small fragment of extrachromosomic DNA contained in the bacterium: the plasmid in fact curls up separately from the main chromosome of the bacterium, includes but a

few thousand bases (compared with a million for the main chromosome), and reproduces itself independently of the bacterium. In vivo, such a method of transmission is rather unusual: when a male bacterium couples with a female bacterium, one of the two strands of the double helix of the male separates and is transferred to the female – the two complementary strands regenerating themselves respectively by duplication in the two bacteria.[1]

Another vector for the future, and one more stable than the plasmid, is the transposon. Transposons are small segments of DNA that can be transposed from one bacterium to another and are often carriers of a gene resistant to an antibiotic.

This characteristic of contingent insertion could be used to transfer a gene, to determine bacteria that have acquired the transposon by the test of resistance to an antibiotic, and to select the bacterium expressing the gene far faster than by traditional methods.

To clone large fragments of DNA (over 20,000 pairs of bases, or about 20 genes), a not too successful attempt has been made since 1978 to use cosmids, the artificial plasmids developed by Barbara Hohn in Basel and John Collins in Braunschweig: cosmids are built up from a plasmid to which has been added a fragment of the lambda bacteriophage (the cos site, or "cohesive end site"). This fragment is made up of a DNA sequence containing a signal to wrap up this DNA in the capsid (outer protein shell) of the virus, which makes it possible for the artificial plasmid to be wrapped up in a protein capsid. It is thus possible to build up large-scale banks of genes "wrapped up" in cosmids, like packages ready for shipment into bacteria but also to animal cells in culture: this is known as the "cosmidic shuttle."

The introduction of a foreign gene into a eucaryote cell and the study of its expression have been

(1) Extrachromosomic DNA is also found in the breathing organites of chloroplasts. The idea of extrachromosomic genetic inheritance is reminiscent of Darwin's gemmules, De Vries's pangenes, and Weissmann's biophores.

55

the source of many difficulties. Most of the time the foreign DNA is integrated by chance into the genome of the cell, the genes are altered, and regulating their expression is virtually impossible. How can they be shipped to the right address, the right street, the right number?

H.E. Varmus's experiments in 1979 showed that the integration of viral genes in precise locations in the genome of the cell is possible thanks to the sequences of the extremities of the genome of retroviruses – the RNA of the retrovirus being copied as DNA, which becomes integrated with the DNA of the cell in the form of a stable "provirus." If the sequences of the provirus are grafted on to the two ends of a gene, they may permit the integration of the gene to the DNA of the cell and the regulation of its expression; it was this that was done for the gene of an enzyme, thymidine kinase, of herpes. The disadvantage of this method is obviously the inevitable presence of viral genes and the possibility of the cell becoming infected.

Another process, developed in 1977 at Columbia University by Richard Axel, Michael Wigler, and Saul Silverstein, reduces this disadvantage: this is the so-called cotransformation process, for which a patent was granted Columbia University on 16 August 1983 (applicable not only to the process itself but also to the products obtained in this way). This is pregnant with consequences, inasmuch as the process is a broad technique, permitting the introduction of any foreign gene whatsoever into a eucaryote cell. It is based on the genetic transformation of eucaryote cells by means of two (or more) foreign genes, one acting as an identification marker of cells capable of surviving under certain culture conditions, the other coding for a protein – hormone, interferon, etc. – not synthesized by the cell. One of the disadvantages of bacteria is their inability to make the required modifications on proteins: in this case, however, the precursor is converted

into active protein within the cell, which can then induce these changes. This procedure facilitates gene amplification.

Correcting Nature?

Shall we one day be able to correct for certain deficiencies that are the source of hereditary diseases? The failure of Martin Cline, who tried to substitute a new gene for a deficient gene in patients suffering from a form of anemia (beta-thalassemia) raised a storm in the national press and cast a cloud on the value of current research. It was blown up out of proportion and taken by the enemies of genetic engineering as an example of genetic "manipulation"; the testimony of the scientific community seems to indicate that it was simply an inexcusable mistake, being in too great a hurry in view of our present knowledge of the mechanisms of genetic regulation.

F. Ruddler grafted onto a plasmid a piece of the virus responsible for the multiplication of viral DNA and a gene of the herpes virus and injected the whole into a mouse's egg, which was then reimplanted in a mouse's uterus. In two cases out of seventy-eight, the cells possessed all or part of the foreign material. In 1982, the scientific journal *Nature* published the work of Professor Kan's team in San Francisco, which had been working on beta-thalassemia. The sufferers from this illness, due to an anomaly in one of the four chains making up hemoglobin, beta-globin, show a deficiency of the gene responsible for the synthesis of its 438 amino acids. The team noted that certain cases of the illness were due to a mutation in the DNA, changing the order of the letters of a codon corresponding to an amino acid (lysine) – which should be written AAG on the DNA and UUC on the RNA, but was written TAG on the former and AUG on the latter. But AUG is an illegible codon, a nonsense codon for

E. coli
The star bacterium of genetic engineering.

transfer RNA, which refuses to translate it, and the lysine is not incorporated into the chain. Instead of directly attacking the gene coding for protein, as Cline did, Kan's team acted on the gene-synthesizing transfer RNA so that it would understand the aberrant codon. But other questions remained, including: Is there not a chance that the transfer RNA may subsequently badly interpret the other codons?

Research is being continued on transgenic mice, that is to say, mice into which a foreign gene, sometimes from a different species, is introduced into an egg before reimplantation of the egg into an adoptive mother. In December 1982, R. Brinster in Pennsylvania and R. Palmiter in California published a spectacular result: they had succeeded in introducing the gene of the growth hormone of a rat into mice eggs, producing mice of twice normal weight. A few months earlier, the same teams had published the expression in manipulated mice of the gene of the viral enzyme thymidine kinase (from herpes): in both cases, the gene it produced was a hybrid gene, composed of a gene regulating expression and of the growth hormone of a rat or of thymidine kinase; in both cases, too, the gene may have been modified in the descendants of the mice. Such work is generally hazardous, as the gene may be destroyed by the organism or silenced (R. Jaenisch's team in Hamburg is concentrating on this phenomenon of inactivation).

Nuclear grafts – that is to say, grafting all of the DNA in the nucleus of a cell – into an organism by means of a reimplanted egg, has been practiced on amphibians since 1952, when Robert Briggs and Thomas J. King did this with frog eggs in Philadelphia. They had little success, as the baby clones did not reach adulthood. In January 1981, an article published in the journal *Cell* raised a furor. It related the success of Karl Ilmensee and Peter Hoppe in Geneva with two mice. But their experiment was immediately contested, and the scientific community remains highly skeptical as to its reality – even though it has never been disproved.

Major problems are apparently raised by what still remains the great enigma to molecular biologists: cellular differentiation. In only a few days, a single cell in an embryo gives birth to thousands of other cells organized for specific functions; these cells all have the same genes, but depending on the function they are to fulfill within the body, some of these genes are activated and others remain inactivated. When one grafts an egg that is starting cellular differentiation, one "brutalizes" the cells that are already differentiated and stable, and this sudden upset undoubtedly causes a break in the chromosomes. And perhaps, as activation of the genes according to the future functions of the cells is well under way, a system close to what would be called "counter-measures" in electronic warfare is initiated.

Rather than recreating chimeras, in both the genetic and the general senses of the term, researchers are turning more and more toward and understanding of the molecular switching system that makes it possible to start or stop biological processes. This could have far-reaching consequences in understanding, among other things, immunological defenses and aging. For example, only recently Mark Ptashne, Alexander Johnson, and

1

2

1. *Stars*
Some of the leading figures in genetic engineering, from left to right: Nobel Prize winner Paul Berg (below); N. Zinder; Nobel Prize winner David Baltimore; Sidney Brenner, famous pioneer of the Medical Research Council in Great Britain; R. Novick; R. Roblin; M. Singer.
2. *Recombining*
The messenger RNA of a human cell is microinjected into the ovocyte (or female gamete) of a frog, which subsequently produces a specific protein of the human cell.
3. *Genetic engineering*
John Kopchick at Merck uses a microinjection procedure to insert DNA, modified by genetic engineering, into a human cell. He thus tests the potential of genes to produce biologically interesting substances and observes the process on a screen.

3

1

1, 2, 3, 4. The gene bank
Clones of what are thought
to be gamma interferon
were cultivated in a petri
dish. The bacteria of each
colony are transferred into
"wells" (containing a
nutritive media) of plates (1).
The following morning,
inocula taken from each
"well" are sucked up
through the end of an
inoculation apparatus with
96 holes (2). The contents
of each little hole are then
expressed onto a nitro-
cellulose filter covering
plate of gelose (3). After
another night of incubation
at a temperature of 37°C, a
colony of bacteria develops
(4). Each colony represents
a clone – which is hoped to
be gamma interferon. From
this bank of 8,000 clones,
completed by Meloy
Laboratories, a hundred
clones of gamma interferon
were identified.

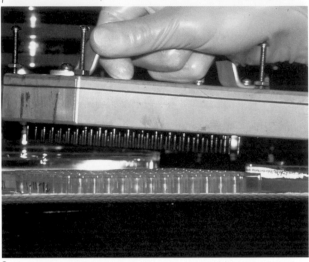

2

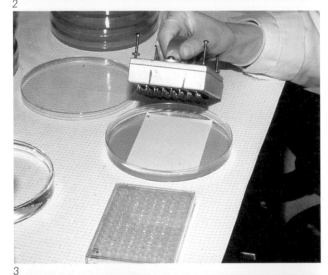

3

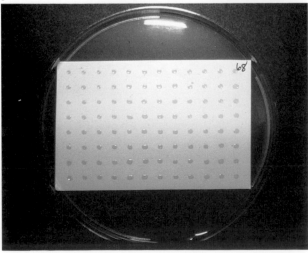

4

Carl Pabo at Harvard investigated the activation-inactivation phenomenon of certain genes of a viral bacterium, the lambda bacteriophage, present in the latent state in certain strains of *Escherichia coli.* This virus becomes virulent on brief exposure to ultraviolet radiation. Thus the genes of the virus are suddenly activated if environmental conditions are changed. The researchers determined that the switch was carried out by means of two regulating proteins (the repressor and the cro protein – an abbreviation for "control of repressor and other things") that act on specific segments of the viral DNA. The hypothesis of the existence of a repressor was advanced as early as 1961 by A. Lwoff, F. Jacob, and J. Monod in France. Its existence was confirmed by biochemical techniques in 1968 by two Harvard teams working simultaneously, one led by Mark Ptashne, the other by Walter Gilbert. The repressor, contrary to what its name might suggest, acts equally to block the expression of one gene as well as to stimulate that of another gene. This discovery was of capital importance for better understanding the mechanisms of molecular regulation.

The study of the regulation of genes within a given cell has just started. It may provide the beginning of an answer to the enigma of cellular differentiation: Regulation operates on different levels – the choice of the DNA segment to be transcribed into RNA, the choice of the transcribed sequences to be "carried" by the messenger RNA, and the rate of translation into proteins.

Genetic-engineering techniques are not limited to those of recombinant DNA (gene isolation and amplification, regulation of their expression, mutagenesis of coding DNA segments). They also include a host of other techniques without which recombinant DNA would not exist: automatic sequencing and synthesis of DNA, automated sequencing and synthesis of peptides, monoclonal antibodies, microbiological engineering, analysis and simulation data-processing systems, etc. For the moment, these techniques are used almost exclusively as research tools to increase understanding of living matter. Unfortunately, however, a certain confusion has arisen in the public's mind as the result of the commercial aspect of these techniques, a confusion further enhanced by scientists who have exchange their lab coats for business suits.

DNA Probe

There is still another instrument of genetic engineering to which we alluded but that today is assuming increasing importance not only for "tinkering" but for conducting clinical tests. This is so-called probe DNA, which is used like a hook to fish among a school of strains (containing all of the genetic material) for the strand complementary to the probe – that is to say, to isolate the bacterium carrying the gene sought for, to extract the fragment of DNA from it, and to study it. In 1960-1961, Marmur, Doty, and Schildkraut showed that the two strands of DNA could be separated by thermal or chemical treatment, and then, using a variety of techniques, once again hybridized. In 1968, David E. Kohne became the first, using hybridization techniques, to isolate a pure gene (that which coded for ribosomic RNA in *E. coli*); this, in particular, made it possible to

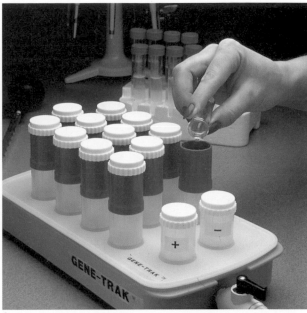

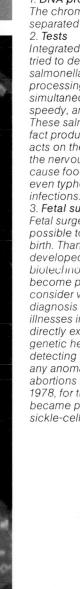

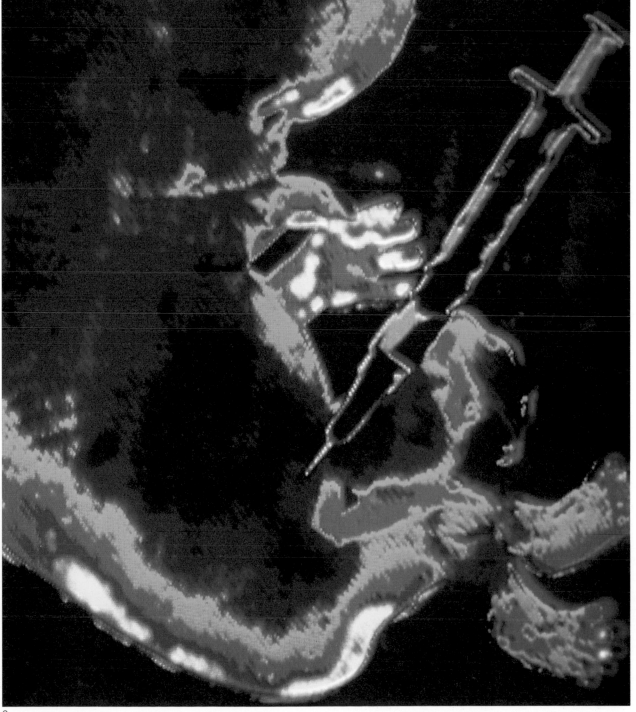

1. DNA probe
The chromosomic DNA is separated on a gel.

2. Tests
Integrated Genetics has tried to develop a test for salmonella for the food-processing industry that is simultaneously reliable, speedy, and inexpensive. These salmonella bacteria in fact produce a toxin that acts on the intestines and the nervous system that can cause food poisoning and even typhoid or paratyphoid infections.

3. Fetal surgery
Fetal surgery makes it possible to operate before birth. Thanks to the tests developed by the new biotechnologies, it has become possible to consider very early diagnosis of hereditary illnesses in fetuses by directly examining their genetic heritage and thereby detecting the existence of any anomaly. The risk of late abortions is thus avoided. In 1978, for the first time, it became possible to detect sickle-cell anemia.

A continent called Life

Imagine a giant continent called Life. To the north lies the Microbe Coast, and to the east the Plants-Flora Coast. From the west to the south stretches the Animals-Fauna Coast. The northern half of the continent is a unicellular state, while the southern half consists of multicellular states from the Tissue to the Living Body. At the center of the continent lies the high Gene-DNA Mountain, while to the south extends the Brain Mountain Chain. The coastal regions of the continent were opened up little by little over several thousand years and settled by mankind: by wine, cheese, soy sauce, and other fermentation industries in the north, by agriculture in the east, and by animal husbandry, the marine industry, and other traditional biotechnologies in the west and south. Starting from the beginning of this century, mankind began to feel that all the secrets of the continent were hidden in the middle of

the central mountain. The roads to the mountain, however, were steep and dangerous, refusing easy access. Finally, in 1953, Watson and Crick descended to the mountaintop by parachute and set up a small camp there for the first time. The subsequent explosive advances in research enabled the construction of several loghouses on the mountaintop and the opening of a trail from the coasts.
For example, a little village of rDNA technology of microorganisms, such as *E. coli* or yeast, was first opened in 1973 by Cohen, Boyer, et al. Today, it has grown to the scale of a town, through which a road has been broadened and paved, almost into a highway, from the coasts to the DNA Mountain.
In the map just sketched, the size of the frontier was deliberately distorted so as to facilitate viewing. Mankind, it is true, has made inroads into the depth of the

continent of life in these past ten years. However, vast unexplored areas offering untold riches still remain. To the southwest of the continent of Life, for example, lies the giant mountain chain of the brain. Almost no roads have been made into this area.
Another important point the map tells us is that the new biotechnology can only be put to practical use in conjunction with conventional biotechnology. Silicon chips have replaced vacuum tubes in electronics, but such substitutions never occurs in biotechnology.
(Adapted from W. Yamaha, Director General of Mitsubishi Chemical Industries, "Biotechnology – Present and Future," presented at the Biotechnology Seminars, sponsored by JETRO [Japan External Trade Organization] and held in 1984 in Stockholm, Lisbon, Milan, Paris, and Bonn.)

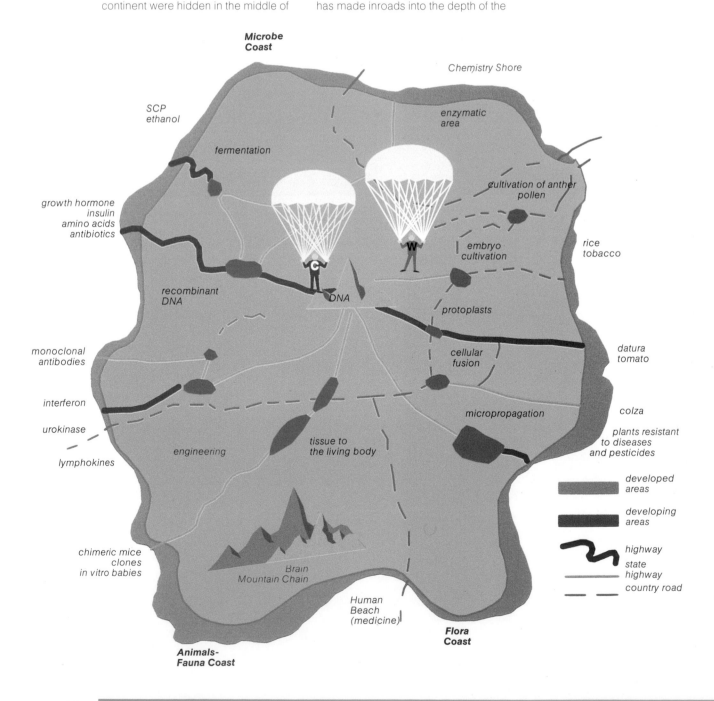

estimate the quantity of genes in a higher organism transcribed into RNA. 1971 marked the beginning of the use of the molecular probe "marked" by radioactive substances or fluorescent "labels" to detect the presence of a virus or infectious organism in the tissues, blood, or urine of an organism. It was also thanks to the DNA probe that when in 1972 Kohne, Hoyer, and Chison compared the DNA of humans with that of gorillas and chimpanzees, they found that the bases of the former differ from those of the latter by only 1% and obtained an initial estimate of the changes in the sequence of DNA during the evolution of the primates.

Thus a new industry came into being toward the beginning of the 1980s: medical diagnosis using genetic probes. Major companies, including Abbott and Johnson and Johnson (through Ortho Diagnostics) have taken an interest in this. Small specialized companies have also entered the fray: Amgen works with Abbott on tests involving infectious diseases and cancer; Cetus prepares tests for intestinal and venereal diseases and does prenatal tests for hereditary illnesses.

One of the most advanced companies in this field seems to be Enzo Biochem, which has developed tests for mononucleosis, cytomegalovirus, and herpes, and is preparing others for hepatitis B and a venereal disease, chlamydia. In the longer term, it is hoped that a test will be developed to detect certain tendencies in adults to hereditary illnesses. But will these carriers of latent genes for hereditary illness want to know this: Will it influence their decisions to have children? And perhaps equally important, will an insurance company assume the risk of insuring such a gene carrier (even if early detection makes prevention possible)?

It is difficult indeed to predict the consequences of the upheaval now affecting not only the world of medicine but also such major and well-established organizations as insurance companies as the result of the development of tests using probe DNA (or, as we shall see, monoclonal antibodies).

Until the present time, an essential hormone in the production of red cells, EPO (erythropoietin), was only produced in such infinitesimally small quantities that its use in therapy was impossible. This is now no longer true thanks to the work of an Amgen researcher, Fu-Kuen Lin, who, using probe DNA techniques, discovered the gene that codes for this hormone. Henceforth, the administration of EPO to certain patients may reduce the risks of blood transfusions (the transmission of impurities, hepatitis B, AIDS, etc.). Amgen has signed a contract with Kirin in Japan for world production rights to this hormone. This could cause serious upheavals among blood banks, which have already been disturbed by the use of monoclonal antibodies at Stanford University that have made it possible to eliminate certain contagenous donors.

"Magic bullets" or "silver bullets": these magical munitions undoubtedly will not be able to eliminate the great mythical diseases of contemporary civilization – cancer, thrombosis, arthritis, etc. – in the near future, but they will make it possible to improve our understanding of how they work. As always, the real upheaval will come from an unexpected quarter: in the very foundations of our social structure, in how we look on diseases and on mankind. If this is true, then we may well speak of a "revolution."

The Story of a Snail

by Richard Axel, Acting Director of the Cancer Research Institute of the United States and Professor of Biochemistry and Pathology, Howard Hugues Institute, Columbia University.

How, in as simple an organism as the marine snail Aplysia, do genes modify behavior? Such was the question to which an answer was sought by researchers led by Richard Axel and the neurophysiologist Eric Kandel at Columbia University. In his investigations, Axel made use of "reverse genetics" (going from the genotype to its expression, the phenotype, contrary to the general rules of genetics), a technique made possible by recombinant DNA. Was there a combinatory mechanism economizing genes that would explain how the human brain could register so many millions of bits of information and then process them?

In speaking of himself, Axel humorously declares, "First of all I worked in the medical department and my professor told me I would undoubtedly never make a good doctor for living beings. I then went into pathology and found I wasn't much better with corpses, and that's how I got involved with molecules."

As molecular biologists studying the brain, our objectives are to identify which genes control specific behaviors and how they do so. We wish to learn how information encoded in the chromosome is expressed and filtered through the complex network of nerve cells that comprise the brain, to generate the rich variety of behaviors that characterize the individual species. Molecular biologists today naturally rely heavily on the technology of recombinant DNA. The application of recombinant DNA to the nervous system may therefore permit us to address questions in behavioral biology in molecular detail.

This problem – how genes control certain behaviors – cannot be analyzed by genetic techniques alone. Genetics attempts to relate observable properties, the organism's phenotype, with specific genes – the genotype. Strengthening such a relationship relies heavily on mutations. In many instances, as in relating red cell dysfunction with mutant globin genes, it is possible to move from outward appearance to changes in specific genes. This procedures, however, frequently breaks down in the study of behavior, since specific behaviors, innate or otherwise, do not directly reflect the state of the gene. Behavior reflects the interaction of a vast network of structures, the central nervous system, which integrates and filters the dictates of the genes in a manner that for the most part is experimentally inaccessible.

The more complex the nervous system, the more elusive the relationship between genotype and behavioral phenotype. The system must therefore be simplified. At the outset, we must choose an organism sufficiently sophisticated that it exhibits complex repertoires of behavior, but sufficiently simple that we can attribute that behavior to identifiable cells. One organism that displays these properties is the marine organism *Aplysia*, a shell-less marine snail that may grow to a rather large size up to 5-10 pounds.

The Simple Nervous System of the Snail

The nervous system of the snail is numerically simple: it contains only about 20,000 central nerve cells collected into four pairs of symmetrical ganglia and one asymmetrical ganglion. This is in striking contrast to the brain of mammals, which contains 1 million times this number of individual neurons. In addition to being few in number, the neurons of *Aplysia* can be quite large, up to 1 mm in diameter, over 1,000 times the size of an average human brain cell. The cells of one such ganglion, the abdominal ganglion, are seen in figure 1. Most of these huge cells contain large quantities of DNA, as much as 2 µg of DNA per cell, a value 100,000 times that of a liver cell or an average mammalian neuron. The function of large collections of related neurons in other nervous systems seem to be carried out in *Aplysia* by single cells of enormous dimensions.

These unusual characteristics greatly facilitate the study of gene expression of a single cell. It is possible to dissect out single neurons manually and examine the activity of individual genes in a single cell. This numerical simplification has made this organism attractive for study by a number of neurobiologists. The elegant work of Eric Kandel and his colleagues at Columbia University has now made it possible to relate the function of particular cells to specific patterns of behavior, and this information may now permit us to attribute behavior patterns to the activity of specific genes.

The Behavior of Snails

It is important to ask first what an organism with so simple a nervous system can do: How sophisticated is its behavioral repertoire? During adult life, the *Aplysia* is largely occupied with elementary behaviors involving feeding and reproduction. Reproductive behavior, including courtship, mating, and egg deposition, in many species, including *Aplysia*, tends to be highly ritualistic, involving a coordinated series of stereotyped behavioral patterns reproduced by all individuals of a given species and excluding the participation of members of other species. One particularly clear example of an innate behavioral repertoire in *Aplysia* involves an elegant array of coordinated behaviors. We shall consider the process of egg laying in greater detail.

Aplysia is a true hermaphrodite: it may serve equally well as male or female. In fact, a single organism frequently experiences the pleasure of both sexes simultaneoulsy. *Aplysia* copulates in long chains of over a half-dozen animals: the chain may also link up to generate an undulating circle. Fertilization occurs internally within a reproductive organ, the large hermaphroditic duct, while development of the fertilized egg occurs externally in the sea. The process of egglaying consists of an elaborate but totally stereotyped behavioral pattern. The egg string is a long, 3-foot ribbon, containing over 1 million fertilized eggs. As the egg string emerges via contraction of the muscles of the duct, the animal stops walking and eating and its heart and respiratory rate go up. The egg string is then caught in the mouth and, with a series of characteristic head waving motions, it is helped out of the organism and wound into a tight irregular mass. A small gland in the mouth then secretes a sticky mucoid substance that attaches to the string, and with one forceful head wave the entire mass of eggs is stuck to a solid substrate, such as a rock. Thus, a series of discrete and seemingly unrelated behaviors come together in a rigidly coordinated sequence that together serve a common function, the deposition and protection of fertilized eggs.

This stereotyped behavioral array has been termed a fixed-action pattern. The innate behavior is thought to be generated by a "central nervous system program," in an all-

or-none fashion, such that the individual elements either occur together in a coordinated sequence or do not occur at all. The characteristic form of such behaviors is often not modified by experience or learning. Thus, each animal inherits a repertoire of fixed action patterns characteristic of the species.

How does an animal inherit such a behavioral repertoire? What does an animal inherit? It inherits DNA. The genes of the DNA can specify stereotyped behavior in two ways. First of all, they can specify a precise network of interconnected nerve and muscle cells that are put in place and "wired" together in the course of the animal's development. A stereotyped behavior is elicited, however, only in particular situations or at particular stages of an animal's life cycle, and only by the coordinated activity of particular parts of the network of neurons and muscle cells. In addition to the network, then, the genes must specify control elements: substances that excite specific preexisting connections in a rigidly determined way to generate a fixed-action pattern at the right time.

The problem of genetic control of egglaying behavior, at least at one level, now reduces to the identification of the controlling elements and ultimately to the genes encoding them. It was first necessary to identify the site of synthesis of these elements. Eric Kandel and Irving Kupferman several years ago identified a cluster of nerve cells sitting on top of the abdominal ganglion. This collection of cells, known as the bag cells, is apparent in the photograph of the abdominal ganglia in figure 1. When extracts of these cells were injected into animals, the full repertoire of behaviors associated with egglaying was observed.

These experiments identified the site of synthesis of the controlling elements, but not their chemical nature. Subsequent studies by Steven Arch of Reed College and Felix Strumwasser of the California Institute of Technology identified a small protein within these cells, a peptide, consisting of only 36 amino acids. This peptide, called ELH for egglaying hormone, when administered to animals, could generate some but not all of the component behaviors associated with egglaying. Thus ELH represented one of perhaps several elements controlling the egglaying repertoire. Recalling the linear relationship between gene or DNA and its protein product, the observation that at least one controlling element was a protein immediately placed the problem in behavioral biology well within the realm of molecular genetics.

A Gene Encoding Multiple Behavioral Neuropeptides
In collaboration with Linda McAllister, James Jackson, Eric Kandel, and James Schwartz at Columbia, we therefore set out to clone, to isolate the gene for ELH as a first step in identifying the genetic ele-

ments modulating egglaying behavior. The result provided a fortuitous surprise: the detailed molecular analysis of a single gene, the gene for ELH, provided a possible explanation for how the entire repertoire of egglaying behaviors may the mediated.

Using the techniques of recombinant DNA, we first isolated the gene for ELH from the chromosome, from the DNA of an *Aplysia* cell. Once we isolated the ELH gene, we sequenced it; that is, we determined its linear order of bases. This was performed using a combination of chemical and enzymatic techniques developed by Allan Maxam and Walter Gilbert at Harvard University. Having obtained the sequence of bases in the ELH gene, we could deduce the sequence of the protein it encoded. The egglaying hormone gene dictates the synthesis of a protein consisting of 270 amino acids. However, the egglaying hormone itself is only 36 amino acids. This created a paradox, the solution of which had important implications.

A schematic of the 270-amino-acid protein deduced from the DNA sequence is shown in figure 2. Indeed, the amino acids of the small ELH peptide can be found within the longer molecule. This presents a problem, however: How does the cell generate intact ELH as a 36-amino-acid peptide, its biologically active form? The cell solved this prob-

Fig. 1
Abdominal ganglion of Aplysia, enlarged about 40 diameters in this photomicrograph. The two clusters of bag cells, where ELH is synthesized, are visible lying athwart the large nerve bundles at the top left and top right. Within the main body of the ganglion one can discern a number of very large neurons, or nerve cells, many of which have been individually identified and found to be invariant in all members of the species. ELH and its companion peptides have been shown to have specific effects on the firing of certain neurons.

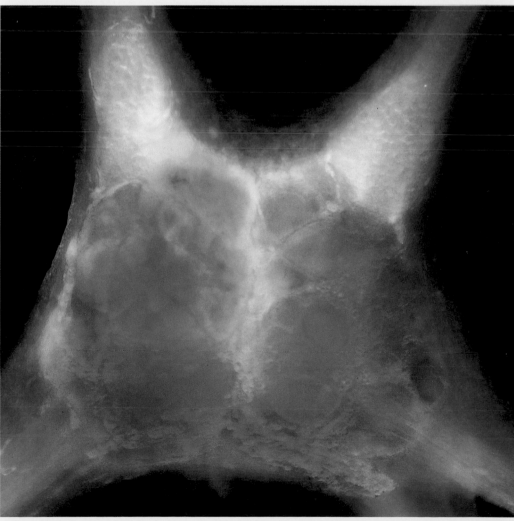

lem by cutting up the larger protein into small pieces after it is synthesized. The cutting process is carried out by specific enzymes: the process is not haphazard, but is quite precise. Signals exist within the protein sequence dictating the cutting sites. The 36-amino-acid sequence of ELH is flanked at its amino-terminal and carboxy-terminal ends by a basic pair of amino acids, lysine and arginine. Previous work on similar precursor proteins in vertebrate systems has identified this pair of amino acids as the most frequent site of cleavage by specific enzymes known as endopeptidases. It was therefore reasonable to assume that similar enzymes exist within the neurons of *Aplysia* capable of recognizing and cleaving at these sites.

ELH, however, comprises only 10% of the protein precursor molecule. We next asked what happens to the remaining protein. Is it merely cast off as a now useless companion to the smaller peptide ELH, or does it also contain other biologically active peptide sequences? Since we knew how ELH was cut from the larger molecule, that is, we knew the cutting signals, we searched the rest of the protein for additional cutting signals. As shown in figure 3, the precursor protein contains 10 such signals. If each signal were recognized and cut, we would therefore generate 11 discrete peptides from a single precursor.

This possibility excited us, since it was now possible that these peptides could perhaps be the other elements controlling egglaying behavior. However, the mere presence of cutting signals does not assure that they will be recognized. Further, we had no indication that the small peptides that could be derived from this large precursor would be biologically significant. Earl Mayeri and his colleagues at the University of California at San Francisco have been investigating the physiological properties of a family of neuropeptides released by the bag cells. We therefore collaborated with these investigators to determine whether any of the peptide molecules predicted from the sequence are present within extracts of the bag cells, and if so whether these protein molecules have any biological activity in the nervous system.

Earl Mayeri and Barry Rothman have identified four small peptides, alpha and beta bag cell factors, ELH, and acidic peptide, within the cluster of bag cell neurons. Each of these peptides is encoded within the ELH gene we have isolated and each is bounded by cleavage sites. Further, they have shown in electrophysiological experiments that three of the four peptides interact with specific identified neurons in the abdominal ganglion, where they act as neurotransmitters, molecules released by one neuron that alter the activity of other cells in a specific manner. These peptides and the neurons with which they interact are shown in figure 3. Let us first consider the neuropeptide, ELH. ELH acts as an excitatory

transmitter augmenting the firing of a specific neuron, R15. At the same time, ELH diffuses into the circulatory system and excites the smooth muscle cells responsible for the contraction of the hermaphroditic duct, expelling the egg string. A second, small peptide, beta bag cell factor, causes the transient excitation of two symmetric neurons, L1 and R1. Finally, a third neuropeptide bag cell factor inhibits the firing of a defined cluster of neurons of unknown function in the left upper quadrant of the abdominal ganglion. Interestingly, this peptide released by the bag cells also is capable of exciting the bag cells.

The specific association of these individual peptides, each derived from a single gene, with the activity of individual neurons suggests a possible mechanism for the generation of the complex pattern of behaviors associated with egglaying. A single gene generates a long protein, which is cut into several small biologically active peptides. Although far from proved, it is possible that individual components of the behavioral array may be mediated by individual peptides or small groups of peptides encoded by only one gene. Since all the behavioral peptides must therefore be synthesized as part of a single precursor protein, the appearance of individual behaviors would be rigidly coordinated. Further, since one peptide cannot be synthesized in the absence of the others, the repertoire of behaviors would occur in an all-or-none fashion. In this manner, a single gene encoding multiple neuroactive peptides could possibly dictate a complex repertoire of innate behaviors: a fixed action pattern.

The Initiation of Egglaying

These observations suggest that ELH, and perhaps additional peptides derived from a single precursor polyprotein, mediates the egglaying process. But what controls the release of these neuropeptides to assure the initiation of the behavioral array at an appropriate time in development? The neurons within the cluster of bag cells are all electrically interconnected. Release of the egglaying peptides occurs following prolonged electrical excitation of the bag cells. Felix Strumwasser and his colleagues at the California Institute of Technology have isolated and sequenced two discrete peptides, as A peptide and B peptide, which, when injected into *Aplysia*, excite the bag cell clusters, causing release of ELH and its companion peptides. The role of these peptides in vivo in initiating the egglaying process has not yet been determined. These two peptides are synthesized by the atrial gland, a secretory organ within the reproductive tract.

We have demonstrated that the ELH precursor, expressed by the bag cells, contains short blocks of amino acids that are homologous to amino acid sequences contained within the A and B peptides. Since the gene encoding ELH is a member of a small multigene family, we anticipated that

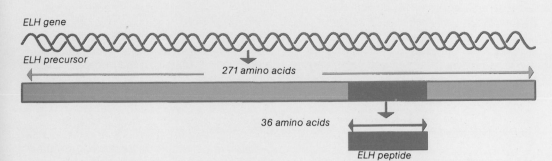

Fig. 2
The ELH gene encodes an ELH precursor protein 271 amino acids long. Since the ELH hormone itself contains only 36 amino acids, it must be cleaved out of the longer precursor protein.

ELH gene

ELH precursor

271 amino acids

36 amino acids

ELH peptide

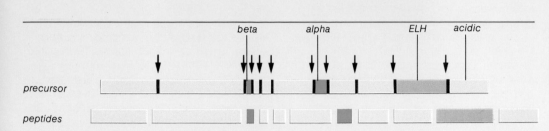

beta   alpha   ELH   acidic

precursor

peptides

Fig. 3
ELH precursor was found to be a polyprotein containing a number of active peptides. The precursor (top) is studded with 10 sites (arrows) at which a protein chain is cleaved by enzymes called endopeptidases. Cleavage at all the sites would release 11 peptides (second from top). Four of those peptides are known to be released by the bag cells: the beta and alpha bag cell factors, ELH, and acidic peptide. Three of them (colored peptides) have been shown to act as neurotransmitters, altering the activity of specific abdominal-ganglion neurons (colored cells) in specific ways (bottom). The beta factor excites cells L1 and R1. The alpha factor inhibits cells L2, L3, L4, and L6. ELH augments the firing of cell R15. ELH also enters the circulation and acts as a hormone, causing contraction of the reproductive tract.

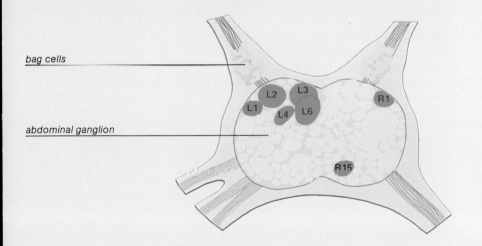

bag cells

abdominal ganglion

L2   L3   R1
L1
L4   L6

R15

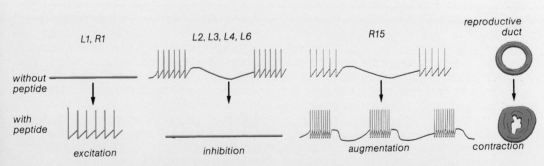

L1, R1   L2, L3, L4, L6   R15   reproductive duct

without peptide

with peptide

excitation   inhibition   augmentation   contraction

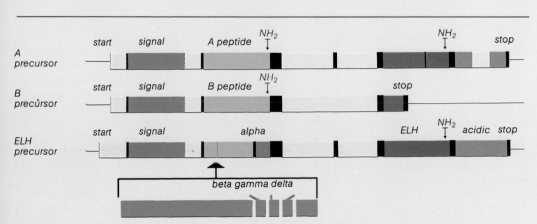

A precursor   start   signal   A peptide   NH2   NH2   stop

B precursor   start   signal   B peptide   NH2   stop

ELH precursor   start   signal   alpha   ELH   NH2   acidic   stop

beta gamma delta

Fig. 4
Three precursor proteins encoded by the three members of the ELH multigene facily are compared. These precursors are cleaved at specific sites (black bars) to give rise to active peptides, some of which undergo amidation (NH2). The three precursors have in common a signal sequence (left) and several regions of homology, or close similarity, that give rise in one precursor or another to the A or B peptide or to ELH and acidic peptide. Single-nucleotide differences alter the A and B genes so that active ELH is not synthesized. An 80-amino-acid insertion in the ELH gene interrupts the A-peptide and B-peptide sequences and gives rise to several different peptides.

perhaps the other genes within the multigene family would encode intact A or B peptide. In a series of hybridization experiments we demonstrated that two members of the multigene family were indeed expressed in the atrial gland. We immediately sequenced these genes and from the nucleotide sequence deduced the protein sequence to demonstrate that the three genes encoding ELH, A peptide, and B peptide comprise a small multigene family of common evolutionary origins.

The similarities and the differences are revealed when the three nucleotide and amino acid sequences are compared in detail (figure 4). Each gene encodes a precursor protein in which lysine-arginine sites (or sometimes a single arginine or two adjacent arginines) delimit the blocks of amino acids that are cleaved to become active peptides. All three precursors begin with a characteristic signal sequence of about 25 amino acids that governs the processing of the protein chain. The newly translated chain enters the lumen of a membrane system called the rough endoplasmic reticulum, where it begins to be modified: the signal sequence is cut off and sugar and phosphate molecules are added to the protein, which proceeds to the organelle called the Golgi apparatus. There the precursor is cleaved and the component peptides are enclosed in small vesicles, or sacs. In response to appropriate stimuli the vesicles fuse with the outer membrane of the secreting cell and release their contents to interact with nearby cells, to diffuse through the ganglion, or to enter the circulation.

The differences between the precursors incorporating the A and B peptides and the precursor incorporating ELH begin after the signal sequence. Let us first describe the A and B precursors. A single-arginine site signals the beginning of either the A or the B peptide, both of which consist of 34 amino acids. At the end of these peptides there is a glycine-lysine-arginine sequence, which serves as a signal not only for cleavage but also for transamidation: the addition of an amino group (NH_2) at the end of the peptide, replacing the usual hydroxyl group (OH). Transamidation "blocks" the end of the peptide, perhaps making it more resistant to degradation. There follows, in both the A and B precursor, a stretch of 47 amino acids unrelated to any known peptide. Then comes another lysine-arginine cleavage site, followed by what looks like the beginning of the ELH peptide.

ELH is not synthesized in the atrial gland, however. Examination of the nucleotide sequence of the atrial-gland genes shows why. In the case of the A-peptide gene, the first 22 amino acids of ELH are encoded correctly. The a single nucleotide difference in the codon for the amino acid at position 23 generates an arginine-arginine-arginine sequence and thus establishes a potential cleavage site that could break up what

would otherwise become the ELH peptide. A different kind of single-nucleotide change is seen in the B-peptide gene. Here one nucleotide in the sixth codon is deleted. The "reading frame" of nucleotide triplets is thereby changed so that a "stop" codon is generated; translation is terminated after only a 6-amino-acid stub of ELH has been synthesized.

Consider now the ELH precursor that is synthesized in the bag cells. The nucleotide sequence of its gene is very similar to that of the gene encoding the A and B peptides, and yet the gene does not specify those peptides; it specifies ELH and a number of other peptides involved in egglaying. It begins with the same signaling sequence seen in the A and B precursors. Then come the first and fifth amino acids of the B peptide. At this point, however, the ELH precursor diverges dramatically from the A and B precursors. The ELH gene contains a 240-nucleotides sequence that is not present in the A or the B gene, encoding 80 amino acids. The insert includes four cleavage signals delimiting three bag cell factors: beta, gamma, and delta. The beta factor, as we have mentioned, is known to have a specific effect on adbominal-ganglion neurons L1 and R1.

After the insert the nucleotide sequence resumes, without any alteration of the reading frame, to encode the sixth amino acid of the B peptide, and then it continues through a sequence that is much like the sequence of the two atrial-gland genes. One divergence in this region is particularly significant. A cleavage site is introduced, generating the 9-amino-acid alpha bag cell factor that, as described above, inhibits the firing of four neurons in the abdominal ganglion.

There follows a stretch of incomplete homology with the A and B genes, after which a cleavage site signals the beginning of the 37-amino-acid ELH peptide. The end of this peptide, as in the case of A and B, is followed by a signal for cleavage plus transamidation. Between the end of the ELH peptide and the stop codon that puts a halt to translation, there remain 27 amino acids: those of acidic peptide, which is also released from the bag cells along with ELH but whose target is not yet known.

Here, then, are three genes that have in common three regions of homology: the A or B region, the ELH region, and the acidic-peptide region. Moreover, within each of these three peptide regions there are near-identities of sequence at fixed positions. The implication is that all three peptides had their origin in a small ancestral peptide whose gene triplicated to generate a larger protein composed of at least three peptides. The gene encoding that larger protein apparently triplicated in turn, giving rise to three independent genes that diverged as they became specialized to satisfy different functional requirements. Then there may have been minor duplications of some regions of one or another gene, as is suggested by the fact that the adjacent beta and gamma peptides are almost identical. Such events presumably allowed for the evolution of variants without significant alteration of the original gene. The various versions of the gene may have been transposed to different sites in the genome, perhaps on different chromosomes.

Peptides as Mediators of Behavior

These three genes encode multiple neuroactive peptides that participate in the neural circuit governing a complex but stereotyped repertoire of behaviors. Several neuroactive peptides have also been identified in the mammalian brain and are thought to mediate specific behaviors. Given the widespread occurrence of neuropeptides, it now seems reasonable to ask, "What properties of the peptides themselves and their mode of synthesis make them particularly suited to their task as mediators of behaviors?" The behavioral potential of an organism is at least in part encoded within a rigid network of connecting nerve cells. Communication between nerve cells is often local and is mediated by the release of neurotransmitter substances as acetylcholine or norepinephrine into the synapse. In this way, specific neurons are called into play by point-to-point contact at synapses. Neuroactive peptides, on the other hand, may act locally as neurotransmitters on neighboring neurons, or they may be secreted into the circulation and act as neurohormones at a distance. Peptides may, therefore, act at several distant points, generating several discrete activities. Peptides therefore provide an additional communication and decision-making network that acts together with the more conventional network of synaptic transmission.

The diverse sites of action of several neuropeptides often permit them to coordinate physiological events with behavior. For example, injection of the octopeptide angiotensin II elicits a behavior in vertebrates analogous to spontaneous drinking. This peptide also acts on the kidney to cause reabsorption of sodium and water. These seemingly disparate actions both serve to rehydrate the organism. ELH also provides a rather nice example. This peptide acts locally to excite specific neurons of the abdominal ganglion, causing one set of behavior at the same time, and acts at a distance on the the hermaphroditic duct, causing contraction and egglaying. This combined effect illustrates an important property of the neuropeptides, the simultaneous coordination of physiological changes with behavioral events to effect a common end.

The synthesis of polyproteins provides a simple mechanism for coordinate control. Several different peptides encoded in a single gene may be simultaneously expressed under the control of a single regulatory element governing where and when a gene is to be expressed. The multiple small peptides contained within a larger polyprotein are not only encoded in the same gene and messenger RNA, but after cleavage from the precursor may be physically contained in the same vesicle until they are released by an action potential. This mechanism of peptide biosynthesis assures not only coordinate synthesis, but coordinate release of companion peptides.

The generation of multiple peptides from a single protein precursor also provides a solution to a numbers problem: the number of individual genes in a chromosome is simply not adequate to describe the diversity of behaviors exhibited by a species. The way in which the protein precursor is cut may vary in different cells or in response to different stimuli. One clear example of alternate processing has been provided by the work of Ed Herbert of the University of Oregon and James Roberts at Columbia University, who demonstrated that a precursor encoding adrenocorticotropin hormone is processed differently in different lobes of the pituitary gland. We have observed one pattern of cutting the ELH precursor in a single cell. In theory, this protein could be cut in 2,000 different ways to generate 2,000 unique combinations of peptides. Each combination could activate different patterns of behavior by modulating the activity of different combinations of neurons. The potential for diversity is even greater, since the ELH genes comprise a small multigene family expressed in an extensive network of neurons. This concept of combinatorial sets of peptides, each derived from a single gene family, greatly expands the informational potential of a chromosome genome.

The organization of genes encoding polyproteins also provides for striking evolutionary flexibility. The interspersion of active peptide sequences in the midst of a larger precursor protein in which other sequences may be nonfunctional allows the evolutionary potential to expand a coordinate set of peptides without interrupting the preexisting set. If the intervening amino acids between two peptide sequences are functionally inert, they may provide a repository for evolutionary change in which one can create additional active peptides. Thus, base changes within the intervening protein sequence may create new processing sites. Alternatively, sequences with their own preexisting cleavage sites may be inserted into this region. Internal duplications within a precursor also provide a mechanism to test new peptide possibilities without destroying the old. The ELH precursor, for example, contains a 240-base-pair stretch not present in the homologous precursor expressed in the atrial gland. This small insertion encodes three peptides, one of which appears to have arisen from a small internal duplication. In this manner, the number of peptides expressed by this precursor has expanded without altering the ability of the polyprotein to express active ELH.

Finally, we have suggested that the combination of peptides that derive from the ELH precursor governs the entire array of egglaying behaviors. If this is indeed true, we may further argue that the individual behaviors within the array may be ascribed to individual peptides or small groups of peptides. Let us consider head waving. The characteristic waving of the head occurs both in egglaying and in feeding as well. Feeding, however, is not associated with the other behavioral component of egglaying. The same peptides may therefore elicit head-waving activity in both feeding and egglaying. This peptide may therefore occur in association with the different companion peptides to generate the two distinct behavioral patterns. In this manner, more complex behaviors may be assembled by combining simple units of behavior, each mediated by one or a small number of peptides.

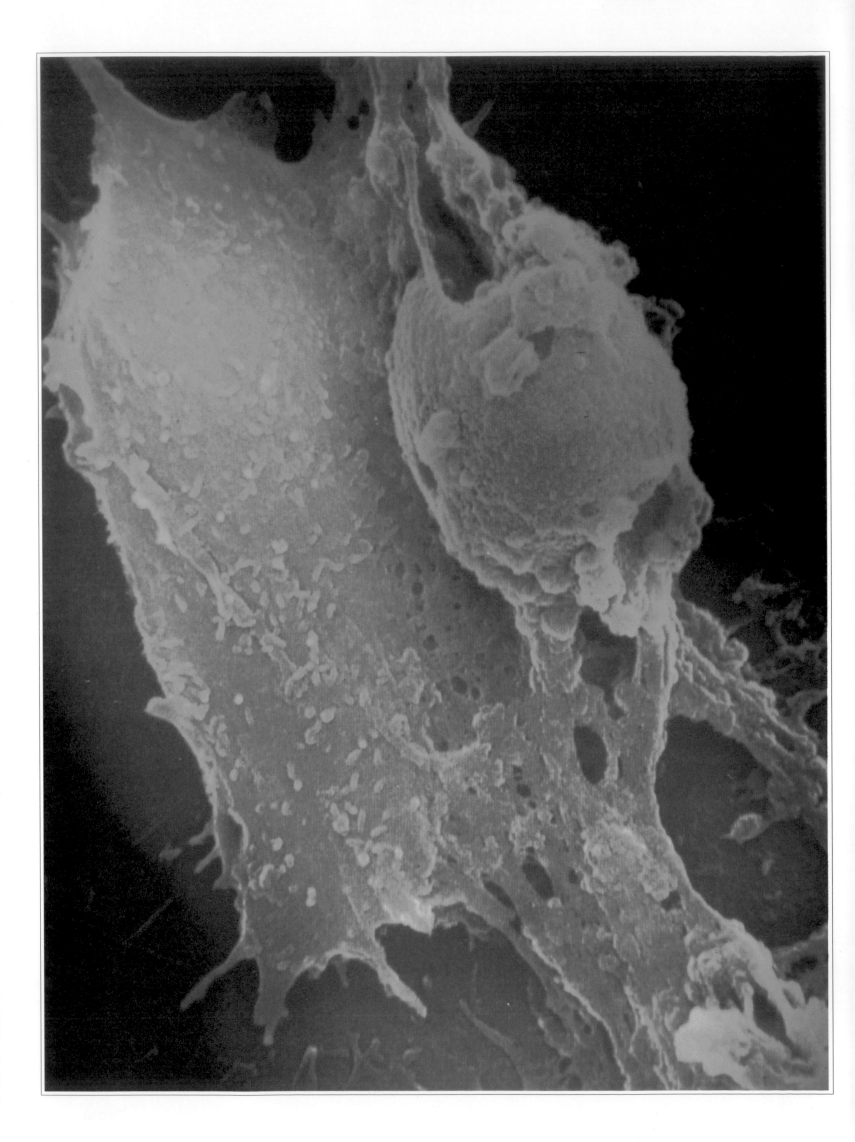

6

Monoclonal Antibodies: Missiles with Homing Warheads

1975: At the British Medical Research Council in Cambridge, England, César Milstein and Georges Kohler were faced with a problem familiar to many other scientists throughout the world: how to obtain a perfectly pure antibody targeted onto a specific viral or bacteriological invader attacking the body.

The antibody was not a new concept in immunology. While the word itself dates back only to 1902, since Jenner's work on vaccination toward the end of the eighteenth century, its existence and underlying principle have been well known: nature had provided man with a defense system capable of protecting him from the onslaughts of his environment. At the end of the nineteenth century, Elie Metchnikoff (Nobel Prize winner in 1908) had already shown that, by phagocytosis, the white blood corpuscles in the blood could absorb microbes in the body (cellular immunity). Later, Emil von Behring (first recipient of the Nobel Prize in Medicine in 1901) showed that there was such a thing as humoral immunity: that the serum from an animal previously suffering from diphtheria injected into another animal would protect the latter against this disease. The antidiphtheria serum developed by von Behring was sold commercially by Hoechst. And between 1898 and 1903, the Belgian immunologist Jules Bordet demonstrated the presence of antibodies in the serum.

Antibodies

With the appearance of an aggressor, a synthesis of proteins – immunoglobulins, or antibodies – starts up within specialized cells, the lymphocyte B cells of the family of white corpuscles found in the spleen, bone marrow, and ganglions.[1] These proteins, excreted into the blood serum, form troops that crowd around the foreign molecule (the antigen) and squeeze it to death. One of their more notable peculiarities is their extreme specificity: an antibody armed against one type of grippe virus will be completely ineffective against another. It is as if we had within our bodies an immense sleeping army of about 18 billions soldiers, each battalion of which only awakes at the approach of an enemy wearing a particular color of uniform. This extreme specialization has for many years puzzled immunologists and geneticists. Would it not take millions of genes to code that many different antibodies? That is impossible. And so?

The basic structure of antibodies was determined by Rodney Porter at Oxford University in 1959, and since then this structure has become even better known: an antibody is Y shaped, the two upper arms corresponding to two specific fixation sites for a given antigen; the specificity of the antibody results from a part of the molecule known as the "variable region." According to the work done by Philip Leder and various other researchers, the immune system produces several billion different proteins by rearranging a few hundred genetic segments. The idea of recombining fragments of DNA and making use of jumping genes that would make possible a multitude of different combinations – somewhat like a child's game of interlocking building blocks – would cast doubt on all of the explanations offered for the synthesis of proteins. By the interplay of introns and exons, the same gene could lead to the synthesis of several different proteins, which would certainly open extraordinary prospects for regulating the expression of the genes. Changes in the three-dimensional structure controlling the specificity of the connection with the antigen would thus result in the replacement in the protein chain of a single amino acid or a modification in the sequence of the amino acids.

Hybridomas and Monoclonal Antibodies

There were two aspects to the problem posed in 1975. In the soup of antibodies derived from the blood of a mouse contaminated by human liver cancer cells, how to distinguish the antibodies characteristic of cancerous cells from those character-

Hybridoma
A tumor cell (the largest), capable of infinite multiplication, fuses with a lymphocyte cell (producing antibodies) to produce a hybridoma, an inexaustible supply of monoclonal antibodies. A new method has been developed by Techniclone International that makes it possible to produce human monoclonal antibodies without employing hybridomas: to immortalize lymphocyte B, it uses an electric field that modifies the membrane of the cell and permits its transfection by a plasmid, the carrier of a cancerous gene. Another (existing) technique permits this immortalization by using, not a cancerous cell but the Epstein-Barr virus: immortalized, the lymphocyte B sometimes no longer produces antibodies.

(1) Lymphocyte T "helpers" control the differentiation of the lymphocyte B cells, while T suppressors can prevent the synthesis of the antibodies in the lymphocyte B cells.

69

1, 2. Tests
*Dr. Bellet explains the test
he developed with Dr. Jack
Wands of the Massachusetts
General Hospital for the
early detection of cancers of
the liver (1). Monoclonal
antibodies, enclosed in
plastic spheres, "capture" a
substance in the blood of
the patient that is secreted
by the tumors (alpha-
foetoprotein). At the same
time, this substance is
recognized by another
monoclonal antibody. More
recently, the same teams
have perfected new tests for
the detection of cancers of
the placenta (during
pregnancy) and cancer of
the testicles. What sets
these tests apart is that they
use monoclonal antibodies
that recognize, not a natural
molecule produced by the
tumor, but a synthetic
"substitute" whose structure
is similar to a very small part
of the natural molecule. The
artificial production of
monoclonal antibodies
against synthetic products
now makes it possible to
recognize better a natural
molecule. Many teams are
following this line of
research today.*

istic of liver cells or other human cells? And how to enable the identified antibody to reproduce endlessly (productive lymphocytes die quickly in a culture)? Would it not be possible to create a chimerical cell that would inherit the potential of both its parents: to produce a specific antibody and to reproduce indefinitely? This is exactly what Milstein and Kohler did by using, in an appropriate environment, a lymphocyte B cell from the spleen of a mouse immunized against an antigen expressed at the surface of a specific type of cell (production of the specific antibody) and a cancer cell from the bone marrow (plasmocytal tumor or myeloma) of another mouse (endless reproduction). The hybridomas (the chimerical cells thus obtained) are distributed among various receptacles and, after two to three weeks in culture, tests (radioactive, enzymatic, or fluorescent) reveal the presence in certain laboratory dishes of the antibody being sought. The accepted cells are then cloned and each hybridoma gives birth to a clone producing a single kind of antibody (monoclonal). We thus have a factory producing a virtualy unlimited amount of pure antibodies. Some recent techniques have even made it possible to increase yields – among them the encapsulation process, invented by Franklin Lim at the Virginia Medical College and exploited since 1978 by Damon Biotech. In this process, the cells are encapsulated in small gelatinous spheres with porous membranes within which they can continue to multiply sheltered from all contamination. This system saves both time and manipulation in the production of monoclonal antibodies: for 1 kilogram of antibody, it takes only 1,000 liters of tissue in culture instead of 100,000 liters, 50,000 mice, and the required purification techniques. This system could also be used to provide diabetics with an insulin secreted by encapsulated human pancreatic cells implanted in the body. Today there are tens of thousands of hybridoma cultures throughout the world (10,000 new lines each year); an international data bank, the Hybridoma Data Bank, was established in 1984.

Ever since the "creation" of monoclonal antibodies by Milstein and Kohler, there has been talk of "magic bullets" aimed at cancerous cells and capable of destroying them – an idea that is hardly new, as it had already been proposed by one of the fathers of immunology, Paul Ehrlich. The origin of cancerous cells remains unknown. It is thought that at a certain moment, "something happens" that overwhelms the body's immune defenses and enables certain "crazy" cells to proliferate. Recent research seems to have shown that there exist on the surfaces of cancerous cells molecular signals, or antigens, against which certain antibodies may be mobilized. Unfortunately, however, cancerous cells carry several tumorous antigens on their surfaces, and a monoclonal antibody, specific to a given antigen, is "deaf" to all other signals. To date, experiments have proved disappointing: the antibody, which can recognize the tumor, cannot by itself eliminate it; some cells avoid its attack and continue to multiply. In 1958, Professor Mathé in France suggested marrying an anticancerous toxic substance to the nonmonoclonal antibody. But the substances tried were too heavy, and the antibodies lost their effectiveness.

Bombarding a Target

With monoclonal antibodies the antibody's target is narrowed down (it must recognize the cancerous cell among about 100,000 healthy neighbors), and so an attempt has been made to have them carry along extremely powerful natural toxins – for example, diphtheria toxins – a single molecule of which can kill a cancerous cell. During tests in vitro, it became apparent that while the toxin did in fact kill its target, it also killed the innocent bystanders, the healthy cells. Genetic engineering was then employed to produce a hybrid molecule, immunotoxin, that would be effective but would lack the property of inappropriate fixation. Toxins are made up of two chains, chain A (toxic) and chain B (responsible for fixation). However, if the relatively unselective chain B is replaced by an antibody, the target is reached selectively. The results of in vitro tests seem promising. But various problems still remain: Is there not a risk that immunotoxins may be neutralized by the antitoxin antibodies mobilized by the body or by tumor antigens circulating in the blood-

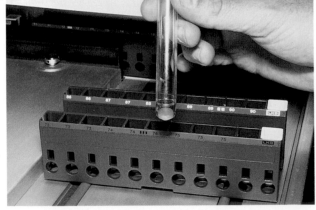

(2) Thanks to liposomes, it is also possible to provide sick cells with an enzyme or protein they may be lacking. Thus in 1970, G. Gregoriadis's team in London was able to insert an enzyme into cells in culture and have the enzyme work. It is also possible to use liposomes to carry glycopeptides or other substances capable of stimulating the production of antibodies by the lymphocytes, thereby increasing the immunity responses of guinea pigs. May it not be possible to carry the gene governing the synthesis of an enzyme throughout the body? Liposomes thus appear to be an interesting therapeutic tool for the future.

stream? Which tumor antigens are truly characteristic? Until now, we have been unable to identify the strictly typical antigens of cancer cells; we only know that certain antigens are expressed strongly by certain cancers, and are probably not by normal cells.

Another way of attacking the tumor would be to bombard it with a "bag" containing anticancer medicines by means of a monoclonal antibody. Such bags, synthesized in the laboratory since 1965, have a diameter of less than 2.5 microns; they are fatty walled vesicles, the liposomes.[2] The problem of coupling (that is to say, of forming a strong covalent bond between the monoclonal antibody and the liposome) has recently been solved. There remains the problem of liberating the medicines once they have reached their destination. These methods are now being tested on animals.

Mention sould also be made of the existence in Great Britain of a treatment for children suffering from cancer of the bone marrow in which the tumorous cells have been eliminated from the patient's own marrow. Purified antibodies are used for this purpose, combined with microspheres of polystyrene filled with iron filings: the antibodies plus filings combine with the cancerous antigenes, and then, by means of a magnetic field, the healthy cells are separated from the tumorous cells.

In vivo experimentation raises many questions. What would be the accessibility of the "killers" to their target? What would be the secondary effects or immune reactions of the human body were the monoclonal antibodies to be based on mouse cells? In 1980, two Stanford University biologists, Lennart Olson and Henry Kaplan, announced the production of human monoclonal antibodies. It was timely news as in 1982 researchers at Sloan-Kettering Institute for Cancer Research established the danger of using monoclonal antibodies from mice in humans. Without the possibility of producing human monoclonal antibodies, experiments could only be carried out in vitro if the subject's health were not to be endangered (a problem that does not arise in the case of mice).

A valuable tool for detection and selection can be made by binding a radioactive isotope or fluorescent tag to a monoclonal antibody, and this was recently done at Standord University, using them, as we shall see, with a cytofluorometer.

Tools for Research

Monoclonal antibodies have become extraordinary research instruments, and they have made it possible to study little-known molecules, such as those that play a part in the parasitic diseases against which we are so unarmed. During its life cycle, and depending upon the host at whose expense it is living, a parasite assumes different proteiform coats, and it is extremely difficult to identify these different coats as its molecular structure, insofar as the membrane is concerned, changes during the cycle. This wardrobe is stored in the chromosomic regions in the form of silent genes. Each of the genes may manufacture a copy that "jumps" into the expression site and codes for a new "coat." However, we have discovered certain monoclonal antibodies that react with the parasite responsible for malaria, in certain stages of its evolution, and this may in time result in the development of a vaccine, something that has been impossible until now.

1

2

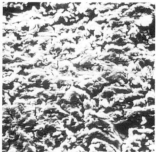

3

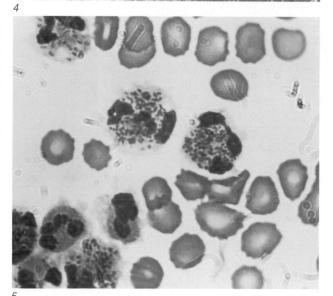

4

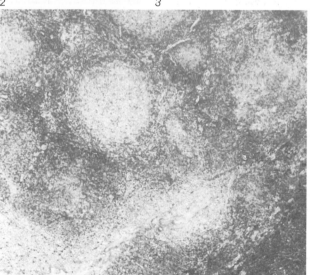

5

1. **Nobel**
Hoffmann-La Roche celebrated the Nobel Prize in Medicine awarded César Milstein (absent), G. Kohler (second from left), and N. Jerne (third from left) in the presence of Nobel Prize winner Linus Pauling (second from right).

2. **Sifting**
Microfilter plate with its 96 "wells" containing independent cultures, which makes it possible to sift monoclonal antibodies according to their activities.

3. **Dosing**
The granular surface on which dosing of immunological measures is carried out, as seen through a scanning microscope. This process, developed by Hybritech, permits extremely fine measurements.

4. **The label**
Section of a lymph ganglion marked by a monoclonal antibody specific to T cells. As can be seen, the "label" attaches itself to regions depending on the lymphocyte T.

5. **Treatment**
Monoclonal antibodies (which are in fact the homing warheads of missiles) are used in diagnosis, but also in the treatment of diseases. For example, Genetic Systems has developed treatments based on monoclonal antibodies for infectious diseases (in conjunction with Miles Laboratories and Cutter Biologicals). In this photograph, the monoclonal antibodies detect and "bombard" Pseudomonas, which in particular affects serious burn cases and leads to their death.

71

By means of radioimmunodetection techniques (marking with a radioactive tracer visible to a scintillation camera), monoclonal antibodies may also make it possible to detect cancer in humans. The use of these tumor-seeking antibodies started in 1949 with the work of David Pressman. But the discovery of monoclonal antibodies in 1975 gave researchers an even more selective detecting device. Today, two teams lead the research in this field: Jean-Pierre Mach's team at the Ludwig Institute in Switzerland, and that led by Steve Larson and Jorge Carrasquillo at the Veterans Administration Hospital in Washington. In 1984, teams of biologists at the Institut Gustave-Roussy in France and of gastroenterologists at Massachusetts General Hospital in Boston developed a test, using monoclonal antibodies, capable of detecting cancer of the liver at a very early stage (an 80% success rate on 3,000 samples of human blood), a test that will be sold commercially by Centocor.

Immunology is at the heart of research and discussions in molecular biology. Today, work is done essentially on three types of receptors in the immune system: the genes at the origin of antibodies, the receptor genes of T cells, and the genes coding for the principal histocompatibility complexes.

An Immunological Identity Card

What are histocompatibility antigens? They are the antigens responsible for rejecting grafts, and if they were known would provide an extremely accurate and valuable "identity card" of an individual's cellular biochemistry.

When a graft occurs, the patient's immune system becomes mobilized against foreign molecules, in exactly the same way it would for viral or bacterial aggression. These foreign molecules – the histocompatibility antigens – in fact enable the body to defend itself against its own cells whenever they exhibit an anomaly – a cancer or an infection. As soon as the body's internal police corps (the lymphocytes) detects a defective cell, it sends the alarm to the lymph ganglions, which sets off the mobilization of other lymphocytes, the K or "killer" lymphocytes, which destroy the aforesaid cells. These antigens differ from one individual to another, and it is precisely this genetic diversity that lies behind each person's biological originality: for example, an antigen will not induce the same immunological response in the recipient of a graft.

The antigens that erect the most powerful wall have been identified: in mice this is the H2 system (described for the first time in 1937 by Peter Gorer and studied for a period of twenty years by Gorer and George Snell at the Jackson Laboratory in Bar Harbor, Maine). The histocompatibility antigens of mice were isolated in 1981 by the Philippe Kourilsky team at the Institut Pasteur and by B. Dobberstein's team in Heidelberg, and the sequence of one of the genes was established for the first time by Leroy Hood. In humans, they are the HLA antigens, discovered in the early 1950s (for which Jean Daus-set received the Nobel Prize in 1980), and the first sequence of the human histocompatibility gene was established in France by Bertrand Jordan and Bernard and Marie Malissen. Furthermore, it was discovered that the organizations of these genes are similar.

With monoclonal antibodies, it has now become possible to test for compatibility between the donors and recipients of grafted organs, as we now have monoclonal antibodies capable of recognizing the most widespread antigens of histocompatibility – much as one determines the compatibility of blood groups.

Diagnostic "Kits"

Such are some of the prospects opened up by the immense field of immunology in which the discovery of monoclonal antibodies must be situated. The first market for monoclonal antibodies is that of biological equipment and reagents, which may revolutionize the entire concept of laboratory examinations: tests for allergies, anemia, leukemia (Hybritech), pregnancy (Organon Technika, Monoclonal Antibodies), and diabetes (Hybridoma Sciences, Sanofi); determining blood groups; diagnosing venereal diseases (Genetic Systems) or hepatitis B (Centocor); the early detection of genetic defects in the fetus; the prevention of diarrhea in newborns (Molecular Genetics); the detection of certain breast cancers (Abbott); etc. Some companies are even thinking of one day using monoclonal antibodies for purifying foodstuffs.

A quick system for determining blood groups has been successfully tested by C. Milstein's team and is in the process of being marketed. At the end of August 1984, Genetic Systems announced the production of human monoclonal antibodies on a large scale for a bacterium of the Pseudomonas family. In September 1984, two French teams (Charles Salmon and Dominique Goosens at the Centre de Transfusion Sanguine and Philippe Rouger at the Pierre and Marie Curie University) announced the production of human monoclonal antibodies capable of protecting a rhesus positive fetus from the immune defenses of a rhesus negative mother (the mother's antibodies destroy the red blood cells of the newborn). The method had also been the subject of experiments by a British team as early as February 1983. This test makes it possible not only to prevent the hemolytic illness of the newborn but also to diagnose allergies by the blood's content of histamines (involved in allergic and inflammatory phenomena). Nelson Teng and J. Sklar at Stanford Medical Research announced a similar discovery simultaneously.

This market for the diagnostic "kit" is less a competitor than a complement to that for "DNA probes." Hybritech, leader among the small companies involved in monoclonal antibodies, is behind Gene Probe, which has taken on the task of exploiting this second market.

Many small companies have specialized in developing and producing monoclonal antibodies: Hybritech, Centocor, Clonal Research, Monoclonal Antibodies, Hybridoma Sciences, Sera Laboratories, Ortho Diagnostics (a subsidiary of Johnson and Johnson); and many large companies have entered the competition: Hoffmann-La Roche, Abbott, Du Pont de Nemours, Mallinckrodt, Wellcome.

One generally overlooked fact should be noted in this competitive jungle: César Milstein and Georges Kohler never applied for a patent on their invention. But they were awarded the Nobel Prize in Medicine for 1984.

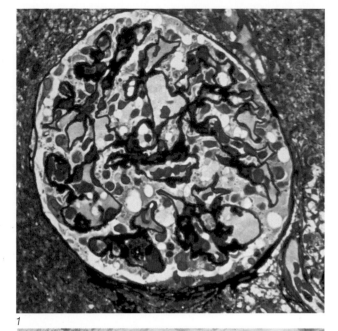

1

2

1. Grafts
Chronic rejection of a grafted kidney: a glomerule is isolated within an internal interstitial sclerosis.
2. Prostate
A section of the prostate "marked" by monoclonal antibodies.
3. Prostate
Test: a researcher at Cetus observes the color reaction caused by the presence of an antigen of cancer of the prostate.

4. Tests
One of the first large-scale uses of monoclonal antibodies was a test developed by Du Pont de Nemours to measure the level of a drug in the body.
5. Pregnancy
Hybritech's visual pregnancy test measures the presence in a woman's blood or urine of a hormone (HCG) to determine whether she is pregnant. This "kit" requires neither radioisotopes nor special equipment, is stable, and has a long shelf life. The tube in the center and the white "marble" represent the negative control. The tube on the right is that of a pregnant woman. The tube on the left is the testtube, and the blue color indicates to the naked eye that this woman is pregnant even before she may notice any change in her menstrual pattern.

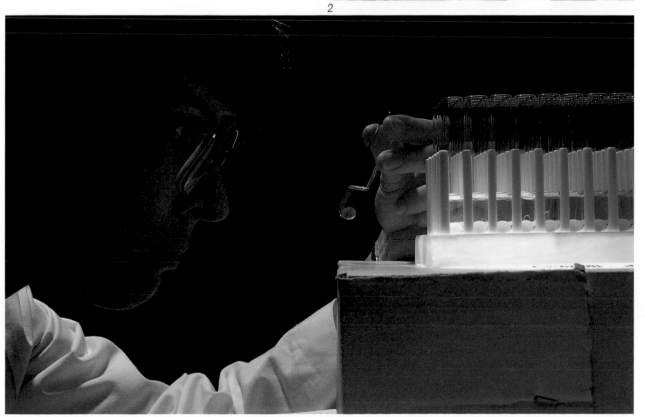

3

4

5

Monoclonal Antibodies: A Key Technique

by Kenneth Mitchell, Principal Investigator Du Pont de Nemours.

Kenneth Mitchell, a Biochemist and Doctor in Immunology, was born in London and came to the United States in 1969. It was there that he worked at the University of Pennsylvania and later at the Wistar Institute. In his own words, Kenneth Mitchell has a "long-term interest in genetics."
He is continuing his basic research on monoclonal antibodies at the Glenolden Laboratory at Du Pont where he specializes in the molecular genetics of immunology and cancer.

Introduction

The basis of the discovery and the development of monoclonal antibodies lies in the clonal selection theory of Burnett (1). In contrast to earlier notions, Burnett held that the heterogeneity of antibody responses arose from the fact that a preexisting set of cells, each with its own predetermined specificity, was present within the immune systems of vertebrates. When an antigen was encountered, only those cells with an appropriate specificity would be called into action, and the progeny of these cells would constitute the immunological memory of the animal and the protein products of the same cells would be the active components of the specific antiserum.

This theory, then, clearly suggested that if a strategy could be devised to isolate and grow individual cells from the immunological repertoire of an animal, it would be possible to produce individual antibodies of a quality and in a quantity not approachable by normal means. The pioneering work of Littlefield (2) and others set the stage for Kohler and Milstein (3), who were the first to describe a procedure that gave rise to hybrid cells secreting antibody of a defined specificity. The essence of the method can be simply stated: Two cells, of which one contributes the capacity to secrete specific antibody and the other the immortality inherent in tumor cells, could be fused by the action of an agent that impairs the integrity of the cell membrane, and the resulting hybrid cell would retain certain of the properties of both parents. Hybrid cells of this kind, growing continuously and secreting antibody, have now been produced by numerous researchers, who have applied a bewildering array of monoclonal antibodies to an enormous range of antigens.

The advent of monoclonal antibodies has revealed an antigenic complexity in the immunological world that was largely unexpected. Notion of antibody heterogeneity were based on the assumption that the specificity of an antiserum reflected the specificity of the individual antibodies contained therein. While this is true in principle, the real nature of specificity deserves a little discussion. If one considers a monoclonal antibody, then the essence of specificity lies in the ability of the antibody to bind to one antigen and not to bind to a second under the conditions of the assay being performed. This binding, then, is clearly related to the relative affinities that the antibody shows for the antigens in question. If one now considers two antibodies, both of which bind to two similar but not identical determinants, then it is clear that one antibody could have a higher affinity for the first determinant and the other a higher affinity for the second. Affinity in these terms implies that at low concentrations the two antibodies will appear to be specific for different determinants, but at high concentrations will clearly show the cross-reactivity of the system. If both these antibodies are present in a antiserum, then the specificity of the antiserum will be dependent on their concentrations, whereas in monoclonal form their individual reactivity patterns can easily be defined.

The specificity of an antiserum, unlike that of a monoclonal antibody, also results from the action of the phenomenon of immune dominance. Immune dominance arises from the tendency of particular molecular species and of particular antigenic regions to elicit a much more substantial response than other molecules or regions. The specificity of an antiserum, therefore, reflects the quantity of antibody present, the number of different antibodies reactive with each determinant, their relative concentrations and affinities, the number of different determinants present on the antigen, and the nature of the system being used to assay for specificity. The specificity of a monoclonal antibody, in contrast, depends solely on the amino acid sequence of the protein encoded by the immunoglobulin genes being expressed by the lymphoid cell member of the hybrid.

Monoclonal antibodies, in general, are specific for small regions (epitopes) of antigens or antigenic mixtures. They, thus, show an extraordinary discriminatory capacity far in excess of that exhibited by the most highly purified antiserum. Studies with monoclonal antibodies directed to influenza (4) or to rabies viruses (5) have shown that strains within the serological groups frequently differ from one another in their detailed antigenic structure. Monoclonal antibodies, therefore, are able to provide an enormous differentiative power, which may, on occasion, prevent their use for the grouping and, hence, classification of antigens.

Hybridoma Methodology

The formation of a hybridoma is achieved by the fusion of two cells, one of which is an immune mouse cell and the other an antibody-secreting tumor cell. Details of the methods and techniques can be found in many books and publications (6); however, a brief description of the principal points is given below. An essential preliminary to the production of a hybridoma is the immunization of the animal that will be the donor of the immune cells. The laboratory mouse, most commonly BALB/c, has been the animal used by most researchers. A variety of immunization strategies have been used, but an initial priming immunization followed by one or several booster shots are usually given. In any case, the final dose of antigen is administered by an intravenous route to promote the accumulation of the antigen and the responding cells in the spleen.

The myeloma cell that is most commonly used as the other parent of the hybridoma is derived from a BALB/c tumor that has been adapted to grow in tissue culture. The original tumor cell line was known as MOPC 21 (MOPC is an acronym for Mineral Oil PlasmaCytoma). Drug-resistant variants of the myeloma cell line have been selected.

(1) Numbers in parentheses refer to "Works Cited" at the back of the book.

74

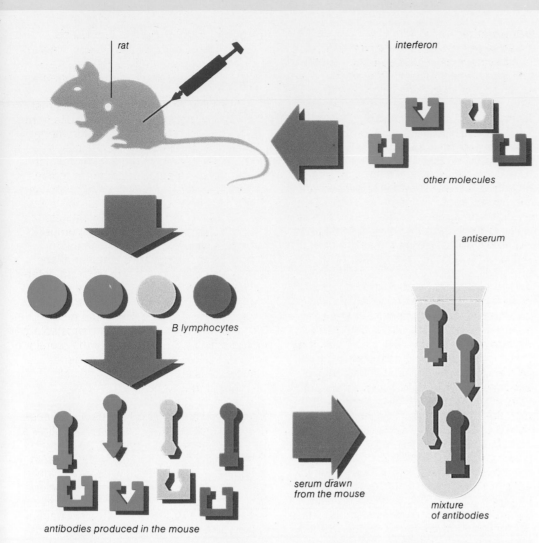

Antibodies

An antigen (interferon, for instance) is injected into the spleen of a mouse – or another mammal – thus alerting the corresponding antibodies. When the animal's blood is drawn off, a certain quantity of the specific antibodies being sought is obtained, accompanied by a mixture of other unwished-for and undesirable molecules.

rat

interferon

other molecules

antiserum

B lymphocytes

serum drawn
from the mouse

mixture
of antibodies

antibodies produced in the mouse

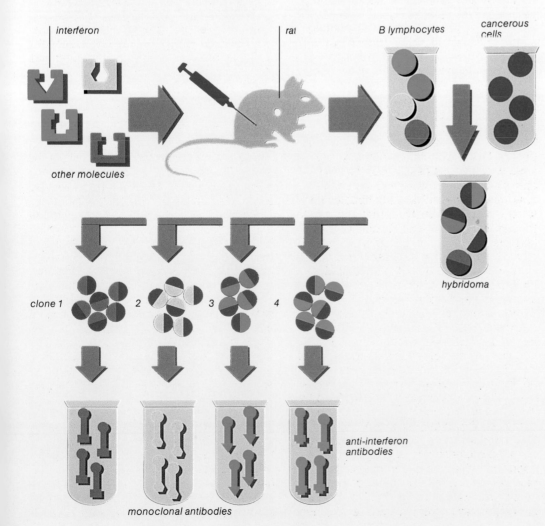

Monoclonal antibodies

An antigen (interferon, for instance) is injected into the spleen of mouse – or another mammal – thus alerting the corresponding antibodies. A few days after the injection, the lymphocyte Bs (the source of these antibodies) are extracted from the spleen, but have the unfortunate characteristic of no longer multiplying in vitro. By using an agent, such as Sendai virus or polyethylene glycol, the lymphocyte Bs can be fused with cancer cells (myeloma), thus forming a hybridoma. The clones made from the hybridoma are separated and tested to determine which produce the "good" antibodies, that is, those capable of recognizing the interferon. But, as Mr. Raees of Hayashibara says, "It's a little like the story of the gorgeous actress who told Bernard Shaw, 'Our child would have my beauty and your brain,' to which the latter replied, 'What a tragedy it would be, Madame, if the contrary ware true.' The tragedy for us is that we never know what's going to come out of the hybridomas."

interferon

rat

B lymphocytes

cancerous cells

other molecules

hybridoma

clone 1   2   3   4

anti-interferon
antibodies

monoclonal antibodies

75

Fusion of the two types of cells was achieved in early work by the action of Sendai virus. The growth of this virus in mammalian cells causes a change in the structure of the cell membranes, which results in the formation of syncytia (multinucleate cells) through a process of cell fusion. A more practical method of cell fusion was developed when it was realized that substances such as polyethylene glycol had the ability to modify the structure of cell membranes in a manner that promoted the fusion of cells that were in close contact. The general strategy has been to mix together a suspension of the two cell types to be fused and to centrifuge the mixture into the bottom of a centrifuge tube. The supernatant medium is then removed, the pellet broken up by gentle tapping, and suspended in a small volume of warm polyethylene glycol solution for several minutes, during which time the cells are gently centrifuged to bring them into intimate contact. The cells are then diluted, first into a medium without serum and then with a serum-containing medium. After a further centrifugation, the pellet is resuspended into a selective medium and pipetted out into a number of wells of a tissue-culture plate. Growth of cells as discrete colonies can frequently be seen within a week or ten days.

Assay for the presence of active immunoglobulin is generally carried out at an early stage, prior to extensive subcloning of the growing colonies. This is carried out for reasons of economy, since there is little reason to grow and nurture colonies that do not secrete an antibody of interest. Binding assays are usually performed by radioimmunoassay or by use of an enzyme-linked method.

Although the initial culture of hybridomas requires the use of a selective medium to suppress the growth of the parental myeloma, once the hybrids have been cloned and subcloned to ensure that the cells are all the progeny of a single cell, there is no longer a requirement for the selective conditions. Subculture and cloning is usually achieved by the limiting dilution method, in which cells are diluted to a concentration where the probability that more than a single cell will be found in each of multiple small subculture is very low. An alternative method is to suspend the cells in semisolid agar in a conditioned medium and, after colonies have appeared, to pick these from the gel and to grow them individually in liquid culture.

Significant amounts of antibody can be obtained by the growth of hybridomas in tissue culture, and concentrations of antibody in the range of 10 micrograms per milliliter can be obtained. In order to get larger amounts of antibody, the hybridomas can be grown as solid or ascitic tumors in a compatible host.

International Activity In the Hybridoma Field

An enormous amount of research and activity has gone into the hybridomas filed over the past ten years. Literally thousands of hybridomas have been generated by researchers in universities and commercial establishments throughout the world. Regarding the international, commercial, viewpoint, most of the major chemical and pharmaceutical companies are mounting substantial research efforts involving monoclonal antibodies. These biological reagents are viewed as providing the ultimate in specificity, since antibodies with minimal cross-reactivity can often be selected by appropriate assays. Small, entrepreneurial, companies are also springing up like weeds; a recent report suggested that some 150 small companies were in existence in the United States, some 20 in the United Kingdom, and many others in Europe. Many of these companies have a major stake in the production of hybridomas directed to antigens on human cells, either for the identification of tumor cells, the classification of human lymphocytes, or the typing of the various pathogens that affect humans. Monoclonal antibodies are also being produced for use as tools in the analysis of the differences between normal and tumor cells and for the study of the actions of various molecules now being found to be actively involved in the metabolism of tumors (namely, oncogenes). Another area that is being actively pursued is in the development of assays for known biological substances: for example, insulin, human chorionic gonadotropin (HCG), growth hormone, and other normal and disease-related entities. Monoclonal antibodies are also being developed for use in purification methods. By coupling antibodies to solid supports, columns can be prepared for the purification of any soluble antigenic substance – for example, interferons and other lymphokines, hormones, and trace proteins of many kinds. Therapeutic applications of monoclonal antibodies can also be envisaged, from methods designed to detoxify plasma in an extravascular situation, to the neutralization of viruses, to the treatment of tumors and the modulation of the immunological systems of patients themselves by the application of antiidiotypic monoclonal antibodies.

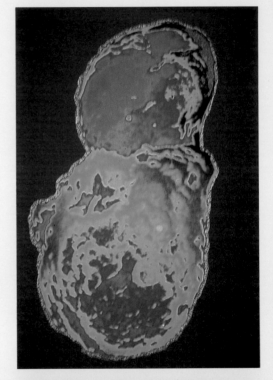

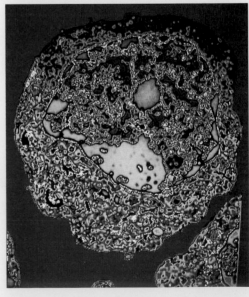

Immunotoxins and Antitumoral Therapy

by P. Gros, Director of Research in Immunology and Oncology, F.K. Jansen, Chief of the Immunotoxins Project, and R. Roncucci, Director of Research and Development, Sanofi.

Many therapeutic approaches are being tried to cope with the twentieth century's most devastating disease – cancer. One of these – bombarding the tumor with killer substances – has been developed in several laboratories, among them those of Sanofi, the Elf-Aquitaine subsidiary.

Sanofi was the first company in the world to file patents in this field, joined in 1980 by various other laboratories: Teijin in Japan and the NIH (National Institutes of Health), the University of Houston, Cetus, and Xoma in the United States.

Roméo Roncucci, doctor of sciences, pharmacologist, and toxicologist, has been directing Sanofi's Research and Development Department since 1979. Pierre Gros joined the Clin-Byla Research Center in 1959, then part of Sanofi, where he now directs research programs on antitumoral agents. F.K. Jansen, doctor of medicine, first worked on the immunology of diabetes; joining the Clin-Byla Research Center in 1975, he initiated the immunotoxin program, which he has directed ever since.

The three of them together describe the current state of the art, the hopes and uncertainties of research, and the prospects of developing an effective product for the future.

There is a general agreement among cancer specialists that the main drawback of available drugs for chemotherapy of neoplastic diseases is more their lack of specificity than their lack of efficacy. Although these drugs, the use of which is now well codified, allowed considerable progress in oncology, they proved more or less incapable of acceptable discrimination between tumor cells to be destroyed and the much greater number of normal cells, with a great variety of properties, which must be protected. This fact often leads to a situation in which the efficacy of the existing drugs is not fully exploited in order to stay within tolerance limits acceptable to the patient.

This lack of selectivity could be overriden if it was possible to provide the cytotoxic molecule (which kills the cells toward which it is targeted) with a seeking device that would permit the toxic effector to bind only to the target cells and leave all other cells unaffected. It must be stressed that, in theory, the achievement of the anticipated action depends upon two mutually related conditions:
• the toxic effector must be extremely potent, so that a very small number of molecules of this drug may kill the target cell once it has been detected by the seeking device;
• the toxic effector-seeking device assembly must not express its toxicity as a free entity in the blood circulation, but only when bound to the target cell.

Historically, this working hypothesis is not new. At the end of the last century, Paul Ehrlich already stated it implicitly in a rather prophetic proposal. However, it took science three-fourths of a century to acquire the tools necessary to put it in a concrete form. Mathé in 1958, then Moolten in 1970, were the first to achieve credible attempts; but these did not result in therapeutic applications. It is only since 1980 that several teams, including ours,[1] have been able to take advantage of recent discoveries, allowing this old dream to become a reality.

Indeed, several studies in the 1970s demonstrated special properties of a class of extremely potent toxic proteins of bacterial and plant origin, the most important of which is diphtheria toxin (produced by *Corynebacterium diphteriae*) and ricin (extracted from the seed of the castor oil plant). These toxins have the following two fundamental properties in common:
• they kill cells by destroying the intracellular protein synthesis machinery located in the ribosomes, a most vital function of the cell;
• both types of proteins include two parts, with fully differentiated functions in their chemical structure: (1) one part carries the toxic capability entirely within itself, for it is the one that inactivates ribosomes (toxic subunit); (2) the other part ensures the cell binding of the toxin molecule and facilitates penetration of the toxic subunit through the cell membrane (recognition subunit).

The cytotoxic subunit is extremely potent, since a single molecule in the cytoplasm of a cell suffices to elicit profound alterations, leading to cell death. Moreover, since the recognition subunit may bind the toxic subunit to practically all animal cells, all conditions are met to make these molecules the most potent known toxins (besides nervous toxins).

From this naturally occurring model, the idea was born to create synthetic hybride molecules retaining the useful properties of these toxins without the undesirable properties.

Thus, if the toxic subunit alone of the toxin is retained (whereas the recognition subunit is rigorously eliminated), the resulting molecule subunit
• has a considerable cytotoxic potential;
• is, however, incapable of expressing this potential for lack of association with a structure suitable to ensure cell binding.

Then, if this toxic subunit is chemically bound to a monoclonal antibody that is an ideal seeking device for a missile, the result is a so-called immunotoxin, that is, a half-antibody, half-toxin synthetic hybrid molecule.

In such a molecule, the antibody moiety would bring out the specificity for a particular marker (antigen), ideally present only on cells to be destroyed, thus permitting anchoring of the immunotoxin solely to the membrane of cells carrying the target antigen. This seeking device having fulfilled its function, the toxic unit would destroy the target cell the same way the natural toxin does.

These so-called A-chain immunotoxins (from the name of the most commonly used toxic subunit extracted from ricin) are not the only ones developed after this principle.

In an effort to avoid any loss in the extreme potency of the natural toxin, one may elect to couple it as such to the antibody capable of providing the chosen recognition specificity. The products obtained are, indeed, very potent, but their specificity is hampered by the conflict that occurs between the antibody that recognizes certain cells and the recognition subunit of the toxin that recognizes other cells. This difficulty may be overriden in laboratory conditions, but may not yet have been solved in vivo in patients. Moreover, there are serious presumptions of general toxicity with such products. In any case, they need to be improved before being subjected to clinical evaluations. As will be seen, the present situation is rather different with A-chain immunotoxins. Owing to the work of our team, among others, preparation, biological characterization, and optimization of conditions of use have permitted
• on the one hand, a first product up to the stage of clinical evaluation in two particular ex vivo treatment models;

(1) H. Blythman, D. Carrière, P. Casellas, D. Dussossoy, O. Gros, G. Laurent, J.C. Laurent, M.C. Liance, B. Pau, P. Poncelet, G. Richer, H. Vidal.

• on the other hand, identification and solution of at least partial difficulties involved in generalizing the clinical use of A chain immunotoxins in in vivo treatment.

Preparation

Preparation is based on conventional techniques adapted for the purpose and consists in the isolation of the A chain of ricin from ricin itself, which is purified from soluble proteins of the seed of the castor oil plant. The process, now in operation in a pilot automatized for safety reasons, permits purification of batches of several tens of grams, enough to cover the present needs.

Concurrently, the chosen monoclonal antibody is obtained by development of producing cells (hybridomas) either in mice or in special reactors using appropriate synthetic media. It is then purified and controlled very carefully to avoid contaminants, especially viral contaminants.

Then the two constituent proteins of the immunotoxin are bound chemically. After a new set of controls, the product is packaged for use in a lyophilized form.

Biological Properties

The molecules studied are expected to carry the following two fundamental properties:
• very high cytotoxic potency;
• selecticity of action strictly imposed by the presence of the chosen antigen on cells to be destroyed.

Different laboratory models have been developed that permit the evaluation of the compounds with respect to these properties on cell cultures particularly suitable for precise measurements. Thus it appears that synthetic immunotoxin molecules are generally slightly less cytotoxic than the corresponding natural cytotoxin (ricin), but have very high selectivity ratios (of the order of 1,000 -10,000). This means that concentrations of immunotoxin 1,000-10,000 times higher than those needed to kill antigen-carrying cells are required to kill cells that do not carry the corresponding antigen. Such a ratio, the in vitro equivalent of the therapeutic index (an evaluation of the safety margin), is reached in pharmacology only rarely, and never with conventional anticancer drugs.

Moreover, it has been shown that the biological properties of A-chain immunotoxins could be further enhanced by adjuvants acting as potentiating agents. Generally, the latter substances are amines or ionophores (agents capable of binding ions and carrying them across biological membranes). Thus, in the most favorable conditions, and in the presence of potentiating agents used in optimal conditions:
• concentrations of immunotoxin required to kill target cells are extraordinarily small (less than 10^{-13} mole, or a few nanograms per liter), that is to say, 10-100 times lower

concentrations than that of ricin alone;
• specificity ratios reach values up to 1 million;
• finally, the death of target cells is obtained almost as rapidly as with ricin itself.

A very extensive, original toxicology study was carried out over a period of more than a year to assess the consequences of a single or repeated administration of an A-chain immunotoxin to three animal species routinely used in such a work: the mouse, the rat, and the rhesus monkey.

As expected, results demonstrated the excellent tolerance of the product on all functions of treated animals up to doses of several milligrams per kilogram (mg/kg)– sufficient to warrant the safety of clinical evaluations.

As for all immunotoxins, the one selected for advanced studies of preclinical development is characterized, as far as its field of application is concerned, by the specificity of its constituent antibody, in this case, the antibody with code name T101, which recognizes an antigen found exlusively on all normal human T lymphocytes (a particular population of white blood cells) and on malignant cells of leukemias and T lymphomas. The corresponding immunotoxin (or IT-T101) should in all instances be suitable for at least four types of therapeutic applications, which will now be considered.

Potential Clinical Applications of IT-T101

1. The most ambitious application is obviously the direct use in intravenous infusion in leukemias or lymphomas where tumoral cells carry the antigen matched by the T101 antibody. It is also the most difficult one, and is not yet available, for several obstacles still prevent its clinical evaluation with reasonable chances of success. Briefly, the obstacles are essentially due to the necessity
• to devise conditions permitting increase of the life span of the molecule in the blood of patients;
• to extend to in vivo conditions in the patient's organism the decisive effect of potentiating agents, achieved in vitro.

Laboratory solutions do exist. Il remains to demonstrate that they may actually be extended to clinical situations.

2. A second application, also in vivo and thus requiring, as for the first application, a certain time before it may be put at work, consists in using IT-T101 as an immunosuppressor capable of lowering the level of immunological defenses. In this case, the immunotoxin will have to destroy in vivo T lymphocytes, its target cells, whose key role in the immunological balance is well known. This situation is limited to particular, none the less important, therapeutic applications, such as organ transplants (kidneys, heart, liver, etc.), for which the temporary decrease in the immune response of the recipient is always favorable for a success-

ful graft during the critical period that follows transplant.

3. Closer to us, and actually already in progress, are applications to the field of bone marrow transplant, a practice with obvious extension to oncology. Two types of applications are found. Despite their similarity they show profound differences in conception.

a. In the so-called autologous graft (or autograft), the treated subject is also the donor of the bone marrow used. Let us suppose a patient with acute lymphoblastic T leukemia. When the disease first becomes apparent, it is easy to obtain complete remission with conventional treatments. The patient no longer shows clinical evidence of the disease, although his body still hosts an undetectable but actual number of residual tumor cells. It is possible to sample bone marrow for subsequent transplant. This bone marrow is by necessity suspected of containing a few undetectable tumor cells. It can be treated in vitro, outside the body, by IT-T101 for "decontamination." It is then stored in liquid nitrogen. At an appropriate time (for instance, during a relapse), the patient will be treated by conventional means (chemotherapy and radiotherapy) at maximal dosage to destroy beyond any doubt any tumor cell present in the body. However, the drawback of this very drastic treatment is that it completely destroys the patient's bone marrow, and the patient would surely die from his inability to regenerate his blood cell lines if left without further treatment. Autotransplant of the preserved decontaminated bone marrow is carried out by simple infusion. It brings back enough viable stem cells to regenerate all of the destroyed bone marrow.

b. In the so-called allogenous graft (or allograft), the difference is that the donor is not the same person as the recipient. In this case, the donor is obviously healthy. There is no need to decontaminate the graft from tumor cells. On the other hand, a healthy bone marrow graft has the capability of setting up a very effective immunological defense, and if the immunological compatibility between the donor and the recipient is not perfect, the graft will recognize its new host as foreign and will react against it. This will result in a serious disease known as graft versus host reaction, which is often fatal. The mechanism of such a reaction is not fully understood. However, the T lymphocytes of the host are known to be the main factor. Specific suppression of this particular cell population considerably weakens the adverse reaction and makes it tolerable, or even suppresses it altogether. Such a suppression may be obtained with IT-T101 by taking advantage of the presence on normal T lymphocytes of the antigen recognized by the T101 antibody.

Both applications to bone marrow transplants have already been subjected to pre-liminary clinical evaluations. From the first studies, the following facts have already been shown:
• general tolerance of the immunotoxin reinjected together with the bone marrow is excellent – this is an essential point for further development;
• until now the rate of graft success has been 100%;
• restoration of blood cells lines in patients treated with worked-out grafts is indistinguishable from that resulting from normal grafts, thus confirming that the bone marrow stem cells are not affected by the immunotoxin – an essential fact;
• in the case of allograft, actual depletion of the graft in T lymphocytes protects effectively against graft versus host reaction;
• on the other hand, in the case of autograft, no information on efficacy will be available for several years.

This preliminary information is very encouraging and is an incentive to pursuing and widening the evaluation undertaken.

Future and Limits of Immunotoxins
Studies now in progress will be crucial for categorizing the immunotoxins, which are a totally new class in the anticancer drug inventory. In this respect, present studies are unquestionably at the research stage in this area. However, besides the studies pertaining to bone marrow transplant, the next step will be the evaluation of the potential of immunotoxins in real in vivo conditions, which have been shown to pose problems of a different nature from those encountered in vitro.

Indeed, when the case of solid tumors comes to be investigated, the problem will be raised of the accessibility of all the cells of the tumor to the immunotoxin molecule because this molecule is larger than those of conventional drugs used in chemotherapy. Moreover, there are reasons to believe that a tumor is not homogeneous as far as the antigens of its cells are concerned. Some cells carry more of them than others or have different antigens. If the whole tumor is to be reached, not one immunotoxin but rather a mixture of immunotoxins carefully optimized and capable of recognizing all tumor cells of a given patient will have to be used. This implies that a large enough number of antigens adapted to the identification of possible targets will have been identified and that the corresponding antibodies and immunotoxins derived from them will be available. All this represents a considerable amount of future study, for, if the antigens of lymphocytes (also found in leukemias and lymphomas) are relatively well known, the specific markers of solid tumors are far less well known, and the presently available monoclonal antibodies with adapted specificities are still very few. The hope that these tumoral markers will be identified still holds, and for the time being monoclonal antibodies specific of antigens found in greater numbers on tumor cells than on normal cells may be used effi-ciently to facilitate the destruction of tumor cells.

Another type of obstacle that is bound to appear during in vivo treatment is linked to the fact that the constituent toxin molecule of the immunotoxin is totally foreign to the human body and thus will elicit formation of antibodies against itself (immunization). These antibodies will neutralize the activity of the immunotoxin, which will progressively lose its efficacy if prolonged or repeated treatments are to be applied. Fortunately, the large number of known toxins usable as immunotoxins may help to solve this problem.

So, if immunotoxins fulfill their promise, they should constitute a new and important weapon in the fight against tumors. Of course, they are not meant to be the absolute weapon against cancer. One hopes, however, that these complex molecules, born from an innovative concept and involving an original mechanism of action, may be the decisive contribution when combined with conventional therapeutic techniques (especially for the eradication of tumors resisting these treatments) as well as techniques expected to result from other ongoing research studies.

Immunotoxins
These monoclonal antibodies with fluorescent labels (the green areas), which bind specifically with the antigens of cancer of the breast cells, are used by Cetus to study immunotoxins.

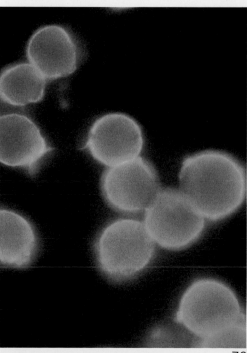

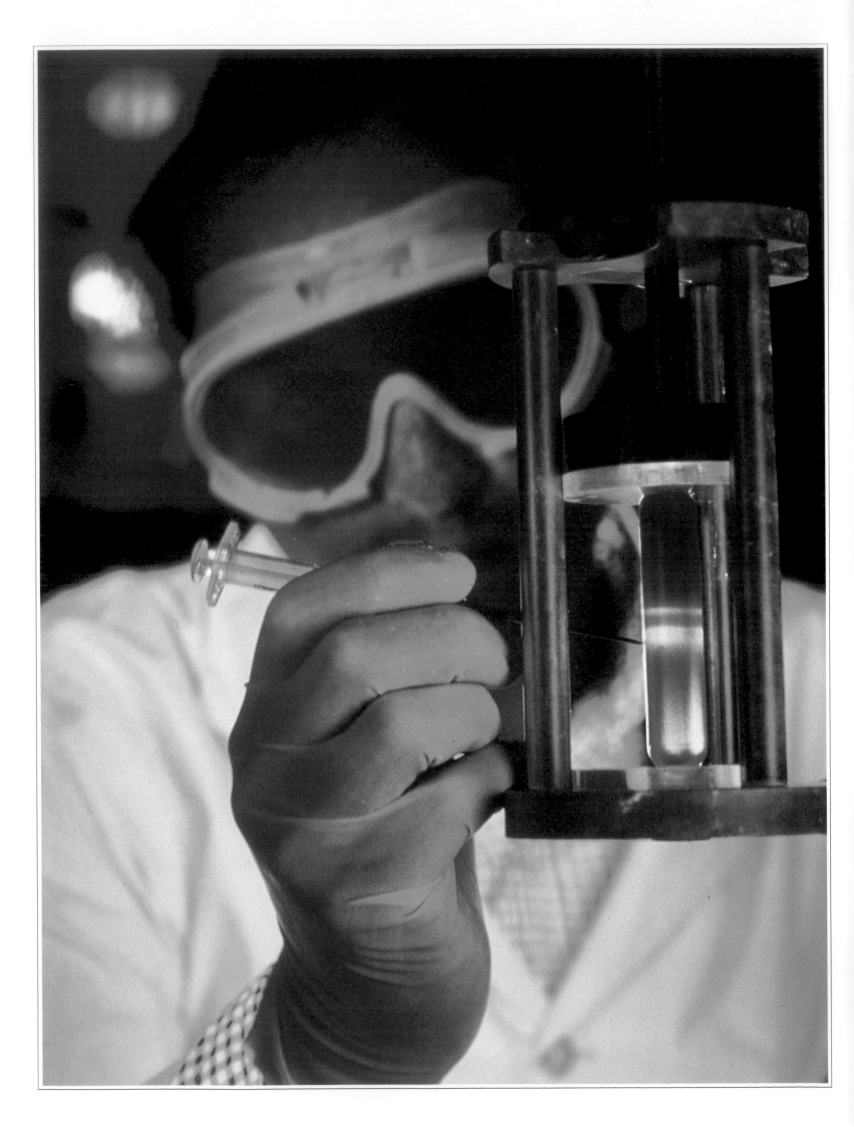

7

Immunology:
Headquarters and Battalions

The human body is an enclosed field within which there is an army constantly on the alert. We saw some aspects of this passive defense, which can be transformed into an active defense, when we looked at antibodies. Today, laboratories throughout the world are engaged in studying these chemical messengers that sound the alarm in the body and launch an attack against any invader – whether viral, infectious, or tumoral – and that have been discovered thanks to genetic engineering. These are the lymphokines, about fifty of which have so far been identified, and which are responsible for orchestrating the movements of an army not all of whose deployment procedures are as yet understood.

The Interferons

The real "star" among the lymphokines, the supreme "magic bullet" of them all, is interferon, even if the enthusiasme that marked its beginnings have dimmed somewhat.

Summer 1956, tea-time at Mill Hill in north London. At the National Institute for Medical Research, two men are talking about viruses. Both are scientists about thirty years old, one a Swiss, Jean Lindenmann, the other a Briton, Alick Isaacs. Lindenmann is speaking about interference, that strange phenomenon by which a person suffering from one viral infection is, in a way, immunized against any other virus. He is starting to talk about a method of deactivating viruses perfected by two Australians, Isaacs and Edney. Suddenly, Lindenmann stops, realizing that the famed biologist of whom he is talking is not Australian but British, and that he is talking to none other than Isaacs himself.

Isaacs had, in fact, worked in Australia, where he had studied the grippe virus. But he had always been obsessed with the same question: How can one virus interfere with another and bar its way? The two men resumed their study of this question together, and little by little realized that what they had

jokingly called "interferon" was not the reaction of one virus to another but a substance secreted by the infected cell as a result of the virus's activity. But it remained to be proved that such a substance really existed. They heated the virus to deactivate it (using the famous method developed by the "two Australian scientists"), then put pieces of chicken membrane in a neutral culture from the medium, added fresh membranes, and exposed them to a new virus: they resisted. In 1957, Isaacs and Lindenmann published their discovery: interferon existed.

Across the world, in Japan, Yasuiti Nagano, who had been studying immunity mechanisms against viruses since 1940 at the Kitasato Institute, observed this very same interference phenomenon, whose existence he postulated in a 1954 article and whose existence he actually demonstrated in 1958, naming it "the inhibitory factor." At that time, apparently, Isaacs and Lindenmann were unaware of the work being done in Japan, just as Nagano knew nothing of the research being carried out in England.

From the time of its birth, two guardian angels watched over the cradle of interferon: skepticism and enthusiasm.

16 January 1980, in a ballroom at Boston's Park Plaza Hotel, Charles Weissmann, a biologist from the University of Zurich, and Walter Gilbert, a biologist from Harvard (who was to receive the Nobel Prize that same year), annouced that Weissmann had succeeded in cloning and expressing alpha interferon. For in the interval, it had become apparent that there was not a single form of interferon but several (alpha, beta, gamma), each with its own different chemical and physical properties and mode of action. In order to exploit this discovery commercially, Walter Gilbert left academia for industry and founded Biogen, whose first laboratory was located in Geneva, in an abandoned watch factory – a sign of the times, perhaps. Today one of the most international of all genetic-engineering firms, it has labora-

Purification
Research on alpha interferon at Schering Plough: clinical tests started in 1981. A production plant has been built in Ireland.

(1) The announcement was made in
Japan on 27 November 1979.

tories in Cambridge, Massachusetts, and Ghent in Belgium, subsidiaries in Belgium and Delaware, and headquarters in... Curaçao! This press conference in 1980 marked a turning point, for now commercial exploitation followed close on the heels of scientific discovery, and instead of appearing in a scientific article in the specialized press, the announcement appeared the following day in the economics columns of newspapers.

That same year, the DNA responsible for the production of beta interferon was isolated by two teams simultaneously, one led by T. Taniguchi collaborating with Kyowa Hakko in Japan,[1] the other by Fiers in Belgium. D.V. Goeddel then cloned gamma interferon, and finally B.A.F. Markham at ICI in Great Britain, collaborating with researchers at the University of Leicester and the Curie Institute, synthesized an artificial gene that coded for alpha interferon 1.

Immediately after the Boston press conference, Biogen stock made a paper gain of $50 million. Interferon existed; it could even be synthesized. But would it be effective?

A few months later, on 12 April 1980, a press conference held by a Glasgow doctor seemed to confirm this: a child suffering from cancer of the ear and given up for dead who had been treated with interferon was improving rapidly. Unfortunately, however, the child died in July. But the damage had been done. Doctors were besieged with requests from patients or parents who suddenly saw a ray of hope. How to tell them that the dose required for a therapeutic trial on a single patient was estimated to cost over $20,000, that it needed betwen 600 million and 1 billion units, that adequate amounts of interferon were not available, or that basic problems in purification had been encountered?

But what had happened between Isaacs and Lindenmann's discovery and the Boston press conference?

First of all, it became apparent that interferon from chickens had no effect on mice or on human beings. For the latter, human interferon had to be obtained. At the beginning of the 1960s, a young Finn, Karl Cantell, working in Philadephia with Kurt Paucker, thought he had detected an antitumor effect caused by interferon. Cantell had, in any case, heard the idea put forth by Ion Gresser, that human leucocytes could provide a higher-quality interferon. Back in Helsinki and working for the Central Public Health Laboratory, Cantell was separating the blood of donors that had to be purified before being sent to the hospitals. The red cells and the plasma were the only things of interest to the doctors: the intermediary layer of white cells separated by the centrifuges were thrown away. Cantell recovered these, injected the leucocytes into eggs, added a virus, and incubated the whole for twenty-four hours at a temperature slightly below 100 °C. He then separated the leucocytes from the media in which they were assumed to have left their interferons.

The problem that arose was which virus to introduce into the egg. After dozens of tests and several years of patient research, Cantell fell on "Sendaï," which appeared active enough to induce the cell to produce an interesting quantity of interferon. "Cantell's soup," as it was known, fantastic in itself, was nonetheless a disappointment. It contained but 1 part per 1,000 of interferon for 999 parts of completely useless organic matter. But the amount of work involved was equally fantastic: at that time, to produce a scant 0.4 grams of interferon required about 51,300 liters of blood!

In the early 1970s, a young Swedish doctor, Hans Strander, tried Cantell's interferon at the Karolinska Hospital on his patients who had contracted bone cancer, and the results seemed encouraging. In New York, Mathilde Krim, an oncologist at the Sloan-Kettering Institute for Cancer Research, convinced the National Cancer Institute to hold a conference in 1975. Three years later, Jordan Gutterman, an immunologist at the M. D. Anderson Hospital in Houston, Texas, received a $2 million grant from the American Cancer Society. Enthusiasm even reached such a peak that a Manhattan discotheque was called Interferon. Later, the American Cancer Society's support rose to $7 million, while the National Cancer Institute's contribution exceeded $10 million. The supreme accolade came on 31 March 1980, when interferon made the cover of *Time*.

The problem of purifying the interferon has been temporarily solved by the use of a monoclonal antibody that recognizes interferon selectively. D. S. Secker and D. Burke were the first to obtain monoclonal antibodies that could detect interferon: immobilized on an insoluble support, the antibody becomes attached to the interferon present in the mixture, and they only need to be separated to obtain the biologically active interferon.

Some pharmaceutical concerns, such as Calbiochem-Behring (the American division of Hoechst), are beginning to produce interferon in substantial quantities using the Cantell method, while MIT is developing a simpler method. The Institut Mérieux in France, South Africa's National Institute of Virology, the government of Taiwan, and the Immunological Institute of the Belgrade Academy of Sciences have all joined in.

Small firms are rushing into the field (Cetus, Genentech, Biogen, Ares Serono, Cytotech), as are large corporations (Searle, Eli Lilly, Miles Laboratories, Merck, Schering Plough, Sanofi, Hoffmann-La Roche, Wellcome, Lederle, Upjohn, Roussel-Uclaf, Hoechst, etc.).

With genetic-engineering techniques, the problems of purification assume a new form: derived from human material, Cantell's interferon could do no harm, even if it did little good. But the interferon expressed by a bacterium is subject to far stricter rules.

Subsequent to the work of Weissmann in Switzerland, but also that of K. C. Zoon and R. Wetzel in California, as well as others, it was determined that there existed at least ten distinct molecular types of interferon alpha, two to five of interferon beta, and three or four of interferon gamma, which have different spectra of activity. Interferon (IFN) alpha is secreted by white corpuscles stimulated by a virus; IFN beta by the cells of conjunctive or fibroplastic tissue; IFN alpha-beta by lymphoblastoid cells, derived from the cells of a Burkitt's lymphoma (cancer of the lymph nodes); and IFN gamma by white corpuscles brought to the state of immunocompetent cells. Alpha and beta interferons would seem to have an antiviral activity, IFN gamma a more pronounced antitumoral activity.

1

1. **Interferon**
Frozen E. coli bacteria
containing genetically
engineered interferon
before the purification
process (this is still alpha
interferon, in this instance at
Hoffmann-La Roche).
2. **Analysis**
Electrophoresis on a gel
testifies to the purity of the
interferon.

3. **Interferon crystal**
4. **Simulation**
Hypothetical structure of a
partial sequence of the
amino acids of interferon.
5. **Production**
The production of gamma
interferon at Biogen for
clinical tests in the United
States.

2

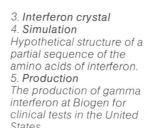

3

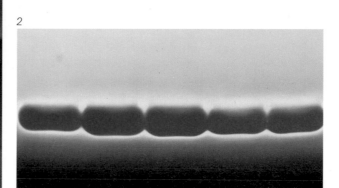

5

4

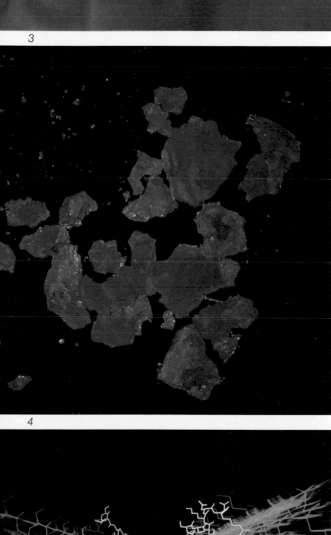

The genes governing the synthesis of these interferons have almost all been located, isolated, and expressed in bacteria or yeasts. But while their structure has been virtually known since the early 1980s, scientists have just begun to study the relationships between their structure and their function.

Interferons are proteins provided with a lateral sugared chain (glycoproteins) of considerable length, as they include chains of from 146 to 166 amino acids. Their three-dimensional structure is not yet known: a sequence of amino acids can thus assume several structures in space, only one of which would be compatible with antiviral activity. As the integrity of the molecule is not, it would seem, essential to its biological activity, one may well ask what part of the molecule is necessary for any expected activity and what is its appropriate spatial configuration? The functional specialization of the different areas of the molecule has been demonstrated by Weissmann (Biogen) and Goeddel (Genentech). In comparing the sequences of amino acids in human IFN alpha and beta, about 20 of them are always found in the same positions, mainly in the two areas of the molecule where there are two sulfuric amino acids, which play an essential role in the shrinkage in space of the proteins (they form a bridge between two distant areas on the molecule). At present, attempts are being made to synthesize only the one chain of amino acids necessary to the molecule's activity.

Research is simultaneously being conducted on the receivers located on all the cellular membranes to which IFN secreted by other cells become attached (there are from 1,000 to 10,000 receivers per cell, it is thought). But we are completely ignorant concerning the nature of the signal that, emanating from the receiver, reaches the DNA of the nucleus, and under whose effect are expressed the genes that are normally repressed. These genes then produce the enzymes that block the translation of the messenger RNA of the virus. It should perhaps be recalled that the virus acts a little like an outside program that short-circuits the cell's own and makes it change its behavior. A virus is, in fact, made up of a nucleic acid (DNA or RNA) enclosed in a layer of protein, the capsid. When it penetrates a cell, it sheds its capsid and, using its own DNA or RNA, forces the cell to produce new viruses. It is then that the cell begins to secrete a substance (interferon) and releases it at the moment that it bursts (cell lysis). When another cell receives this interferon, which becomes attached to a receiver on the membrane, it produces changes in the membrane and in the cell. And there lies the quandary: Just what is the structure that goes as far as the nucleus? In any case, there is "something" that inhibits the synthesis of the proteins and can slow down the proliferation of cancer cells or the multiplication of a virus. Thus IFN acts as an immunoregulator, a sort of biological rheostat.

The current balance sheet on the experimental use of interferon is noncommittal: positive, it would seem, as concerns certain viral diseases (colds, hepatitis, herpes, and rabies), which require small quantities, inconclusive as concerns cancerous tumors. Treatment with IFN seems to act on bone cancer (osteosarcoma), but, as this type of cancer is rare, statistical studies become difficult. For the more frequent types of cancer (of the breast or bone marrow), there seems to be improvement in 10-30% of cases (but no cures), accompanied by painful side effects, such as a significant upsurge of fever. The trend is increasingly toward the combined use of interferon, chemiotherapy, and radiotherapy.

Glycoproteins, like interferon, cannot be obtained by genetic-engineering techniques. The cultivation of cells in large tanks, as practiced by Sumitomo Chemical for alpha interferon (in a pilot plant that opened in 1985 and is expected to produce 5 billion units per year), is still extremely costly (and liable to contamination). Hayashibara has perfected an original method that may make it possible to "harvest" a thousand times more.

Hayashibara is a family firm founded in 1883 by Katsutaro Hayashibara (and that continues to be managed by one of his descendants). Expert first in the chemistry of starch and then in enzymatic techniques (maltose syrup, the conversion of dextrose), the company initiated its immunological researches in the early 1980s as the result of a meeting between Ken Hayashibara and Tsunataro Kishida, a researcher at the University of Medicine of Kyoto, who had taken an interest in IFN since Nagano's initial work.

Transplanting human cells to animals to preserve live cells (for toxicity tests) has been carried out industrially for over forty years now. But, generally speaking, this is a passive method of protection rather than an active method of cultivation. Why not use it to produce IFN? This was the origin of the Okayama hamster farm after a selection had been made of noncannibalistic hamsters so that, contrary to custom, the mothers did not eat their offspring. In 1981, construction began of the Fugisaki Institute, whose purpose is to produce 300 billion units of IFN annually.

The improvement in the culture of cells producing IFN, the identification of the monoclonal antibodies capable of recognizing them specifically, and the production of IFN by genetic engineering may make it possible to obtain interferons in substantial quantities far more cheaply, but several problems remain to be solved. Which interferon works best on a specific disease? Should it be employed alone, or in conjunction with other things? What doses should be given, how, and for how long? Why do reactions vary from one patient to another?

Meanwhile, the stakes are rising fast: in 1980, having supported Cetus's program for the industrial production of IFN, Shell Oil wagered $2 million. Biogen became allied with Schering Plough for the production of IFN alpha and IFN beta (particularly for IFN alpha 2 against the common cold, which will be marketed in Japan by Yamanouchi). Tests on skin cancer began at the San Francisco General Hospital in 1982 under the auspices of Paul Volderbing (University of California at San Francisco) and Michael Gottlieb (University of California at Los Angeles). Genentech has joined with Hoffmann-La Roche for research on IFN beta and the synthesis of IFN alpha, which will be sold by Takeda Chemical Industries in Japan. Purification of Hoffmann-La Roche IFNs will be done in agreement with Damon Biotech, which produces antiinterferon antibodies. The development of IFN alpha and beta was also the purpose of the agreement between Genex and Bristol Myers. Where IFN gamma is concerned, Genentech has become allied with Boehringer, and contracts have been signed in Japan with Toray and Daiichi Seijaku. In Japan itself, about fifteen different companies are competing, including Toray (IFN beta), Mitsui Toatsu Chemicals, Kyowa Hakko (IFN beta and gamma), and Suntory, where the cloning and expression of the gene of IFN gamma was achieved at the Suntory Institute for Biomedical Research by the team led by Dr. Shoji Tanaka. Not to mention Meiji Seika, which obtained the Searle license for IFN gamma, Toyo Jozo (IFN alpha), which signed a five-year contract with Johns Hopkins University Medical School, and Shionoji Pharmaceutical, which obtained Japanese sales rights for Biogen's interferon gamma. As for Hayashibara Biochemical Laboratories and Mochida Pharmaceuticals, they are conducting a joint lymphokines-interferons program.

Such a list is mind boggling. All seems ready for the birth of the magic child. But what form will he assume, and what will his behavior be? That no one can yet tell. "We are talking about the planet Andromeda," says Wataru Yamaya, General Manager of Mitsubishi Chemical, "but we haven't the least idea of what it may be – we've only got as far as the moon. But the important thing is that we're already in space."

Interleukins

Other stars in the immunological firmament are the interleukin 2s. A certain group of lymphocyte Ts, the "helper" T cells start secreting *interleukin 2* – the headquarters staff – which in turn activates an entire alarm, defense, and extermination system. The interleukin 2s become attached to the surface receivers of the lymphocyte Ts, which results in the rapid multiplication of "helper" cells and, consequently, an increase in the number of interleukin 2s.

At the same time, the interleukin 2s become attached to the surface receivers of another group of T cells, the *"killer" T cells*. These "killers" start dividing, secreting substances that destroy the invader of the tumorous cell.

In the case of interleukins alone, for they are the key to the immune system, there are nearly fifteen companies now engaged in clinical tests. Leading the field at present are Cetus, Biogen, and a small newcomer that calls itself quite simply Interleukin 2. How is interleukin 2 formed? The alarm signal, on the one hand, leads to the mobilization of the lymphocyte Ts (alert but not active) and, on the other hand, to the secretion of interleukin 1 in the macrophages, thus pushing certain types of lymphocyte Ts to produce interleukin 2, which moves toward the alert lymphocyte Ts and gives them the order to intervene. The interleukin 2s are thus immunomodulating hormones acting on cells with specific receivers and exciting these (the word hormone comes from the Greek verb "to excite"). They exercise their activity on the lymphocyte Ts, which are differentiated in the thymus (a gland near the heart toward the front of the rib cage whose role in immunology has been appreciated for almost twenty years now).

The cloning and expression of interleukin 2 was successfully accomplished by four teams of researchers simultaneously in 1983: one led by Prof. Tadatsugu Taniguchi of the Cancer Institute in Tokyo, one led by Prof. Fiers in Belgium (in conjunc-

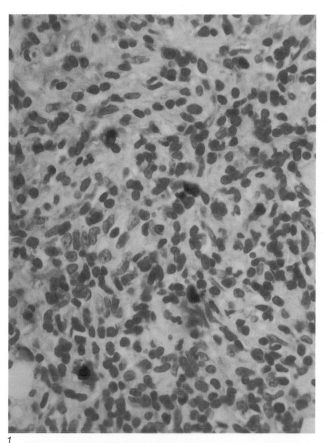

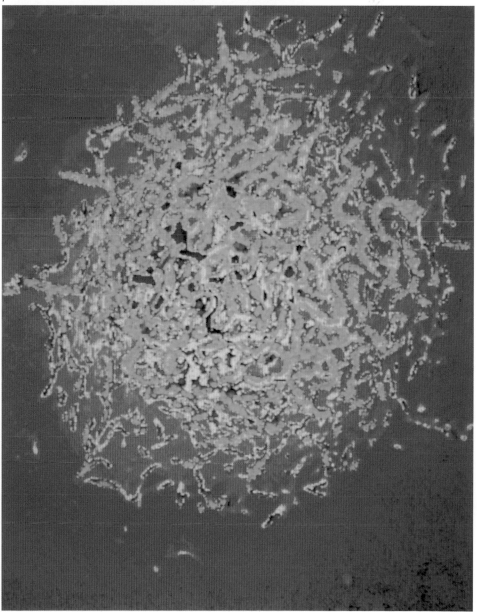

1. *Immunoglobulins*
Here, antibodies, or immunoglobulins, "detect" a destructive lymphoma.
2. *Macrophage*
The "ogre" of immunology. In contrast to the killer cell, the macrophage does not spit venom at an intruder, but simply swallows and digests it inside the cell.

1

2

tion with Biogen), Prof. Robert Gallo's team at the National Institutes of Health in Bethesda, Maryland, and Prof. Robb's team in Wilmington, Delaware (Du Pont de Nemours). For the initial research, use was made of interleukin 2 produced in vitro by a tumor-ous strain of lymphocyte Ts. Immunochemical puri-fication was carried out at Du Pont de Nemours thanks to a monoclonal antibody that, for unknown reasons, is difficult to obtain in other laboratories. It became apparent that the interleukin 2 from the tumorous strain is slightly different from the natural interleukin 2 produced by normal lymphocytes, which may reflect a functional difference. The inter-leukin 2 produced by cloning and expression in a bacterium is thus being studied, and it is thought that it will be possible to produce the pure hormone in large quantities. During the summer of 1984, vari-ous research organizations, including the National Cancer Institute and the National Institute of Allergy and Infectious Diseases, began their first large-scale clinical tests. Different tests have indicated that the insertion of interleukin 2 into the body increases the production of "natural killers" (NK) in AIDS cases. As interleukin 2 may also be consi-dered a growth factor, since it permits the prolifera-tion of lymphocyte Ts, it may eventually be used against cancer (where, as we have already seen, growth factors play an important role). The adminis-tration in vivo of interleukin 2 has already made it possible to reestablish the immune systems of some mice whose thymuses were removed. All such signs are encouraging, but most researchers still remain cautious as to the outcome of the tests.

And the Others...

While interleukin 2 and interferon are the best known of the lymphokines, they are not the only ones being investigated by research laboratories. There is also *interleukin 1*, which has just been pro-duced by genetic engineering and whose clinical tests will soon begin. In the running here are SmithKline Beckman, Hoffmann-La Roche, Colla-borative Research, and Immunex working together with Syntex. Interleukin 1 may speed the recovery of patients after surgery and help both to diagnose and to treat chronic inflammatory diseases. There are also various factors, such as the *colony stimu-lating factor*, a hormone that stimulates the bone marrow to produce lymphocytes and that could help in treating anemias and leukemias or in rein-forcing the defenses of those who have had a graft of bone marrow. Clinical tests will soon start using this lymphokine. Involved are SmithKline, Cetus, Immunex, and the Genetic Institute in association with Sandoz. The *macrophage activation factor* is still being produced in such limited quantities as to make clinical tests impossible. Research on this factor is being jointly conducted by Immunex and SmithKline. This lymphokine may be used in the treatment of certain cancers, such as lung cancer. The *B cell growth factor*, developed by Biotech Research Laboratories, Hoffmann-La Roche, and Biogen, may increase the effectiveness of vaccines and be used in the treatment of leukemia, but it too is still produced in insufficient qualities. In addition to these, the *tumor necrosis factors*, secreted by B cells and macrophages, which cause tumors to explode, have been added to Genentech's research program.

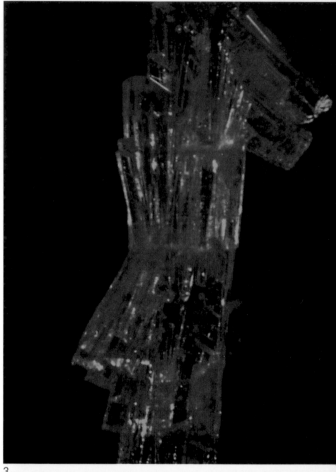

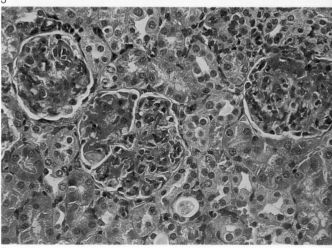

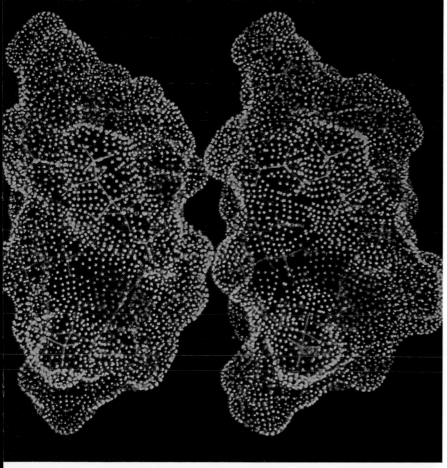

Cyclosporin an immuno-depressant

Immunodepression, or immunosuppression, makes it possible to eliminate some of the unwanted reactions of the immune system. It has once again become popular since the introduction in human clinical treatment of antilymphocyte T monoclonal antibodies and cyclosporin and makes it possible to consider treating such autoimmune diseases as myasthenia, hyperthyroid conditions, diabetes, and multiple sclerosis. It is in this way, for instance, that test doses of cyclosporin, administered during the first weeks following the onset of diabetes, seem to have slowed down the development of this disease. Treatments with immunodepressants have also made it possible to make progress with organ transplants. Cyclosporin was isolated from a Norwegian fungus *(Tolypocladium inflatum)* whose spore cells can be seen here (1). In 1975, its structure was determined at the Sandoz Laboratories by Roland Wenger, and since 1983 it has been sold commercially. (It is produced by the massive cultivation of the fungus.) But what part of the structure of the cyclosporin is responsible for its activity as an immunodepressant? If an answer can be found to this question, it may become possible to develop derivatives, modified to improve the immunosuppression phenomenon or reduce secondary risks. To study the relationship between a substance's structure and its biological activity, the total synthesis of the substance (crystals of synthetic cyclosporin: 3) is a particularly interesting investigatory technique. Thus researchers at Sandoz have been able to determine that this activity undoubtedly stems from a single amino acid – here in crystalline form and photographed in red light (2). To study the active area, a comparison is made on a computer between the Van der Waals surfaces (reached by water) of natural cyclosporin (left) and of modified cyclosporin (right), and it becomes apparent that a simple modification of the density of spatial compression on the right is enough to block the activity of the molecule (4).

Cyclosporin acts on a disease that affects mice, glomerulenephritis, caused by the formation of immune complexes that are deposited in the renal glomerules (5). The appearance of the glomerules on the mouse undergoing treatment is once again almost normal (6) – there are only minor residual lesions.

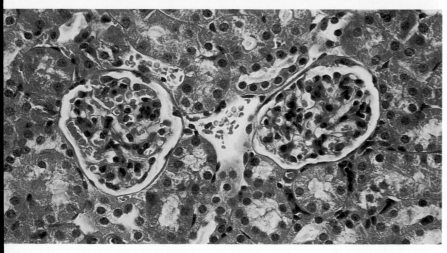

That all of this research is of considerable interest is obvious to everyone, but the use of lymphokines in the treatment of diseases still remains a distant prospect. Perhaps they will have to blended into a subtle cocktail of carefully balanced ingredients. Researchers at the National Cancer Institute, for instance, recently showed that giving a mouse with lung cancer interleukin 2 and "killer cells" in separate doses had no effect, but that a combination of the two destroyed the tumor.

In the course of their research, scientists have discovered more and more about how autoimmune responses work. In the case of rheumatic arthritis, for instance, it would appear that the immune mechanism can run away with itself. In Texas, Marguerite B. Kay discovered that certain cells, having become old, commit suicide, so to speak, by inducing the immune system to come and destroy them. It is almost as if they put on an enemy uniform (by producing an antigen) so that the soldiers in the same regiment of antibodies would detect them and finger them to the voracious macrophages. In addition to the implications that this discovery may have on research into the aging of cells, it may also mean that the immortality of cancer cells is in fact due to the absence of this suicide antigen. Work is also being concentrated on the receivers in an effort to "unhook" them and thus parry the responses rejecting grafts or those involved in autoimmune illnesses.

Such is, essentially, the panoply of arms available to immunology. If we eventually succeed in the industrial production of lymphokines, it will be possible to reinforce a faltering immune system or to give back to a patient suffering from AIDS or receiving anticancer chemotherapy the defenses he had lost. It may also become possible to encourage the body to put up a fight against the so-called autoimmune diseases (that is to say, those in which the chemical soldiers of immunology no longer know where to stop, but, overflowing the line of defense, attack the body itself), such as rheumatoid arthritis and multiple sclerosis.

Once again, large pharmaceutical concerns (Hoffmann-La Roche, Du Pont de Nemours, Schering Plough) and small companies (Genentech, Cetus, Biogen) are competing. But have not certain economists predicted that by the year 2000 the market for lymphokines in America alone will equal the $1.7 billion now paid for antibiotics?

"The thing that should be remembered is that when the day comes that these techniques are perfected, a single 1,000 liter fermenter, in the case of certain lymphokines, would be enough to supply the entire world. That is why competition is so stiff and why there will be so many losers," concludes Hiromu Sugama, Director of Bio-industrial Development at Mitsui Toatsu Chemicals.

This fact-finding trip we have made into the army ever on the alert to defend the body, but an army that may at times lose some of its shock troops and even be cut off from its headquarters and call for reinforcements, shows that the situation is far less certain than the initial cries of victory might have indicated. But it also describes one of the most fascinating adventures of those modern-day explorers who try to map the mysterious ways of an unknown world: the researchers in molecular biology.

Immuno-genetics

by Jean Dausset, Professor of Experimental Medicine
at the Collège de France, member of the Académie
des Sciences and the Académie de Médecine, Nobel
Prize in Medicine in 1980.

*Born in Toulouse in 1916, Jean Dausset has
devoted his life to the immunogenetics of
blood cells. In 1958, as chief of the Immu-
nohemological Laboratory of the Centre
National de Transfusion Sanguine, he described the first leucocytal antigen, which
became HLA-A2. At the same time, with
Professor Robert Debré, he undertook a
reform of the university and hospital structures in France, which made it possible for
biology to flourish and led to a renewal of
medical research. It was in 1965 at the
Hôpital Saint Louis, where he was in charge
of the Immunological Department, that he
described the first tissue group system, later
to be called HLA.*

*In 1977, the Collège de France named Jean
Dausset to the Chair of Experimental
Medicine, a position held from 1855 to 1878
by Claude Bernard. In 1980, along with the
Americans B. Benacerraf and G. Snell,
J. Dausset was awarded the Nobel Prize in
Medicine for their discoveries "concerning
the genetically determined structures of the
surface of the cell which determine immu-
nological reactions."*

Immunogenetics is the study of genetics
using immunological tools, essentially specific antibodies, but also cells specifically
immunized against a particular antigen.
These methods are used to detect a "specificity" present in a few individuals of a species that is absent from other individuals of
the same species. Researchers thus identify "groups" of individuals that are "carriers"
of this specificity and other "groups" that
are not carriers. The most commonly known
groups are the blood groups A, B, and 0.

But immunogenetics is not limited solely to
the study of "gene products," as these specificities are called. It has also begun to
study the genes that code for them. These
genes are found in the large deoxyribonucleic acid molecule (DNA), clustered in the
nucleus of every cell in the organism.
Immunogenetics is concerned with finding
the exact location of genes on each chromosome and determining the nucleotide
sequence and mode of expression.

The history of immunogenetics began in
1900 with Karl Landsteiner's discovery of
the A, B, and 0 blood groups. By separating
the serum or liquid part of the blood from
the red corpuscles, and mixing serum from
one individual with red corpuscles from
another, he observed for the first time that
red corpuscles tended to agglutinate in
some combinations but not in others.
Landsteiner then divided the population
into "groups," and with his pupils went on to
demonstrate the existence of two specificities or antigens, A and B, in human red
corpuscles, or erythrocytes. Depending on
whether these antigens are present simultaneously, in isolation, or are absent from
erythrocytes, four blood groups can be
defined: A, B, AB, and 0 (standing for zero
and not the letter O, to show that there are
no antigens in this group).

In 1910, Von Dungern and Hirszfeld published a theory according to which heredity
of A and B specificities depended on two
totally independent pairs of genes. This
theory was proved to be false, and it was
not until 1924, twenty-four years after
Landsteiner's discovery, that Bernstein proposed the theory that is now universally
accepted: A and B specificity or absence of
specificity is conditioned by a single series
of genes, which all occupy the same site or
locus on the same chromosome. These
three genes (gene A conditioning specificity A, gene B conditioning specificity B,
and gene 0 conditioning absence of either A
or B on red corspuscles) are, by definition,
alleles, or variations of a single gene, occupy-
ing a single locus in the genome. These three
alleles define a system, i.e., the AB0 system.

Moreover, we may also observe that the
two genes A and B can coexist in one individual. We say then that they are codomin-
ant, i.e., that neither one inhibits or dimi-
nishes expressions of the other.

To arrive at a correct understanding of this

allelic system, which has served as a basic
for immunogenetics, we must also grasp
the following points:
- that the DNA molecule in man at the
moment of cell division contains 23 pairs of
chromosomes;
- that at the moment of cell division leading
to germinal cells (spermatozoa and ova),
these pairs split up and form two new cells;
- that each individual, created by the fusion
of a spermatozoon and an ovum, therefore
possesses two homologous genes, i.e., two
genes corresponding to the same locus,
one from his father and the other from his
mother.

We can thus make up the following table
for the AB0 system:

		Spermatozoon		
		A	B	0
Ovum	A	AA	AB	A0
	B	BA	BB	B0
	0	0A	0B	00

Phenotype is the name given to the
expressed characteristics (in this case
those found on the surface of red corpuscles), while genotype refers to the fundamental consitution of the two homologous genes.

Since the 0 gene is not expressed, the
phenotype, or group A, is found in all individuals of genotype AA or 0A; likewise,
group B is found in all individuals with
genotype BB, B0, or 0B. Phenotype AB is
found in individuals of genotype AB and
phenotype 0 in individuals with genotype 00.

If we know that anti-A antibodies are found
naturally in B individuals, and anti-B
antibodies in 0 individuals, while AB individuals have no natural antibodies, we can
begin to understand the rules of blood
transfusion:
- 0 individuals are universal donors since
their red corpuscles do not agglutinate in
any individuals;
- AB individuals are universal receivers
since they have neither anti-A nor anti-B
antibodies.

These properties provide the basis for the
everyday practice of blood transfusion
according to the following diagram:

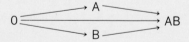

Moreover, certain populations around the
world have characteristic blood groups. For
example, the inhabitants of Siberia are
largely group B, while American Indians
belong almost exclusively to group 0.

Immunogenetics can thus be a remarkably
accurate tool for anthropologists.

We have emphasized the AB0 system here
since it was the first and most important

discovery for medical practice. But there are many other blood group systems, which are also inherited according to Mendelian laws but which differ from the AB0 system in that there are never any natural antibodies in the serum of nonimmunized individuals. Immunization always occurs as a result of transfusion or pregnancies incompatible with this system.

The most well-known such group is obviously the Rhesus system. Here one of the alleles is referred to as Rhesus negative because of the simultaneous absence of the three factors C, D, and E on the Rhesus molecule it carries. These factors are replaced by three other factors: c, d, and e. Rh negative individuals have received the cde factors from both parents, and can therefore be easily immunized against the CDE factors, which they do not possess (whereas the reverse is far more difficult, or even impossible).

Rh negative women can therefore easily become immunized against the red corpuscles of the fetus they are carrying, but the process of immunization can have a disastrous effect on the fetus, destroying its red corpuscles, or even causing death in utero.

The following diagram shows a family at risk from this problem:

$$\text{Father} \frac{\text{Rh CDE}}{\text{Rh cde}} \times \text{Mother} \frac{\text{Rh cde}}{\text{Rh cde}}$$

First child	
$\frac{\text{CDE}}{\text{cde}}$	Usually not affected since there is no preimmunization
Second child	
$\frac{\text{cde}}{\text{cde}}$	Not affected
Third child	
$\frac{\text{CDE}}{\text{cde}}$	Affected because the mother has been immunized by the CDE antigens of the first child

Thus, on average, one in two children in this family (cde/cde) will not be affected, but the disease will strike the other children with increasing severity, since the immunization will be accentuated by further incompatible pregnancies.

The Rhesus system is also important in transfusions: it is obviously essential to give only Rh negative blood transfusions to Rh negative patients to prevent development of anti-Rh antibodies, which could be disastrous for women, but also for men if they accidentally or unavoidably receive Rh positive blood.

P. Gorer, in 1936, and G. Snell, in 1948, discovered tissue groups in mice. The H2 system was extensively studied and has since served as a model for understanding its human equivalent, the HLA system (H for

Histo-Tissues, L for Leucocytes, A for first tissue system).

The discovery of tissue groups in man was independent, however, of the discovery of similar groups in rodents. It was made (Dausset in 1958, J.J. Van Rood in 1962, R. Payne in 1963) in the same way as the discovery of red corpuscle groups, by observation of agglutination of white globules (leucocytes) in one individual as a reaction to the serum of another immunized individual (figure 1).

This system is an extraordinarily complex one. In fact, each individual receives from each of its parents for the same segment of chromosomes (chromosome 6) at least 6 genes, which are expressed on the surface of the cells. Three of them – HLA-A, HLA-B, and HLA-C (referred to as class I) – are expressed on the surface of almost all the organism's cells. The others – HLA-DP, HLA-DQ, and HLA-DR (referred to as class II) – are only expressed under normal conditions on certain subpopulations of leucocytes: B lymphocytes and some monocytes (figure 2).

Thus up to 12 distinct specificities can be expressed on the B lymphocytes of any individual. Where the individual has received the same specificity from both parents, this figure is of course smaller: such individuals are then referred to as homozygotic for the corresponding gene.

Finally, there also exist many variations or alleles for each of these 12 genes. For example, there are around 20 for HLA-A, 40 for HLA-B, 15 for HLA-DR, etc. This gives us an almost infinite number of possible combinations.

The discovery of all these alleles was made collectively by the scientific community, which began organizing international cooperative research on this point in 1965. Had it not been for such exceptional cooperation and the enthusiasm of all involved, this goal may never have been attained, or at least not for many more years.

The HLA system, or what we might more correctly call the HLA complex, is often referred to as the MHC (Major Histocompatibility Complex) or Tissue Matching System, which emphasizes its importance in tissue transplantations. For a tissue transplantation to be tolerated for more than a short period of time, it must be compatible not only in terms of the AB0 system, but also as compatible as possible in terms of the HLA system. This was demonstrated first of all in skin grafts (figure 3), and as the different techniques developed, in kidney, liver, and heart transplants, and finally in bone marrow transplants.

Kidney transplants are now common practice. The best results are obtained when the donor is a brother or sister (sibling) with the same HLA group as the patient. If there is

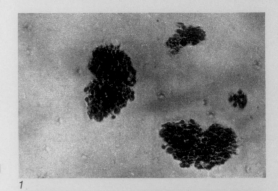
1

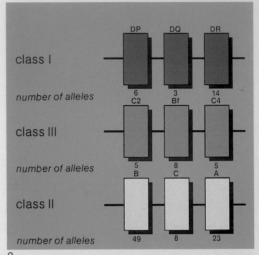

class I

number of alleles

class III

number of alleles

class II

number of alleles

2

3

Fig. 1
Leucoagglutination obtained by mixing normal leucocytes with the serum of a patient containing antileucocyte antibodies developed after a large number of transfusions (observed since 1952).

Fig. 2
Representation in diagram form of the short arm of chromosome 6, on which we find genes coding for molecules HLA class I or II, and complementary factors (class III). Each rectangle represents a gene, or a familly of genes. The distances between rectangles illustrates the supposed distances between genes. The figures correspond to the number of variations or alleles for each gene.

Fig. 3
Two skin grafts, 1 centimeter (cm) in diameter, made to the forearm of a volunteer. On the left is a "white," nonvascularized, graft, rejected immediately; on the right, a "pink," vascularized graft, accepted temporarily, but which will be rejected under normal conditions by around the tenth day.

no sibling or the sibling is unable to be a donor, the kidney must be taken from a body with the closest possible compatibility. This is why national and international organizations have been set up to organize exchanges of kidneys between teams. A central computer contains all tissue groups in its memory along with the essential characteristics of patients waiting for transplants. At the present time, powerful immunosuppressant drugs give remarkable results in long-term grafts.

The second clinical impact of tissue groups was the discovery that certain HLA groups were found with greater frequency in individuals suffering from certain diseases. Thus, antigen HLA-B27 is found in 90% of people afflicted with rheumatoid spondylitis; 80% of people suffering from insulin-dependent diabetes possess either HLA-DR3 or HLA-DR4 or both. The table of main pathologies gives a list of the most striking associations with an evaluation of the relative risk for each. For example, an individual with positive HLA-B27 has 180 times more chance (Relative Risk) of contracting rheumatoid spondylitis than a subject with negative HLA-B27.

It is obvious that the associations can be of great service in clinical practice. They can be used in diagnosis, and sometimes in therapy, but, most important, open up the possibilities of preventive medicine. It becomes possible to identify the individuals most likely to contract a particular disease, which is extremely useful where preventive or early treatment can be prescribed to arrest development of the particular disorder.

Researchers believe that these associations exist because HLA antigens play an essential role in immune response. This has been demonstrated in guinea pigs (B. Benacerraf, 1963) and mice (H. McDevitt, 1969). The antigens of the major histocompatibility complexes constitute the individual's profile, and the organism learns not to develop any immune reaction to them, except in some very rare cases of autoimmune diseases.

In 1974, R. Zinkernagel and Doherty discovered that foreign antigens (such as viruses) modified MHC antigens, making them unrecognizable to the organism. The antibodies or killer cells (cytotoxic cells) then developed a defense reaction against these antigens.

Immunological mechanisms may be described in diagram form as a process involving three essential actors, which researchers have not yet been able to identify completely:
1. MHC molecules (self) function as a "trap" for structures foreign to the organism (X) and form a self +X complex found on the surface of mononuclear cells (especially monocytes). But in some cases, this complex does not form and there is no immune response. The individual does not respond to X, since he does not possess the tissue group needed to form this complex.
2. The "self +X" complex = other presents itself to the lymphocytes, which transmit it to other lymphocytes, some of which (lymphocytes B and plasmocytes) secrete specific antibodies against the foreign antigen, while others (T lymphocytes) lyse or dissolve cells with the "self +X" complex on their surface.
3. A feedback reaction, which also originates in the cell (T lymphocyte suppressors) works to inhibit this process once it has attained its objective.

In this subtle process, we can see the essential role played by the HLA complex in defending man against the aggressions to which he is permanently exposed. It is the seal of his personality, his individual profile distinguishing him from all other beings by its particular composition.

This diversity is useful not for the individual himself, but for the species. In fact, under conditions of outside aggression such as an epidemic, some individuals with a particular HLA profile cannot respond to the agent, and generally do not survive, whereas others do possess the correct profile and respond well. They either do not catch the disease or, if they do, are able to recover.

Thus the species can continue to perpetuate itself despite the wide range of aggressive agents to which it is exposed, particularly since these agents are themselves capable of diversifying.

This brief summary demonstrates the wide range of theoretical knowledge, basic concepts, and practical applications that researchers have harvested from the study of immunogenetics: transfusion, transplants, immune response. An excellent harvest in less than one hundred years.

Main pathologies associated with or linked to HLA

HLA antigens	Pathology	Relative Risk*
A1	Hodgkin's disease	1.4
A3	Idiopathic hemochromatosis	8.2
B5	Behcet's disease	6.3
B14, B47	Congenital suprarenal hyperphasia	15.4
B27	Rheumatoid spondylitis	87.4
B35	Subacute thyroiditis	13.7
Cw6	Psoriasis	13.3
DR2	Multiple sclerosis	4.1
DR3	Systematic lupus erythematosus	5.8
DR3	Addison's disease	6.3
DR3	Basedow's disease	3.7
DR3	Myasthenia	2.5
DR3	Extramembranous glomerulopathy	12.0
DR3 and/or DR4	Insulin-dependent diabetes	6.4
DR3 and/or DR7	Coeliac disease	4.8
DR4	Rheumatoid polyarthritis	4.0
DR4	Pemphigus	14.4
DR5	Bierner's anemia	5.4
DR5	Hashimoto's disease	5.6
DR5, DRw8	Juvenile arthritis	5.2

* Relative Risk: factor by which the risk of contracting a disease is multiplied for individuals carrying the corresponding HLA antigen. For example, individuals in group HLA-Cw6 run 13.3 times more risk of contracting psoriasis than individuals who do not possess this group.

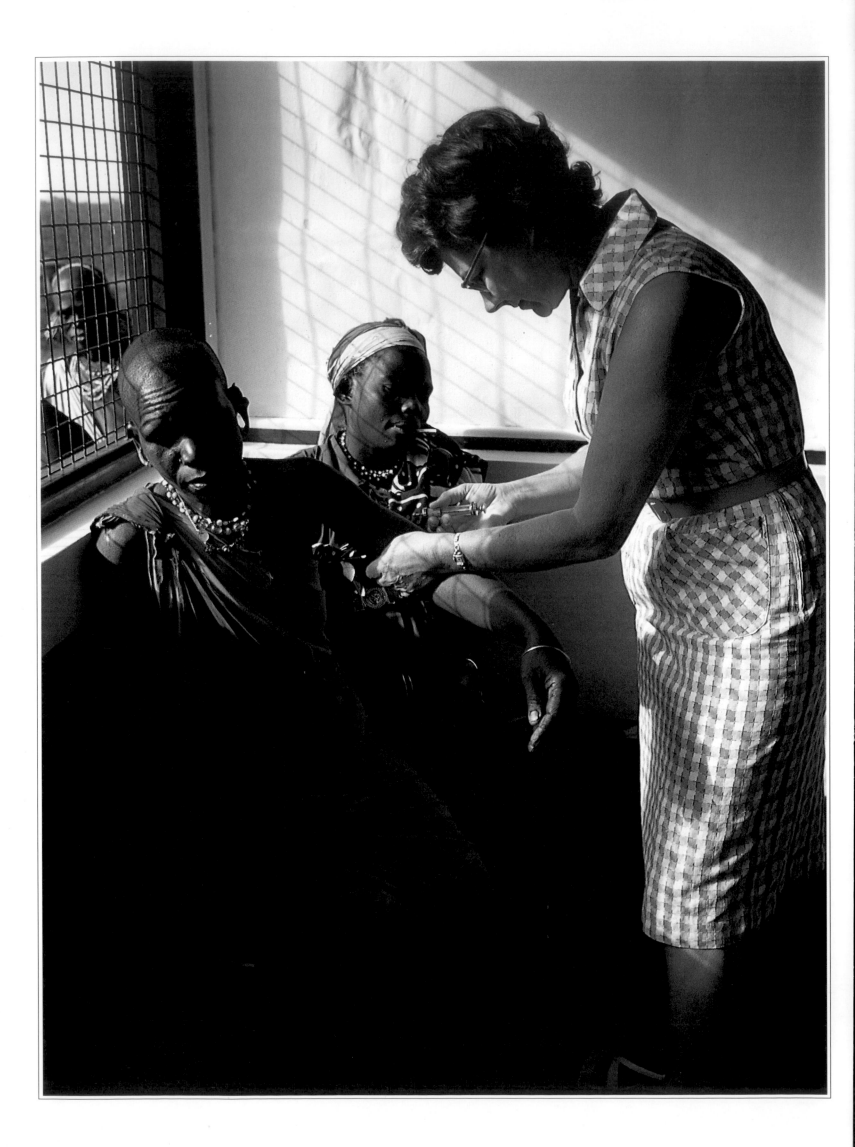

8

Preventing Invasion:
The Vaccines

The first weapon in the armory of immunology was forged in 1797 when an English doctor, Edward Jenner, inoculated a child with the virus of a vaccine to protect him from smallpox. But it was Pasteur who, during the second half of the nineteenth century, established the scientific basis of vaccination. Let us return to that day in 1864 when, before a brilliant audience at the Sorbonne, he declared, "Life is a germ, and a germ is life!": fermentation, decomposition, putrefaction, all are caused by minuscule living organisms. And what about disease? In trying to protect his experimental media against all outside contamination, Pasteur opened the way not only to aseptic conditions and clinical sterilization, but to the "pasteurization" of foods such as milk (thus saving thousands of lives). He also founded the new discipline of bacteriology. From the "disease" of beer and wine, he turned to the infection of animals and men, and the rest was history: first the vaccination of sheep against anthrax, then from 1883 onward, the vaccination of children against rabies.

Since Pasteur's day, the principles of vaccination have remain unchanged: the body is injected with a dead or "stunned" microorganism to alert the sentinels of its immunological defenses. The molecules of the corresponding antibodies recognize the "cloak" of the virus or the microbe and mobilize for action. Then they immediately neutralize the invader by preventing it from becoming attached to the host cell. Unfortunately, present-day vaccines still pose many problems. The "stunned" invader may awaken and become dangerous. Certain vaccines are expensive or risky for those preparing them (hepatitis B); others are relatively ineffective (cholera); still others do not exist yet (malaria). It is difficult to protect the body from parasitical diseases, as the "wardrobe" of the parasite enables it to assume a thousand disguises and evade even the most watchful antibody. To top it all off, such recent phenomena as the resistance developed by certain bacteria (gonorrhea and meningitis) to antibiotics and sulfa drugs and the appearance of new strains in such illnesses as grippe require the elaboration of new vaccines against the diseases that they cause.

With the discoveries of biotechnology, there has been a strong tendency to turn to "second-generation" vaccines produced by genetic engineering, or even, with an eye to the more distant future, "third-generation" or synthesized vaccines.

Where industry was concerned, it was high time. For a variety or reasons, annual sales of vaccines stagnated, then slumped: the market for animal vaccines hovered around the $50 million mark, while that for human vaccines came to about $100 million on a worldwide basis (the annual loss from animal illnesses is estimated to cost some $50 billion!). According to forecasts made by Lance Gordon, head of the Immunology Department of Connaught Laboratories, the market for human vaccines alone could jump to $250 million in 1987, thanks to sales of the new vaccines.

Looking forward to the future are the grand old veterans in the field: the Institut Pasteur and the Institut Mérieux in France, the "big five" in the United States (Merck, Lederle, Connaught, Parke-Davis, and Wyeth) and Wellcome in Great Britain. Following them come the new advanced-technology companies (Genentech, Cetus, Molecular Genetics, Amgen, Biogen, Transgene, and Akzo), and finally the newcomers: Norden Laboratories (of the SmithKline Group), Beecham, Burroughs Wellcome, Anchor, and Pitma-Moore (of the Johnson and Johnson group).

One of the first vaccines obtained using genetic-engineering techniques was that for hepatitis B.

There is, in fact, no effective treatment for viral hepatitis, of which the most dangerous form is hepatitis B, with its long-term risk of cirrhoris and cancer of the liver. The number of chronic carriers in the world (that is, those in whom the surface antigen of the virus persists after the end of the illness) comes to about 200 million. Even with the first attempts to develop a vaccine in the 1970s, a major

Prevention
A vaccination campaign in Ethiopia.

93

obstacle arose: the impossibility of cultivating the virus in cellular strains. For the first time, thoughts turned to a vaccine of human origin: Would it not be possible to extract from the serum of chronic carriers certain nontoxic particles of the surface antigen of this virus whose structure was henceforth well known?

The first commercial vaccine was developed by Philippe Maupas and Institut Pasteur Production in 1980 and M. Hilleman at Merck. Yet, because many people feared becoming contaminated by blood-borne factors, such as those causing AIDS (which had for so long remained a mystery), the sources of human blood began to dry up. In any case, such vaccines can be exceedingly costly, the processes for producing them can be long and difficult, and some of them are even dangerous.

A solution was sought in genetic engineering. In the vaccine developed by Merck, the gene that codes for the surface antigen (the active ingredient) of the virus is separated from the DNA, purified, and reinserted into a yeast, which excretes the vaccine.

1, 2. Vaccine
Nineteenth-century American farm for producing vaccine (1) and an engraving of Gillray at the same period (2).
*3. **Koch, Ehrlich, Behring***
The "founding fathers" of immunology: Robert Koch isolated the bacillus of tuberculosis, the treatment for which, developed by Hoechst, was one of the first immunological preparations. Paul Ehrlich, a disciple of Koch, worked on antibodies and developed a treatment for syphilis. Another disciple of Koch, Emil von Behring, discovered antitoxins and their applications in serotherapy. All three received the Nobel Prize for their work.

(1) Or more accurately the "epitope," that is to say, the specific antigenic determinant.

2

1

3

This shortens the process, which takes only a few weeks instead of several months, and eliminates problems linked to shortage of plasma. The crucial problem at present is how to ensure that the new vaccine is permanently effective.

The virus causing hepatitis A has also been identified; in contrast to that causing hepatitis B, it can be cultivated in cellular strains. The cloning of the A virus in a colibacillus was accomplished in 1981 by Friedrich Deinhardt in Munich, with the result that we now have a detailed genetic map of the virus. M. Hilleman's team is now trying to produce a vaccine.

A microbe can also be made inoffensive by excising from it that part of its DNA that causes the infection. Thus at Juntendo University in Japan, Tatsuo Yamamoto has excised from the bacterium the gene responsible for traveler's diarrhea ("Montezuma's revenge"). Rendered inoffensive, the bacterium subsequently mobilizes the antibodies that, it would seem, provide effective protection.

In Berne, biochemist René Germanier approached the problem differently: he modified the genetic heritage of the bacterium responsible for typhoid in such a way that it absorbs galactose in such quantities that it becomes incapable of metabolizing it, so that the mutant dies by poisoning after having lived long enough to develop an immunity response in the intestine, but not long enough to have obliged the organism to produce toxins. Following similar principles, B. Roizman of the University of Chicago has prepared an antiherpes vaccine.

Herpes – above all, the genital herpes that now affects 5-20 million Americans, with 300,000 new cases each year – is a plague against which, once again, there is no effective treatment. The gene responsible for herpes simplex has been cloned and expressed in *E. coli* by Lynn Enquist of Molecular Genetics, who signed an agreement with Lederle for the commercialization of the new vaccine.

Synthetic Vaccines

But with the increase in knowledge resulting from rapid strides in molecular biology and in genetic-engineering techniques, an increasingly serious attempt is being made to identify the active antigenic areas of the pathogenic microorganism and to copy its structure chemically. This would then make it possible to synthesize the short peptides ("vaccines of the third type") capable of provoking the formation of antibodies. This hypothesis had been germinating in the mind of a researcher at the Rockefeller Institute, Walther Goebel, in the late 1930s, and his experiments showed that the presence of the entire pathogenic agent was not indispensable in order to obtain protection. Various work on the synthesis of peptides, particularly that of R. Merrifield at Rockefeller University in 1960, aided in proving this experimentally.

With the aid of monoclonal antibodies, it has become possible to identify the antigen [1] that sets off the immune response and to study the sequence of amino acids. One may also – and this is frequently the method employed – isolate the fragment of the genome that controls the synthesis of this antigen, study the sequences of the nucleic acid, translate with the help of the genetic code, and

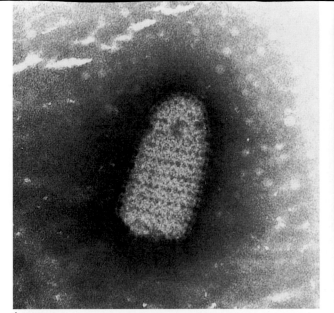

1

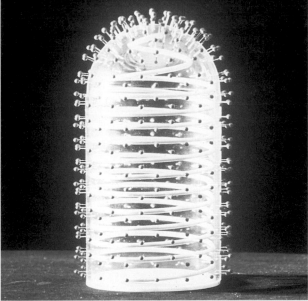

2

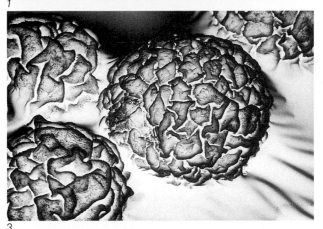

3

4

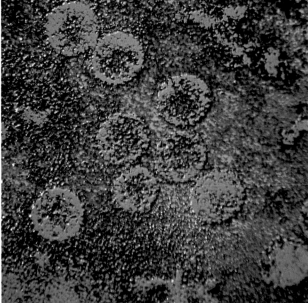

5

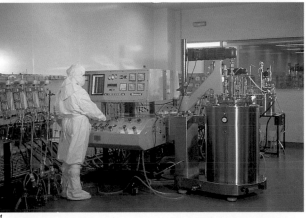

6

1, 2. Rabies
*Rabies virion seen through a
transmission electron
microscope (1), and a
model (2) of the rabies virus.
Hundreds of thousands of
people are given antirabies
treatments annually.*
3, 4. Vero cells
*Vero cells in the course of
division (3), and a bank of
Vero cells (4); The Institut
Mérieux produces polio
viruses on Vero cells carried
on microballs in
biogenerators, and this has
led to a study of this process
on an industrial scale
(1,000-liter biogenerators).
This technology should
revolutionize the production
of inactivated virus vaccines
and vaccination procedures
in the Third World.
Furthermore, monkeys (from
which the cells were
obtained) are no longer
necessary for the
production of this new
vaccine.*
5. Hepatitis B virus.
6. El Salvador
*On 3 February 1985,
undoubtedly for the first time
in the world, the belligerents
in a civil war observed a
cease-fire for an entire day
to enable UNICEF to
vaccinate 400,000 children.*
7. Mérieux, Salk, Leloir
*Charles Mérieux, son of the
founder of the Institut
Mérieux (right) with Jonas
Salk, inventor of the polio
vaccine, and Luis Leloir,
Nobel Prize Winner in
Chemistry in 1970.*

realize a molecule with the same peptide sequence as that of the surface protein of the virus.

Antigenic determinants are areas on the surface of the proteins – as they must be accessible to the antibodies – and that, consequently, are composed of hydrophilic sequences (the exterior of the protein being in contact with the aqueous matter of the body). Thus in studying the structure of the protein, we try to identify the regions rich in lysine, arginine, aspartic acid, and glutamic acid (four extremely hydrophilic amino acids).

Various methods are used to identify these areas. Thus the mutants of the virus against which a neutralizing antibody is available can be tested. When it is noted that it is no longer recognized by the antibody, an attempt is then made to identify the peptide sequence responsible for the change.

If the three-dimensional structure of the protein is known, it then becomes possible to use a computer, creating a model of the protein and rolling a ball representing a molecule of water on its surface. If the points on the surface accessible to the water are connected, a map is obtained of the peptides accessible to the antibodies. Even then it remains necessary to fabricate a fragment, apart from the protein, that retains its initial spatial conformation.

In 1970, R. Arnon and M. Sela at the Weizmann Institute, in Israel, showed that the injection of a peptide can induce the production of antibodies recognizing the entire molecule (in this instance the lysozyme from egg white). In 1982, Louis Chedid's team at the Institut Pasteur and Michael Sela's at the Weizmann Institute demonstrated that synthetic vaccines are no longer just a dream of science fiction: by synthesizing a peptide with fourteen amino acids from a fragment of diphtheria toxin that included the two immunologically active zones of the original toxin, they produced an antidiphtheria vaccine. In this case we already have an effective, low-cost traditional vaccine, but it was the first demonstration that vaccine can be synthesized, and henceforth the path lay bear to making other more useful synthetic vaccines. At the New York University Medical School, Ruth and Victor Nussenzweig are chemically synthesizing the antigen of a stage of the parasite responsible for malaria, and an agreement has been signed with Genentech, which plans on initial clinical testing in 1986. At the Tennessee Medical School, Edwin Beachley is working on a synthetic vaccine against rheumatic fever. Research aimed at developing a synthetic vaccine against hepatitis B has been intensified at Baylor University Medical School (Gordon Dreesman), at Scripps Clinic (Richard Lerner), in Houston (Melnick and Hollinger), and in New York (A.M. Prince). The amino acid sequence of the antigen having been deduced from the viral genome, the peptides corresponding to the antigenic regions of the molecules were synthesized; the antibodies developed without problem, but we still do not have the least idea of whether they neutralize the virus. A the Weizmann Institute, research is being continued on a grippe vaccine capable of acting against several strains at the same time. And so forth and so on.

Generally speaking, the small size of the molecule obtained (less than forty amino acids) precludes stable, strong, or durable protection. Consequently, there is an increasingly strong trend to turn to a tripartite model: "carrier-haptene-amplifier."

The carrier is a neutral protein support. The haptene is not exactly an immunogenic substance in itself, but it is capable of recognizing the antibodies produced against it, providing it has been grafted onto the carrier macromolecule (haptenes were discovered in 1930 by Landsteiner). The amplifier, as its name implies, amplifies the immune reactions directed against the antigen. For the moment, this is a muramyl-dipeptide (MDP), synthesized and tested in France by L. Chedid and E. Lederer collaborating with the Institut Choay.

In the long run, it is hoped that it will be possible to combine several antigenic molecules in a single vaccine, for this would make it possible to provide protection against several diseases at the same time or against several forms of the same illness. One might even be able to vary the structure of the carrier at will, to match the genetic characteristics of each patient. But these are only long-term prospects.

Animal Vaccines

The first vaccine produced by genetic engineering (and the first to be sold starting in 1982) was a veterinary vaccine against diarrhea in pigs. It was marketed simultaneously in the Netherlands (Akzo), the United States (Cetus), and France (Rhône-Mérieux). Undoubtedly the most sophisticated synthetic vaccine for livestock was the one perfected by the Institut Pasteur and on which Sanofi is now conducting tests. Combining the amplifier with the LH-RH brain hormone that controls the liberation of the hypophysis sex hormones, this vaccine in effect castrates animals.

The next veterinary vaccine to be produced by genetic engineering and placed on the market will undoubtedly be that for aphthic fever (foot and mouth disease). Caused by a virus similar to that which causes polio, this devastating disease affects cattle, pigs, sheep, and goats and decimates herds endemically in Asia, Africa, and Latin America. Furthermore, in certain countries without serious inoculation programs, half of the deaths can be attributed to vaccines that have kept their virulence. This is why the market of 3 billion doses sold each year does not include either Australia or the United States, which are relatively unaffected, and in any case prefer to slaughter the animals when an occasional epidemic does occur. An enormous market would thus be opened if it became possible to produce a completely safe vaccine. Research teams at Genentech and the US Department of Agriculture's Plum Island Animal Disease Center have succeeded in synthesizing the gene coding for one of the antigens of this illness (VP1) and have expressed it in a bacterium. Tests have begun on this in Argentina and Colombia. The vaccine will be sold commercially by International Minerals and Chemicals of Chicago. In the Richard Lerner Laboratory of the Scripps Clinic, J.L. Bittle is trying to develop a synthetic vaccine whose results may be even better than those obtained with the entire protein.

Research is being continued on porcine diarrhea, as the vaccine produced genetically by Cetus has one disadvantage: it must be administered to the sow while she is pregnant in order to preserve the piglet. Molecular Genetics has thus adopted an approach of producing and administering the antibodies directly instead of just stimulating them.

Thus far, we have only looked on *Escherichia coli*, that star bacterium of genetic engineering, from its positive side as a microfactory producing all sorts of useful substances for the human body. Nevertheless, it should not be forgotten that in its pathogenic form, it is the cause of various diseases among men and animals. One of the most serious of these is colibacillary diarrhea, which affects calves, lambs, and piglets and results in many deaths. These strains are endowed with three main characteristics: they produce "enterotoxins," which set off the diarrhea; they are not easily detached by the muscular contractions of the intestine, as they are solidly anchored to the mucous membrane by the filaments of their "fur"; and their adherence factors are more often than not associated with a given species.

A vaccine has been developed that acts against enterotoxins in pigs. Administered to the sow while pregnant, a toxin is transmitted to the piglet that neutralizes any enterotoxin that may be produced by alerting the antibodies. But more and more, vaccines are being sought that will neutralize the "fur" of the bacteria, which, though it has no pathogenic power in itself, firmly attaches the toxin-producing bacteria. Thus it is a matter of finding how to unhook it so that it may be eliminated by bodily functions. This "fur" consists of fine filaments (the *pili*) from 3 to 4 nanometers in diameter whose synthesis is coded by a plasmid. Today, vaccines are being sold that associate different adherence genes and the genes responsible for the expression of some of these adherence factors (K88 and K99 in pigs) that have been cloned. Thus Molecular Genetics in the United States has developed a monoclonal antibody aimed at K99 in calves. During the course of tests, it was noticed that the calves treated with this resisted infection. Yet over a million head are killed by this bacterium in the United States. At present, diseased livestock are being treated with antibiotics, and the goal of Molecular Genetics is to perfect a more effective (and less expensive) preventive treatment. Synthetic vaccines are also being produced by Intervet and Cetus. A later procedure, due to the work of Dr. George G. Katchatourians of the University of Saskatchewan in Canada, is being sold by Connaught. This involves a vaccine made of mutant bacteria that are inactive and incapable of reproducing themselves.

Research continues on other veterinary vaccines: against rabies (which results in losses of tens of millions of dollars, particularly in South America), sleeping sickness, etc. As the world's herds of livestock number some 20 billion animals, the potential market is enormous. But it should not be forgotten that tests carried out on animals can teach us much and be a great help in developing the genetic or synthetic vaccines needed for people. Research in this field is far from simple: in contrast to hormones, for instance, whose molecules have a specific structure, which need only be reproduced, the active part of a vaccine is frequently complex and difficult to find. Once again, the various approaches employed are less competitive than complementary and can provide mankind with a whole new range of immunological arms with which to prevent and combat disease.

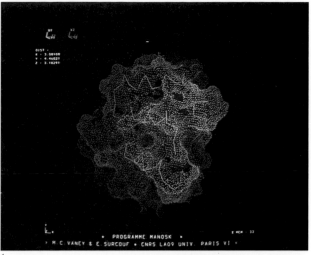

1

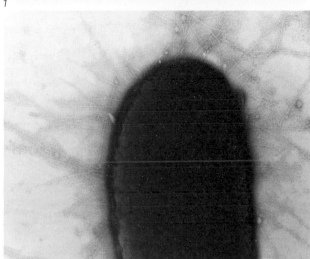

2

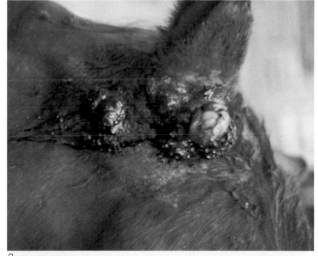

3

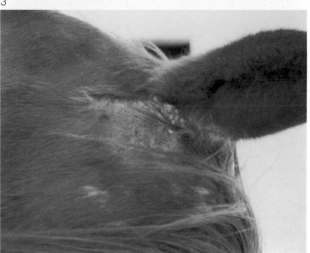

4

1. Synthetic vaccines
To determine active antigenic sites, use is made of the rotation of a molecule of water on the surface of a protein: when the structure of the surface proteins of a virus are well enough known, the most suitable short sequences of polypeptides for recognition by antibodies can be found.

2. Pili
The "fur" of E. coli bacteria responsible for porcine enteritis consists of tiny filaments that enable the bacteria to attach fhemselves to the mucous membrane.

3, 4. Animal health
New products have appeared that make it possible to modify biological response even after the onset of illness. Ribi ImmunoChem is already producing a "detoxified" endotoxin to combat the most frequent tumor in horses, the sarcoid. The cancerous nodule in a horse (3) has disappeared (4) six months after the start of a treatment that consists of activating monocytes and macrophages by injection.

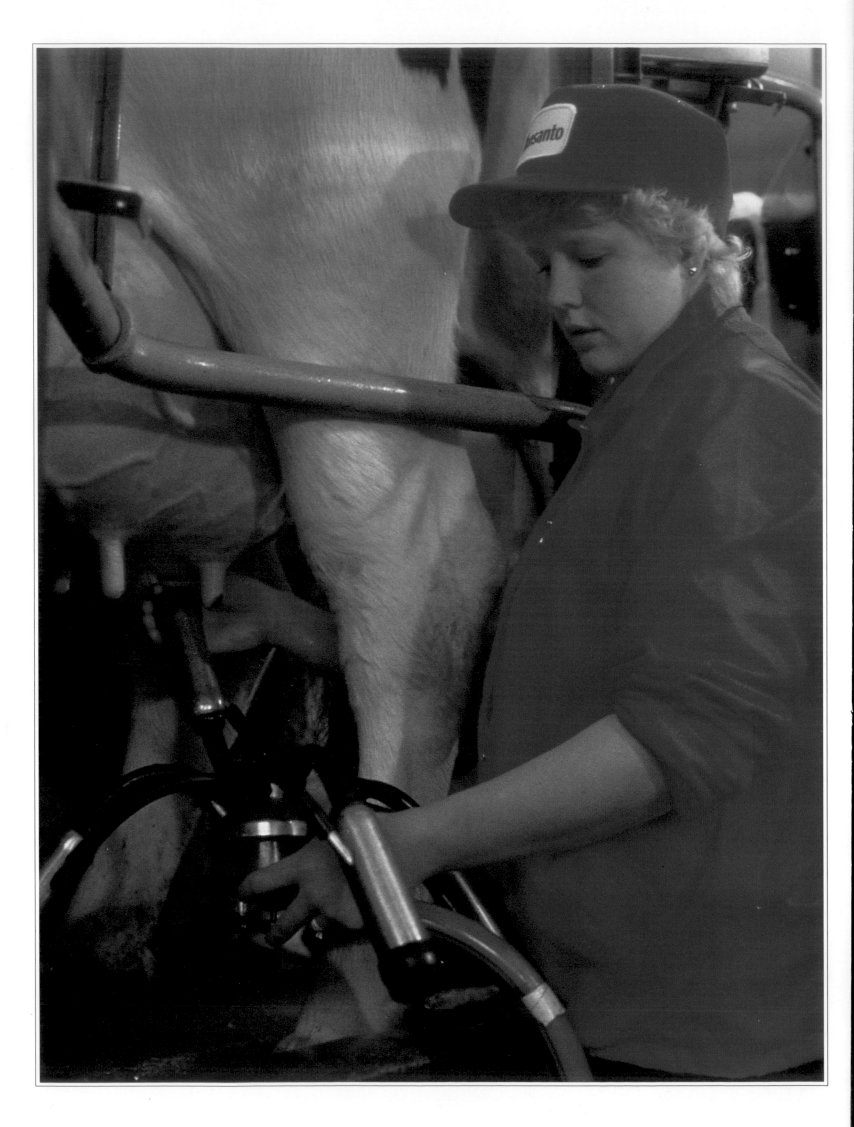

9

Rousing the Troops:
The Hormones

In 1968, within a few months of each other, two research teams, the first led by Roger Guillemin, the second by Andrew Schally, announced the discovery of a hormone in the hypothalamus (the region in the brain above the pituitary gland), revealing it to be the key area in orchestrating hormonal secretions, and thus the link between the nervous and endocrine systems.

In 1972, Roger Guillemin discovered a substance that "slowed down" the secretion of the growth hormone by the pituitary gland. Five years later, the cloning and expression of this peptide of 14 amino acids, *somatostatin,* was a major breakthrough in genetic engineering. Today, antisomatostatin antibodies are being sought to counter this braking action. The fact is that a deficiency of growth hormone can cause a variety of troubles – pituitary dwarfism, problems in mending broken bones or recovering from burns, malnutrition, and obesity.

The growth hormone itself was discovered in 1920 by Evans and Long. Christened *somatotropin,* it stemmed from a precursor with 217 amino acids, of which the first 26 are eliminated during its secretion by the cell. In 1966, C.H. Li determined the sequence of the 191 amino acids, and in 1967, P. Roos isolated somatotropin and obtained it in a relatively pure form. It was subsequently commercialized, but many problems remain to be solved: only 4-6 milligrams (mg) of the hormone can be extracted from a single pituitary gland, representing only two doses in normal therapy. So a very large number of cadavers are needed. In contrast to insulin, animal cadavers cannot be used, as the human body cannot tolerate the growth hormones of other species. In drug costs alone, the annual expense of treating a single child can be $3,500-$7,000. In addition, obtaining a completely pure hormone is a difficult matter, with the result that there is always a danger of initiating immune responses. The chemical synthesis of a polypeptide of this length is extremely complex and would be far too costly. In cell culture, yields are very low, and dete-

riorate rapidly. In 1978, the Swedish firm Kabi Vitrum, which claimed more than half of the world market for human growth hormone, signed a genetic-engineering research contract with Genentech, which, a year later, succeeded in cloning and expressing the gene of the human growth hormone in *E. coli* bacteria. Yet, curiously enough, the hormone thus obtained was slightly different, and at one end contained an additional amino acid, methionine, whose effects on humans are as yet unknown. In Japan, Kabi Vitrum's long-time partner Sumitomo Chemicals has been working on this matter and in the process investigating various other growth factors – somatostatin, IGF (Insulin Growth Factor), and an entirely new factor, discovered in 1981 by Roger Guillemin at the Salk Institute. This is the GRF or Growth Releasing Factor, which, as its name implies, initiates the secretion of the growth hormone.

The questions now arise not only whether it will become possible to eliminate the undesirable methionine from the chain of the growth hormone, but also whether it will be possible to produce this factor, which controls the release of the growth hormone by the hypothalamus, directly. The GRF is, in fact, a far shorter chain (of 44, 40, or 37 amino acids), which could be produced by chemical synthesis (expensive) or by genetic manipulation. Nevertheless, there are several problems still to be overcome with respect to genetic engineering, such as maturing the precursor from which the GRF is derived and the instability of the protein in bacterial media, and it was this that forced the Boyer team to link the somatostatin to a carrier protein.

In September 1983, the Sanofi group in France announced the industrial production of GRF (or *somatocrinin*) and stated that seven other groups of international scope were also in the running in Europe, Japan, and the United States. In collaboration with the Hoffmann-La Roche Research Institute, R. Guillemin's team succeeded in cloning the gene behind the synthesis of the hormone. Clinical

Growth hormone
The bovine growth hormone is ordinarily secreted by the pituitary glands of cows. It has been biosynthesized by Genentech for Monsanto. The hormone was tested at Cornell University on Holstein cows: daily milk production rose from 30 to 188 liters.

tests conducted in San Diego, San Francisco, Lyon, and Tokyo have proved satisfactory. And, for the first time, a method of synthesis in the liquid phase developed by Sanofi seems to have enabled it to outdistance its rivals. Said Roger Guillemin, "This is the first time that a biological molecule of this size has been synthesized chemically to be produced on an industrial scale" (from *Le Monde,* 23 September, 1983).

Sanofi also succeeded experimentally in expressing the gene of the human growth hormone in animal cells (Vero cells, from monkey kidneys). The hormone expressed is apparently similar to the hormone extracted from the pituitary gland.

Still other factors have been discovered, but to date their role remains a mystery. This is true, for instance, of the *somatomedins* (growth mediators), which are secreted by the liver but whose production is stimulated by the growth hormone. These may control the growth of certain components of the body, such as bones.

Pharmaceutical companies are not the only organizations to take an interest in growth factors, for they are equally important to agroindustry. Monsanto, for example, has signed an agreement with Genentech and tests are under way. Injections of bovine growth hormone may increase production of cow's milk by up to 40%, and the weight of large livestock by 10-15%. The potential market in the United States alone is estimated at between $250 and $500 million. Growth factors may also prove of interest where the wool industry is concerned. Sheep's wool, like human hair, is produced by the cells of the skin and is in part controlled by EGF (Epidermal Growth Factor). If EGF is injected, the wool grows faster and becomes so fine that it falls out as soon as it is brushed. Research is now being pursued in Australia on a variety of related problems – above all, concerning the quality of the wool thus obtained.

Similarly, the use of the growth hormone in stock breeding should result in meat that is richer in proteins and has less fat and water. An additional advantage to consumers is the absence of toxic residues. In Japan, livestock is apt to be finny rather than furry: Kyowa Hakko and the University of Kitasato are working on a growth hormone that will accelerate the growth of salmon and increase their weight at maturity.

During the 1970s, the family of neurohormones was enriched by some new and strange members, the "brain morphines." For many years it had been assumed that if such opiates as morphine and heroin acted on the brain, it was because there were corresponding receptors on the nerve cells. In 1971, at Stanford University, Avram Goldstein suggested a way of finding these receptors, which were brought to light soon after by Solomon Snyder and Candace Pert at Johns Hopkins University and almost simultaneously by Eric Simon at the New York University School of Medicine and Lars Terenius at the University of Uppsala.

But why would the brain have such receptors for morphine and heroin? Would not the answer be that the brain itself produces chemical compounds of a similar nature and, above all, of a similar structure? On 2 May 1975, John Hughes and Hans Kosterlitz, working at the University of Aberdeen in Scotland, revealed their discovery of such substances,

which they called "enkephalins" ("in the head"). In 1976, Roger Guillemin detected other substances "closely related" to the enkephalins, endorphins (ENDogenous mORPHINES). Three years later, in 1979, Kenji Kangawa and Hisyauki Matsuo of the Mizayaki Medical College in Japan and Masao Igarashi of the School of Medicine of Gunma University in their turn discovered what they called "alpha-neo-endorphins," which would be assimilated to the dynorphins (DYNamic endORPHINS, so-called because of their great power), named in December of the same year by Avram Goldstein, who discovered that it was, in fact, a peptide that he had identified... four years earlier.

Research subsequently became focused on the precursors of these substances, with various teams taking an interest in the problem: Sid Udenfried's at the Roche Institute, Shigetada Nakanishi's at the University of Kyoto, Robert Crea's at Genentech, that of Ed Herbert and Michael Comb at the University of Oregon, etc. For these neurotransmitters of a special kind ("inhibitory neurotransmitters"), which also function as hormones – sending messages over long as well as very short distances – may act on pain, as certain experiments (including Guillemin's) have shown, as well as on the mechanism of addiction (to drugs, for instance), and indicate the possibility of using chemical or biological means to produce substances many times more powerful than morphine or heroin but, it is hoped, without their pernicious secondary effects. Scores or even hundreds of times more powerful than heroin, these substances have also been found in the intestine, stomach, spinal cord, and other parts of the body and even in species other than mammals or vertebrates.

In addition to GRF, the growth hormone, and (perhaps) later endorphins, genetic engineering is exploring the possibility of producing other hormones, such as the luteinizing hormone or gonadotrophin, which can counter sterility or certain forms of cancer, and which have been available for the past ten years by extracting them from human pituitary glands, urine, or placentas. The genes that govern their production have been isolated and expressed in cellular cultures by researchers at Molecular Genetics. But even now a product of genetic engineering is being sold commercially, and it is a hormone: insulin.

Insulin

On 5 July 1983, the American firm Eli Lilly, which holds 50% of the world market and 88% of the US market for insulin extracted from cadavers, launched Humuline, the first genetically engineered insulin.

By 1921, when the saga of insulin began, the mechanisms of diabetes had already been under scrutiny throughout the nineteenth century. As early as 1916, Sir Edward Sharpy-Shafer thought that the origin of certain forms of diabetes lay in the failure of certain "islands" in the pancreas to secrete a substance, which he called "insulin" (from the Latin word "insula," an island). In 1921, at the University of Toronto, a young surgeon, Frederick G. Banting, and a student, Charles H. Best, isolated insulin from a dog's pancreas and demonstrated its effectiveness against diabetes. Barely two years later,

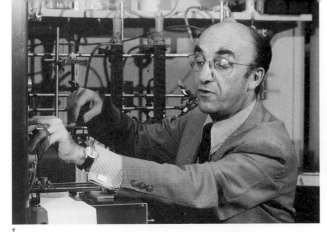

1. *R. Guillemin*
Nobel Prize Winner in Chemistry in 1977, Roger Guillemin was born in Dijon, France. He moved to the United States and now conducts his research at the Salk Institute.

2. **Banting and Best**
Sir E. Banting, who received the Nobel Prize in Medicine in 1923 for his discovery of insulin, with his disciple, C.H. Best.

3. *Insulin crystals*
Crystals of BHI zinc insulin, the first ever formed through genetic engineering. These crystals of human insulin have a high degree of purity; otherwise they would not have crystallized in the presence of zinc.

4. **Purification**
At the Celltech research center in Great Britain, Dr. Sarojani Angal purifies the human growth hormone.

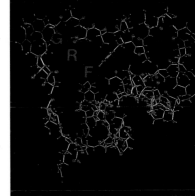

5

5. *GRF*
In 1982, Doctor G. Sassolas of Lyon, in treating a patient suffering from acromegaly, identified a tumor that could secrete the hypothalamic GRF factor, the existence of which had been postulated but which had never been isolated. The tumor was studied at the Salk Institute, and a few months later, R. Guillemin's team published the GRF sequence.

1. Neoendorphin
The structure of "brain
opiums" is being studied in
many laboratories, among
them Suntory's, shown here.
2. Interactions
Two steroid molecules
interact electrostatically with
the phosphate groups of the
DNA and are positioned in
the large trough in the form
of a characteristic pile of
molecules, the structural
details of which remain to
be determined
experimentally. The DNA
may serve as a matrix for
multiple molecular
interactions: as a result of
the molecular asymmetry of
the double helix (alternating
small and large troughs),
molecules the size of a
steroid can penetrate
completely into the space of
the large trough.
3. Corticosteroid
A corticosteroid,
synthesized by Roussel-
Uclaf, has powerful
antiinflammatory and
antiallergenic properties.

102

Banting and J.J.R. MacLeod (the head of the department) received the Nobel Prize in Medicine. Businessmen were even quicker to prick up their ears: in June 1922, an agreement was signed between Eli Lilly (founded in 1907 and in 1922 still a quite modest pharmaceutical concern) and the University of Toronto, and the new product went on the market the following year. On the other side of the Atlantic, Behring immediately joined battle, and in Denmark Harald and Thorvald Pedersen founded Novo in 1925 to produce insulin from pigs. In 1955, the sequence of the polypeptide was determined by Frederic Sanger, and by 1963-1965, synthetic insulin was being produced successfully. In addition to Lilly, Hoechst (Behring), and Novo, other firms were also in the race: Organon, Choay, and Nordisk.

The potential market is relatively limited in size. It is thought there are some 60 million diabetics in the world, of whom only 4-5 million require treatment with insulin. Furthermore, the last word has not been heard from electronics, with the implantation of micropumps stimulating insulin within the body. Neverthess, as T. Beppu says, "Bio-industry is an archipelago of small islands which can lead to the conquest of vast territories."

Steroids

However fascinating recent research may be, it should not obscure the fact that, thanks to biological methods, we already have at our disposal a valuable hormone that is available in appreciable quantities at a reasonable price. Cortisone is a steroid hormone (with a carbon chain). In addition to the corticosteroids (including cortisone), of which Roussel-Uclaf is the world's leading producer, and which are being produced in Japan by Mitsubishi Chemical, the important steroids where industry is concerned are testosterone and estradiol estrogen (used in contraceptives) and spirotolactone (a diuretic). Steroids are used in the treatment of hormonal deficiencies, contraception, inflammation, and certain allergies, and are manufactured from vegetable alcohol (sterols) made from soybean oil residue or from the roots of a Mexican plant, barbasco.

It was in 1938 that cortisone was first isolated by Edward Kendall and Tadeus Reichstein during the course of their research on the steroids secreted by the adrenal glands of mammals. In 1949, Philip Hench demonstrated its therapeutic effect on rheumatoid arthritis. Unfortunately, chemical synthesis based on beef bile proved a long and complex process. Nevertheless, the physiological effect of the molecule is dependent on one crucial step: the attachment of an atom of oxygen at position "eleven" on the carbon chain. In 1951, two researchers at Upjohn, D. Peterson and H. Murray, discovered a way of reducing the stages in the chemical synthesis of cortisone from 37 to 11 by showing that a strain of bread mold exercised an enzymatic action on progesterone, an intermediary in the synthesis of cortisone, in such a way that an atom of oxygen was liberated at position 11. The result was not long in coming, and the price of cortisone suddenly dropped 70%. Since then, different laboratories have succeeded in replacing certain chemical steps by biological processes, so that today corti-

sone can be sold for 400 times less than its original price. The most important of all the products produced today by bioconversion, steroids are a prime example of what biological methods can contribute to the synthesis of certain substances that it would be virtually impossible to obtain in any other way.

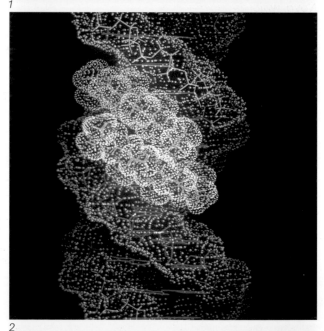

1

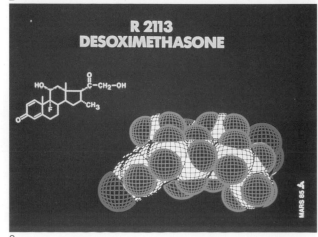

2

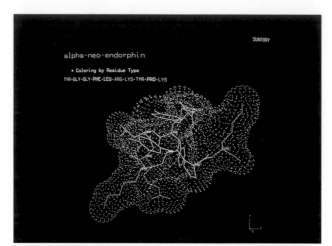

3

Human Insulin by Semisynthesis

by Hans C. Barfoed.

Born in Denmark in 1928, Hans C. Barfoed received his M.Sc. in 1951 in Chemical Engineering from the Technical University of Copenhagen. From 1952 to 1959 he worked for Danish Fermentation Industries Ltd. Joining Novo Industries Ltd. at the time as Research Chemist in the Enzyme Laboratories, Barfoed later rose to manager of Industrial Applications in Enzyme Research and Development in 1969. He was appointed Manager of Scientific and Technical Information for Novo in 1976.

Making insulin by chemical synthesis is a costly and difficult process because the polypeptide is so large. Today, two competing methods are employed: the genetic (or biosynthetic) method, developed by Eli Lilly in collaboration with Genentech, and which is presently being explored by Novo in conjunction with Biogen, and the enzymatic (or semisynthetic) method, presently used by Novo, which is a leader in the field of enzymes.

To understand the difference between these two options, H. Barfoed describes the Novo technique, while I. Johnson describes the production of human insulin by genetic engineering (next article).

Numbers in parentheses refer to "Works Cited" at the back of the book.

Possibilities and Limitations of Insulin Treatment

Before the discovery of insulin in the beginning of the 1920s, the diagnosis of diabetes was in many cases tantamount to a death sentence. Only a few insulin-requiring diabetics survived more than a year after they had contracted the disease, and those who did had to face considerable deterioration in their quality of life, involving rigorous dietary restrictions.

With the introduction of insulin treatment, the direct threat against their lives was removed, and to a large degree their lifestyles could be normalized. However, along with the prolonged life span of diabetics, late complications appeared, involving their circulatory and nervous systems, kidneys, and eyes. Thus, diabetes became the most common cause of blindness. Medical science has been working intensively to disentangle the causes of diabetic late complications in order to find preventive measures, if possible. Although it is still not entirely clear why these complications occur, there are strong indications that the greatest possible normalization of the metabolism in general, and particularly the level of sugar (glucose) in the blood, is essential so as to reduce their incidence and severity.

In spite of these improvements in the treatment, a complete normalization of the metabolism in diabetics is difficult to achieve.

Impurities of Insulin and Species Factors

Impurities in the insulin can contribute to such problems in the treatment as insulin allergy, lipoatrophy (waste of fatty tissue from the skin in the injection areas), and certain forms of insulin resistance. The impurities to be found in insulin are peptides and proteins that are extracted along with the insulin itself. Examples of such impurities are proinsulin, glucagon, pancreatic polypeptide (PP), somatostatin, and vasoactive intestinal peptide (VIP).

As immune reactions caused by impurities in the insulin have become less common

with the introduction of highly purified preparations, interest has shifted toward those immune reactions that may be caused by differences in species, e.g., differences among insulins from pigs, cattle, and man.

The Structure of Insulin

The insulin produced by the human pancreas – and the pancreas of other mammals – is a polypeptide consisting of 51 amino acids arranged in two chains, the A and B chains, with 21 and 30 amino acids, respectively. The chains are interconnected by S-S disulfide bridges formed by a reaction involving a cysteine residue in each of the chains. This sequence of primary structure of human insulin is shown in figure 1.

Besides the primary structure, the insulins are characterized by a spatial configuration (secondary and tertiary structures) as well as a specific association pattern (quaternary structure).

The insulins currently used for the treatment of diabetes are extracted from the pancreas of cattle and pigs. These insulins are remarkably similar to human insulin, but do not have exactly the same amino acid sequence. Bovine insulin differs from human insulin in the A8, A10, and B30 positions, whereas porcine insulin differs only in the B30 position. It should be added that the sequence of the A chain is identical in dogs, rabbits, and certain whales.

Although these differences may seem slight, some research indicates that they are indeed of significance in determining the antigenicity of the insulins, i.e., their tendency to provoke formation of antibodies against insulin in the patients' blood. Thus it has been found that bovine insulin is more antigenic than porcine insulin, corresponding to the greater difference in primary structure compared with human insulin. It would be expected that human insulin itself would be the least immunogenic of them all, and clinical experience does in fact confirm this.

Fig. 1

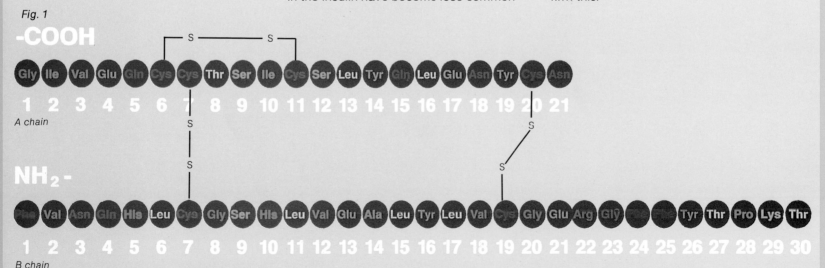

A chain

B chain

103

Enzyme-Catalyzed Synthesis of Peptide Bonds

In principle, any enzymes catalyzing a given reaction will also catalyze the opposite reaction. However, in the case of proteases, the reaction conditions – in particular, the large amount of water present in the reaction mixture – will usually favor hydrolysis, i.e., splitting of peptide bonds. Nevertheless, it has been found that proteases in mixtures of organic solvents and water are able to catalyze the formation of peptide bonds (1). This observation has made the enzymatic conversion of porcine insulin to human insulin possible, and this possibility has been realized by several workers (2-4).

Schmitt and Gattner (1978) were the first to describe the selective removal of the terminal alanine residue in the B chain whereby des (Ala^{B30}) insulin was formed (5). They used an enzyme process based on carboxypeptidase A, an enzyme found in yeasts. By using an ammonium buffer, the enzyme was made specifically to cleave off the alanine (B30) without cleaving off asparagine A21, the terminal amino acid in the A chain. The insulin prepared in this way was used in two methods for producing human insulin that were introduced by Morihara et al. in 1979 and by Gattner et al. in the same year.

In both cases the porcine insulin from which alanine B30 had been removed was made to react with a threonine ester, whereby the corresponding human insulin ester was formed.

The Novo Method

The method developed by Novo for the conversion of porcine into human insulin consists of a single enzymatic step, transpeptidation, catalyzed by porcine trypsin. The mechanism of this reaction is as follows (6):

Initially, porcine insulin forms a complex with the trypsin. Within this complex an ester bond is formed between the carboxyl group of lysine B29 and the hydroxyl group of serine in the active center of the enzyme, and alanine B30 is released. In aqueous solution this des (Ala^{B30}) insulin-trypsin ester (DAI-trypsin) would hydrolyze to yield DAI and trypsin. However, in the presence of an inert organic solvent in high concentration and a threonine ester, the amino group of the ester competes so favorably with water that virtually no hydrolysis takes place. Instead, a human insulin ester (HI-OR) is formed.

The concentration of organic solvent, and thus of water, is limited by considerations concerning the stability of trypsin against denaturation. When the trypsin is stabilized by addition of calcium ions and the temperature is kept below 37°C, the reaction may be carried out in a medium containing less than 20% water.

In Novo's laboratories the transpeptidation reaction was optimized to yield 97% of the theoretical amount of human insulin from porcine insulin.

Furthermore, it was found that under the conditions chosen, certain other components of a relatively crude porcine insulin solution, e.g., proinsulin, were converted to human insulin ester along with the insulin itself. Thereby the yield of human insulin is raised by about 3%, so that actually no loss of insulin activity takes place during the conversion.

Purification

After the human insulin ester has been formed, the remaining operations are in principle the same in both methods, except for a final ion exchange chromatography introduced in the Novo method to remove the last trace of human insulin ester still present after cleavage.

In the first purification step, trypsin is removed by gel filtration at a low pH, where the enzyme is inactive. This step reduces the content of trypsin to below the detection limit of 1 part per million (ppm). From the known efficiency of the subsequent purification steps in separating trypsin and insulin, the trypsin level in the final product (human monocomponent insulin) can be estimated to be less than $1 : 10^{22}$, or 2-3 molecules of trypsin in 1 kilogram (kg) of insulin. The use of trypsin from porcine glands ensures that no pancreas-foreign contaminants are introduced into the purification process.

In the next step, the human insulin ester is separated from any residual unconverted porcine insulin by anion exchange chromatography. The removal of porcine insulin from the final product is controlled by digestion, with trypsin and amino acid analysis of the result of this digestion. As the chromatographic method used very efficiently separates threonine and alanine, the detection limit for the latter is 0.1% of the threonine peak. As no trace of alanine is detectable by this test, any contamination with porcine insulin is below 0.1%. In the following step, the human insulin ester is hydrolyzed, yielding human insulin and alcohol. Finally, any residual unhydrolyzed human insulin ester is removed by ion exchange chromatography.

Purity Specifications

The final product complies with the specifications for Novo's monocomponent (MC) insulins and is designated human monocomponent insulin (HM-insulin). Additional tests have been included to demonstrate the absence of porcine insulin, just as a further purity test by high-performance liquid chromatography (HPLC) has been added. According to the specifications for this test, the insulin peak must represent at least 99% of the total absorbance of the chromatogram.

Identity Tests

Semisynthetic human insulin has been

compared in a number of ways with human insulin prepared from human pancreas, and complete identity has been demonstrated.

1. Amino acid sequence.
The amino acid sequence (primary structure) as determined by classic techniques (7) has been shown to be identical in the two types of human insulin.

2. x-ray diffraction.
Human insulin crystallizes with zinc ions to yield regular rhombohedrons that are indistinguishable from porcine insulin crystals by microscopic examination. By x-ray diffraction spectrography it is possible to show small difference between the crystals of the two insulin species, but the diffraction patterns of human insulin made from human pancreas and by semisynthesis are completely identical. This proves that the primary structure (sequence), secondary structure (helical conformation), tertiary structure (overall spatial conformation), and quaternary structure (association behavior) are identical for both preparations.

It has been shown that biosynthetic human insulin, made by fermentation of genetically engineered bacteria, yields the same diffraction pattern as the two others, so that all three preparations are structurally identical.

3. HPLC.
The more pronounced hydrophilic character of human insulin relative to porcine insulin, due to the presence of a hydroxyl group in threonine, makes it possible to distinguish between the two by liquid chromatography.

4. Immunoreactivity.
In radioimmunoassays (RIA) using ^{125}I-labeled insulin from human pancreas and HM-insulin, identical standard curves were obtained with different sera containing antibodies against insulin (8). HM-insulin can therefore replace insulin of human pancreatic origin as a standard in RIAs for immunoreactive insulin (IRI). Consequently, HM-insulin has been placed at the disposal of the WHO (World Health Organization) Bureau of International Standards.

5. Potency.
In a variety of bioassays (mouse convulsion test, blood glucose assay in mice and rabbits), the potency of HM-insulin has been found identical to that of porcine MC insulin, namely, 185 I.U./mg N, corresponding to 28.5 I.U./mg dry insulin.

6. Immunogenicity.
Immunization of rabbits by the method of Schlichtkrull et al. (9) with HM-insulin and porcine MC insulins showed that the two insulin types had comparable and very low immunogenicities.

Clinical Experience with Semisynthetic Human Insulin
Clinical tests with Novo's HM-insulin were started in 1980. At the time of the first introduction on the market (United Kingdom,

1982) clinical trials had been carried out on over 2,000 persons in more than 20 countries. These studies included normal volunteers, newly diagnosed diabetics, and existing diabetics treated with other preparations. The therapeutic effect, i.e., the lowering of blood glucose, was very similar in all insulin species tested, and the transfer to human insulin caused very few problems. Patients were transferred from a variety of bovine, porcine, and mixed species insulins to human insulin, and no significant adverse effects have been reported.

A few differences have been found between porcine and HM-insulins. Thus, Schernthaner (10) found that human insulin produced lower levels of antibodies than porcine insulin in newly diagnosed diabetics. Soluble human insulin also appears to be absorbed slightly faster than soluble porcine insulin, which may be due to its higher solubility, caused by the more hydrophilic character of the human insulin molecule. A decreased counterregulatory hormonal response to the decrease in blood sugar (particularly glucagon and growth hormone) has been observed in some studies when comparing human with porcine insulin. The clinical significance of these observations is still not clear.

In the small group of patients with manifest allergy or immunological resistance to animal insulins, human insulin has proven very variable. In most of the patients symptoms

have disappeared completely following a change to human insulin. The lower antigenicity found in newly diagnosed patients could, in principle, lead to better metabolic control in these patients, due to the absence or very low concentration of insulin-binding antibodies. Whether this can lead to a reduction in the incidence or severity of late diabetic complications is an open question and will, by its very nature, remain so for another 10-15 years.

Jan Markussen
At Novo, the inventor of the system that made it possible to transform porcine insulin into human insulin.

Insulin by Genetic Engineering

by Dr. Irving S. Johnson, Vice President of Lilly Research Laboratories, a division of Eli Lilly. He is a member of the American Association for the Advancement of Science, the American Association for Cancer Research, and the New York Academy of Sciences.

Dr. Johnson is a native of Colorado. He earned a Ph. D. in experimental biology from the University of Kansas and is also a graduate of the Institute of Management of the School of Business of Northwestern University. Dr. Johnson joined Eli Lilly and Company in 1953 as a bacteriologist and has held many positions within Lilly Research Laboratories since that time. He is currently a Vice President for Lilly Research Laboratories. In that role, Dr. Johnson was responsible for the project team that was successful in developing the process for the commercial production of biosynthetic human insulin. Dr. Johnson continues to be responsible for research programs in the area of recombinant DNA technology, as well as in the areas of immunopathology, encompassing research on connective tissue disease, anti-inflammatories, antiasthma, cancer, and allergies.

Eli Lilly and Company deserves the credit for being the first to take the daring step from the laboratory to commercial production of a recombinant DNA product. Humulin® (human insulin of recombinant DNA origin, Lilly) was approved by the United States Food and Drug Administration in 1982. Today it is being sold in 18 countries. And, in the United States alone, more than half of the newly diagnosed diabetics are starting therapy with Humulin.

Furthermore, the company has developed a new production process for Humulin that involves the biosynthesis of the precursor of insulin – human proinsulin – which requires but a single fermentation operation (an advantage in time and money). The company is working in other areas of recombinant DNA research, including the development and testing of proinsulin as a therapeutic agent, human growth hormone, bovine growth hormone, and another field of biotechnology – monoclonal antibodies for anticancer therapy.

Numbers in parentheses refer to "Works Cited" at the back of the book.

Acknowledgments.
Grateful acknowledgment is made to the scores of Lilly associates who have made significant contributions to this project, particularly B.H. Franck and R.E. Chance. As in any major technological development of this type, the laboratory, engineering, production, and logistic requirements that had to be met involved the close cooperation of all levels of our administrative and technical staffs.

Eli Lilly and Company has been involved in the development and manufacture of insulin and other products for diabetics since 1922. In that year our scientists began working with Frederick G. Banting and his associates at the University of Toronto to develop a standardized and clinically acceptable insulin product.

In the early 1970s we became concerned about a possible shortage of insulin. Until then, the world's insulin needs had been derived almost exclusively from pork and beef pancreas glands, which were collected as by-products from the meat industry. This supply changes with the demand for meat and not the needs of the world's diabetics.

Today, the diabetic population is growing more rapidly than the total population. Because of the uncertainty of the insulin supply and the forecasts of rising insulin requirements, it seemed not only prudent but a responsibility of the scientific community and insulin manufacturers to develop alternatives to animal sources for supplying insulin to the world's diabetics. Lilly established several internal committees of scientists to examine various solutions to the problem, including the potentially new technology called genetic engineering.

The function of DNA in a cell is to serve as a stable repository of coded information that can be replicated at the time of cell division to transmit the genetic information to the progeny cells and to encode the information necessary for synthesizing proteins and other cell components.

In 1972, Jackson, Symons, and Berg (1) described the biochemical methods for cutting DNA molecules from two different organisms, using restriction enzymes, and recombining the fragments to produce biologically functional hybrid DNA molecules. In 1973, Cohen, Chang, and Boyer (2) reported that they could make a hybrid molecule that would express the foreign DNA within it as though it were a part of the original molecule's natural heritage (3). That profoundly significant accomplishment also generated major concern over potential biohazards. Because of this concern, work was halted – including work at my own organization – until after the Asilomar Conference in 1975 (4). In June 1976, the National Institutes of Health (NIH) announced guidelines for recombinant DNA work, marking the end of the two-year moratorium on this type of research. All Lilly research in this area was conducted under the guidelines and with approval of the Recombinant Advisory Committee (RAC).

The physiology and genetics of the bacterium *Escherichia coli (E. coli)* has been studied for many years, and, as a result, most recombinant DNA research has been done with this microorganism. For safety and containment reasons, Eli Lilly chose the K-12 strain of *E. coli* to use in our recombinant DNA research and production, a weakened strain of *E. coli* that cannot colonize the intestinal tract of humans or animals. The functioning of the protein synthesis apparatus of bacterial cells is obviously central for being able to produce human insulin in these cells. Although proteins are synthesized only on the ribosomes in the cytoplasm of the cell, the genetic code for production of proteins resides in both the chromosomal DNA and in the small rings of cytoplasmic DNA called plasmids. Both of these sources of DNA are transcribed into messenger RNA, which is subsequently translated into proteins. The basis of recombinant DNA technology is our ability to manipulate this bacterial plasmic DNA, which is accomplished by isolating the plasmid DNA, cleaving with restriction enzymes, and inserting the desired DNA. The desired DNA is obtained either by synthesis, isolation from natural sources, or a combination of these procedures. In the human insulin work, the A- and B-chain genes were prepared by synthetic nucleotide chemistry, while the human proinsulin gene was semisynthetic – that is, the gene was constructed using a segment of the natural DNA that codes for proinsulin along with a fragment of synthetic DNA. The nucleotide synthesis was performed by Itakura and coworkers at the City of Hope, and by Goeddel and coworkers at the Genentech Corporation (5, 6). After the desired DNA is obtained and inserted, the plasmid DNA is a rejoined using a ligase enzyme and then reintroduced into the host cell through a process called transformation. The cell is then cloned, that is, many copies are made, and, after verifying that the desired and correct gene is present, the material is started from the same seed pool, which has been verified to have the correct gene present.

In order to maximize the production of the desired protein in the *E. coli* cells, the gene message that was inserted into the plasmid also contains a so-called promoter. This promoter determines the rate at which messenger RNA is formed, and consequently there is greater production by the cell of the desired gene product. For human insulin biosynthesis, two promoter systems have been used, originally beta-galactosidase and now tryptophan synthetase, or Trp E. The Trp E promoter yields more of the desired product as compared with the beta-galactosidase promoter system. When the *E. coli* cells are producing the desired gene product, one can observe electron dense bodies (see figure 1), which by immunocytochemical techniques (7) have been shown to be the promoter-linked product (A chain or B chain or proinsulin) – called the chimeric protein. The chimeric product can be schematically represented as Trp E-Met-A Chain (or -B Chain or -Proinsulin). The methionine linkage provides a specific chemical cleavage site for release of the desired polypeptide from the promoter protein Trp E.

Figure 2 indicates the two schemes we

have explored for producing biosynthetic human insulin. The current method is to make the A and B chains in separate *E. coli* fermentations; the second route is the production of proinsulin in a single *E. coli* fermentation. As far as we have been able to determine, both methods yield equivalent preparations of biosynthetic human insulin (8, 9).

Figure 3 illustrates in more detail the current method used to produce biosynthetic human insulin.

The chain combination procedure was developed at the Lilly Research Laboratories (9). This procedure consistently gives

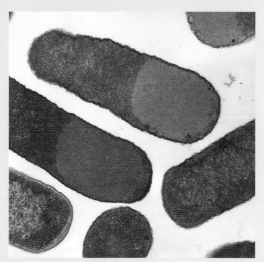

Fig. 1

higher yields of insulin than ever reported for this reaction. The insulin yield is approximately 60% relative to the limiting B chain. The biosynthetic human insulin is purified and isolated by column chromatography and crystallization. The excess chain materials and by-products are then recycled. This chain combination procedure is an extension of studies from several laboratories (10-13) that were conducted during the 1960s when chemically synthesized chains of insulin were combined to yield synthetic insulin preparations. However, the yields are significantly better than in the earlier studies, due in part to the availability of modern analytical techniques, such as high performance liquid chromatography (HPLC). The availability of this technique allowed the examination and optimization of the many variables of this complex series of reactions, and thus the achievement of excellent yields of human insulin.

The preparation of human insulin employing human proinsulin is shown in figure 4.

As indicated earlier, the biosynthetic human insulin procuced via this scheme from human proinsulin (figure 4) is identical to the insulin made by the chain combination method. Before turning to a discussion of the characterization of biosynthetic human insulin, we should point out that one of the exciting aspects of the proinsulin scheme is that we are now able to produce human

proinsulin and C-peptide, which we felt might have interesting activities in their own right and which could now be investigated clinically for their potential usefulness in the treatment of diabetes. Preliminary clinical data indeed suggest that the primary metabolic effect of proinsulin is on the liver, thus differing from insulin, which primarily affects peripheral tissues. Proinsulin may eventually be a superior therapy for type II diabetes or may be used in combination with insulin.

Doctors Guy Dodson and Tom Blundell in England have recently examined BHI crystals by x-ray diffraction analysis and have found them to be identical to crystals of natural pancreatic human insulin, to crystals of human insulin prepared from pork insulin, and to crystals of human insulin derived from biosynthetic human proinsulin. This x-ray diffraction analysis assures us that all of the chemical bonds in BHI are identical to those of pancreatic human insulin (15, 16).

A number of techniques and tests proved to be valuable in monitoring the recombinant human insulin process.

High-performance liquid chromatography (HPLC) techniques developed at Lilly can detect proteins that differ by a single amino acid (17), and HPLC tests showed that human insulin (recombinant DNA) is identi-

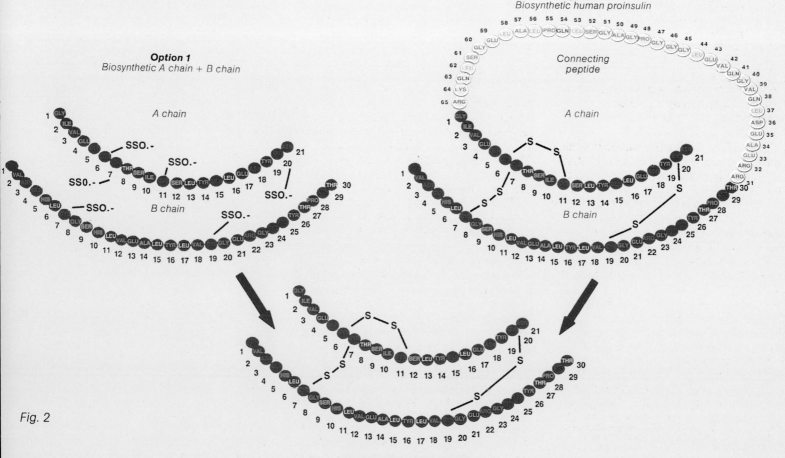

Fig. 2

Option 1
Biosynthetic A chain + B chain

Option 2
Biosynthetic human proinsulin

Biosynthetic human insulin

cal to pancreatic human insulin and that it is close to, but not the same as, pork insulin, which differs from the human insulin by one amino acid; beef insulin, which differs by three amino acids; and sheep insulin, which differs at four residue positions. A chromatogram of human insulin (recombinant DNA), pancreatic human insulin, and a mixture of the two showed that they were superimposable and identical. HPLC has become an important analytical tool for determining structure and purity and is now considered to be a more precise measurement of potency than the rabbit assay, although most government regulatory agencies around the world still emphasize the rabbit potency assay.

A measure of the correct tertiary structure and appropriate folding is the circular dichroic spectra. The spectra for porcine insulin and for human insulin (recombinant DNA) were found to be identical. Crystallographic studies by x ray further revealed the structural integrity of the recombinant molecule (15). We also found the amino acid composition of human insulin (recombinant DNA) and pancreatic human insulin to be identical. In addition, we have compared polyacrylamide gel electrophoresis for human insulin (recombinant DNA), pancreatic human insulin, and pork insulin, as well as isoelectrical focusing gels for these three insulins.

Another technique that we found useful for ensuring that we had the appropriate disulfide bonds and lacked other types of protein or peptide contaminants was HPLC of a specifically degraded sample. There is a staphylococcal protease that cleaves insulin in a specific way at five sites – always next to glutamic acid, except for one site between serine and leucine. After treating the insulin with the protease, we looked for and identified the various peptide fragments by HPLC and found none that were not derived from insulin.

In the end, we employed twelve different tests to establish that what we had produced was human insulin. We believe the correlation among three of the tests was particularly important – the radioreceptor assay, radioimmunoassay, and HPLC. Moreover, the pharmacological activity of human insulin (recombinant DNA), as demonstrated by a rabbit hypoglycemia test, showed a response essentially identical to pancreatic human insulin.

Another serious question remained to be answered – namely, that of the potential contamination of the product with trace amounts of antigenic E. coli peptides. Relevant to this question is the difference in starting materials between human insulin of recombinant DNA origin and pancreatic animal insulins. The glandular tissue is collected in slaughterhouses, with no control over bacterial contamination. The desired gene product is isolated from a few cells of the islets of Langerhans, which make up

less than 1% of the glands; thus more than 99% of the tissue represents tissue contaminants and undesirable materials. The common protein contaminants of the animal insulins are other pancreatic hormones or proteins, many of which are highly immunogenic.

In contrast, with recombinant DNA production of human insulin, almost 100% of the cells (E. coli) produce the desired gene product. Because of the method of manufacture, none of the pancreatic contaminants of the animal insulins are found in the human insulin of recombinant origin. The issue of proteinaceous contamination derived from the bacterial host cell was addressed through a series of experiments that were made possible by running large-scale fermentations of the production strain of E. coli, which contains the production plasmid with the code for the insulin chain sequence deleted. The small quantities of peptides isolated after applying the chain purification and disulfide linking process to the "blank" preparation were shown not to be antigenic except in complete Freund's adjuvant (18); in addition, no changes in amount of antibody to E. coli peptides were detected in serum from patients who had been treated with human insulin for more than a year (19).

In July 1980, we began clinical trials of our human insulin in the United Kingdom. Within weeks, similar tests were under way in West Germany and Greece and, finally, in the United States. Plants, specifically designed for the large-scale commercial production of human insulin (recombinant DNA), were built at Indianopolis and at Liverpool in the United Kingdom. On 14 May 1982, we filed our new-drug application for human insulin with the FDA.

Clinical studies with human insulin (recombinant DNA) indicate its efficacy in hyperglycemic control. It appears to have a slightly quicker onset of action than animal insulins. In double-blind transfer studies with animal insulins, patients previously treated with mixed beef-pork insulin had a 70% decrease in bound insulin in comparison with a base line. Species-specific binding of human, pork, and beef insulins at 6 months decreased by 30% in control subjects treated with pork insulin, and by 51% in patients transferred to human insulin. Species-specific binding of beef and human insulins decreased equally whether patients were maintained on purified pork insulin or switched to human insulin. Species-specific binding for pork insulin, however, remained constant in both groups (20). The clinical importance of these findings remains to be clarified in long-term studies. Occasional patients hypersensitive to animal insulins and semisynthetic human insulin derived from pork insulin tolerated human insulin (recombinant DNA) well. Recombinant technology now permits us to study human proinsulin and mixtures of human proinsulin and insulin much as

they are secreted by the beta cell. These studies may provide an improved modality of therapy in diabetes.

The power of recombinant DNA technology resides in its high degree of specificity, as well as the ability it provides to splice together genes from diverse organisms – organisms that will not normally exchange DNA in nature. With this technology, it is now possible to cause cells to produce molecules they would not normally synthesize, as well as to produce more efficiently molecules that they do normally synthesize. The logistic advantages of synthesizing human insulin, growth hormone, or interferon in rapidly dividing bacteria, as opposed to extracting these from the tissues in which they are normally produced, are obvious.

We have shown the practicality of using recombinant technology to produce proteins of pharmacological interest as fermentation products. This was accomplished without adverse environmental impact or increased risk to workers.

The preparation of human insulin utilizing recombinant DNA technology marks a significant accomplishment in the field of molecular biology and provides a secure source of insulin for the future treatment of the insulin-dependent diabetic.

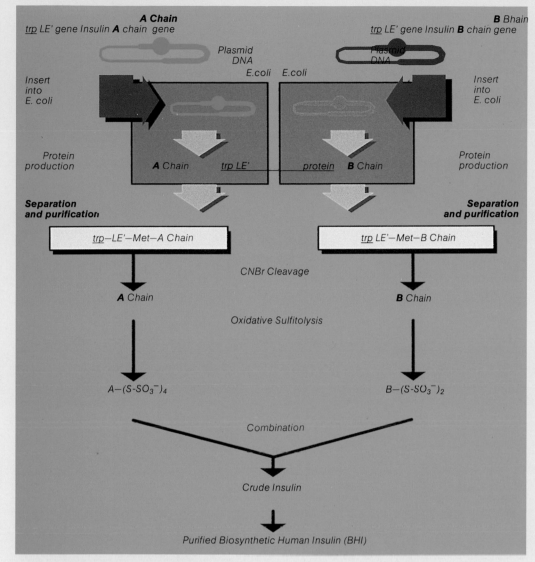

A Chain
trp LE' gene Insulin **A** chain gene

B Bhain
trp LE' gene Insulin **B** chain gene

Plasmid DNA

Plasmid DNA

Insert into E. coli

Insert into E. coli

E.coli   E.coli

Protein production

Protein production

A Chain   *trp* LE'   *protein*   **B** Chain

Separation and purification

Separation and purification

| *trp*–LE'–Met–A Chain | | *trp* LE'–Met–B Chain |

CNBr Cleavage

A Chain   **B** Chain

Oxidative Sulfitolysis

A–(S-SO₃⁻)₄   B–(S-SO₃⁻)₂

Combination

Crude Insulin

Purified Biosynthetic Human Insulin (BHI)

Fig. 3
Illustrating the current method used to produce biosynthetic human insulin. The chimeric protein, Trp E-Met-Chain, is produced in the E. coli cells in separate fermentations. Methionine is used as a cleavage point, since it is sensitive to the chemical cyanogen bromide, or CNBr, and because methionine does not occur either in the A or B chains or proinsulin. After the cyanogen bromide cleavage, the A and B chains are converted to the stable S-sulfonate derivatives, purified, and chemically combined to yield insulin. The insulin is then purified by modern gel filtration and ion exchange chromatographic procedures. At this point, it should be emphasized that all of the biosynthetic human insulin presently being produced by Eli Lilly is derived from this chain combination procedure and that all clinical studies have been conducted with such insulin.

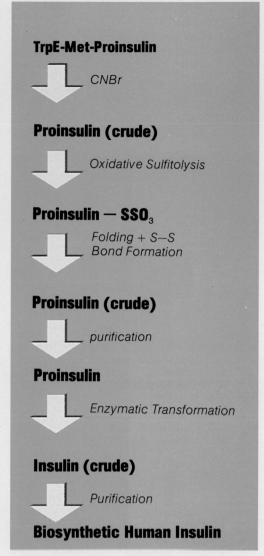

TrpE-Met-Proinsulin

CNBr

Proinsulin (crude)

Oxidative Sulfitolysis

Proinsulin — SSO₃

Folding + S–S Bond Formation

Proinsulin (crude)

purification

Proinsulin

Enzymatic Transformation

Insulin (crude)

Purification

Biosynthetic Human Insulin

Fig. 4
The preparation of human insulin employing human proinsulin. The chimeric protein produced by the E. coli cells is Trp E-Met-Proinsulin. As in the A- and B-chain case, the chimeric protein is cleaved using cyanogen bromide. The proinsulin is subsequently converted to its S-sulfonate derivative by oxidative sulfitolysis and then isolated. Then the proinsulin-S-sulfonate is treated with a thiol reagent, beta-mercaptoethanol, which allows the proinsulin molecule to fold and form the proper disulfide bonds. Yields as high as 70% are achieved in this process, which was also developed in the Lilly Research Laboratories (9). The proinsulin is then purified by ion exchange chromatographic methods, and the by-products of the folding reaction are recycled. The proinsulin is then converted in greater than 95% yield to insulin using a combination of the enzyme trypsin and carboxypeptidase B. This process is a modification of the procedure originally developed by Kemmler and cowokers (14). The biosynthetic human insulin is subsequently purified by gel filtration and ion exchange chromatography, and by crystallization.

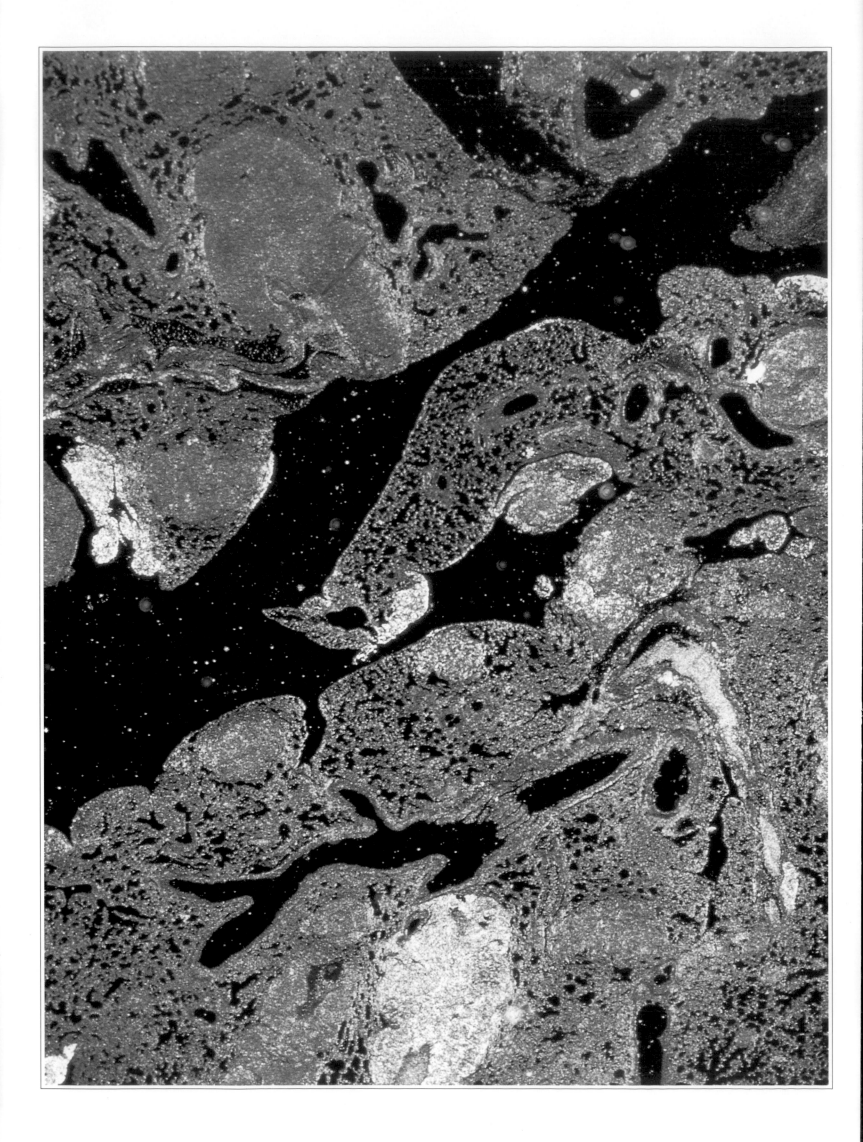

10

When Cells Go Mad: Cancer

"Nature often whispers its secrets in a calm and steady voice which rumbles through the impenetrable jungle of events," said Peyton Rous in 1942 at a time when his research into the viral origins of certain tumors seemed completely stalled. In discovering the first virus to be a source of cancer in 1911, Rous had provided researchers with one of the best tools to study the transformations that occurred in a tumorous cell. At that time, he had tried to establish the viral origin of a sarcoma of the conjunctive tissue of a chicken by injecting this cancerous tissue into healthy chickens, which in turn became affected. He then injected into other chickens a cancerous solution from which all cells (including the tumorous cells) had been filtered out. Once again, the chickens developed cancer. Rous thus concluded that a virus, too small to be detected by the microscopes of that time, was the source of the disease. His idea was met with the greatest of skepticism, and when he was finally awarded the Nobel Prize in Medicine for his work in 1966, Rous was 87 years old!

If viruses have become the privileged instruments for studying the mechanisms of cancer, this is because their genetic structure is simple and the number of their genes quite low. A researcher can thus induce and study the tumors that arise in the same type of cells and that reveal themselves by the same genetic changes, which can be quite easily studied in cultures of tissues.

Like all viruses, tumorous viruses (retroviruses, for example)[1] are constituted of genes enveloped in a "cloak" of proteins, the capsid. To reproduce, they "squat" on a healthy cell and use for their own benefit all its biochemical reproductive apparatus. Tumorous viruses act in a specific way: they integrate their own genetic material into that of the healthy cell; after that, when the cell divides, the viral genes are also duplicated.

When in the early 1960s researchers began to study retroviruses, the entire process of inducing a cancer remained a mystery and was difficult to observe, as it took place inside animal tissues. Renato Dulbecco and Marguerite Vogt were the first to induce a tumor in healthy cells in vitro: they infected the cells of the conjunctive tissues (fibroplasts) of hamster embryos with polyoma virus. The contaminated fibroplasts, injected into rats, proliferated and formed tumors. The infinitesimally small quantity of viral DNA introduced into the healthy cell was thus the origin of the changes in the structure and behavior of the diseased cell. Furthermore, the phenomenon was reversible: when the genes of the viral DNA are inactivated, the cells once again become healthy.

Retroviruses are nomads, endowed with a certain degree of freedom within the cells of higher animals. They are RNA viruses with a particularly original way of duplicating themselves, inasmuch as the RNA of the genome of the retrovirus must first be copied in DNA (cDNA) by an enzyme. This enzyme, reverse transcriptase, which was discovered in the late 1960s almost simultaneously by Howard Temin of the University of Wisconsin and David Baltimore of MIT, caused a revolution in thinking: its discovery cast doubts on the idea – at that time well established – that genetic information could only go in one direction (from DNA to RNA). It made it possible to understand certain strange processes and to analyze the specific viral RNA in the infected cells: thanks to reverse transcriptase, it was henceforth possible to synthesize the DNA complementary to the RNA of the retrovirus and use it as a probe to detect the viral RNA present in the cell, to the total exclusion of (quite abundant) cellular RNA. H. Temin also determined that viral cDNA becomes integrated within cellular DNA and becomes a "provirus" (an obligatory step in the multiplication of the retrovirus).

Not all retroviruses are oncogenic – cancer causing. There are some that will never become virulent, others that act only after a long latent period, and still others that induce a tumor almost immediately. In 1969, to explain this difference,

Metastases
Section of animal lung tissue with multiple tumors (or metastases), shown by the higher areas.

(1) A retrovirus is a virus whose genetic material consists of RNA, in contrast to viruses whose genetic material consists of DNA. While certain tumors are induced by retroviruses (HTLV-1 or HTLV-2, for instance), others are caused by DNA viruses (adenovirus, polyoma, SV40, etc.).

Temin postulated the existence of a precursor, a "protovirus" possessing some of the characteristics of Barbara McClintock's "mobile genetic components." Retroviruses would seem to have evolved from these mobile components... something that Temin was able to prove almost ten years later.

The Oncogene

Parallel research conducted on carcinogenesis since the beginning of the twentieth century had suggested that carcinogenic agents did not seem to bring the cell any new information, but altered its genetic material. Could cancer thus have a genetic basis, even in the case of nonviral cancers? This is just what various research teams began to ask themselves in the late 1970s. Could the aberrant behavior of cancerous cells (their uncontrolled proliferation),[2] haphazard intermingling, lack of respect for territoriality, quite different external membranes, high rate of metabolizing sugar, and the production of atypical substances be controlled by genes that had been altered in one way or another?

Different teams came to the same conclusion: it is a gene – the oncogene (from "onkos," the Greek for tumor) – that is responsible for transforming the healthy cell into a cancerous cell.

In the early 1970s, an oncogene of the virus of Rous's sarcoma had been identified (known as "src," as oncogenes are designated by a three-letter abbreviation). In 1975, however, Dominique Stéhelin, Michael Bishop, and Harold Varmus at the University of California discovered that this src gene was not really a viral gene; in fact, the retrovirus had appropriated another gene (the protooncogene, or cellular oncogene), diverted it from its normal function, and in a way obliged it to provoke a cancer. But just what are the mechanisms behind this "Operation Zombie"? That still remains a mystery.

This surprising idea, that there is a perfectly normal cell corresponding to any given oncogene, was confirmed in 1976 by the simultaneous discoveries of Robert Weinberg, M. Wigler, and Mariano Barbacid: an oncogene used as a probe hybridizes strongly with a DNA sequence of the same size and of almost identical structure; these protooncogenes were subsequently discovered in many mammals, chickens, fruit flies, and even in an archeobacterium. They undoubtedly play an important role in normal metabolism, as they seem to have existed for from 600 million to 1 billion years. Recent research seems to suggest that some of them also play a part in the process of regulating cell growth.

In 1981-1982, at the Robert Weinberg Laboratory at MIT, Chiaho Shih, drawing his inspiration from the famous Avery-MacLeod-McCarty experiment (which had shown that the transmission of the virulence of one strain of bacteria to a nonvirulent strain was due to a transfer of DNA), undertook a transfer of the same kind: Could the cancerous phenotype be transmitted from one cell to another? He isolated the DNA from the cells of a mouse with a tumor and proceeded with a "transfection," in other words, transferring it to healthy cells of the same mammal: two weeks later, the culture showed cells resembling those induced by the infection of a tumorous

virus. These cells, injected into healthy mice, produced tumors. The cancer was thus caused by the transfer of nonviral genes, inasmuch as, in this specific case, it had been originally induced by exposing the cells to a chemical carcinogen. This experiment, repeated with all sorts of other tumorous cells, always produced the same results.

A discovery of any kind invariably raises new questions. In this case, it was, What sequence or sequences were responsible for the change? Was it one or several segments of DNA? Did it act alone? But most important of all was to find the single segment out of thousands of DNA fragments that contained the genome of the cell! A major problem was encountered in using the recently investigated gene-cloning methods, that of finding a specific probe for the sequences being sought (as the only thing known about the DNA in question was the changes it wrought on the cells). Using different techniques, three research teams tried to penetrate this enigma: Geoffrey Cooper's at the Dana-Farber Cancer Research Institute – working in conjunction with the Harvard Medical School, Robert Weinberg and C. Shih's at MIT, and Michael Wigler's at the Cold Spring Harbor Laboratory.

We now know that there are unquestionably two kinds of activating oncogenes: protooncogenes activated by the retrovirus that have been "diverted" and forced to act as oncogenes, but also oncogenes resulting from the mutation of protooncogenes "under the influence" of various carcinogenic agents (chemical, nutritional, hormonal, hereditary, and radioactive), which frequently result in the most deadly forms of cancer (of the lungs, breasts, or colon). Nonetheless, recent experiments in molecular hybridization have shown that these two groups were not incompatible. All of which leads to the conclusion that cancer is not a generic term covering a group of separate illnesses but a single disease whose mechanisms are the same for different types of tumors.

What is the nature of the mutation that transforms a "good" gene (protooncogene) into a "bad" gene? Using rDNA genetic recombination techniques, it has become possible to locate the region where this takes place for a gene "ras": it is on a segment 350 nucleotides long, where a single base changes, the guanine of the protooncogene being replaced by the thymine of the oncogene. The mutation thus involves but a single nucleotide among the roughly 5,000 in a normal human protooncogene.

Still other methods of activating the protooncogenes were also discovered: by translocation or chromosomic rearrangement, either spontaneously or under the effects of outside agents, the protooncogene changes place inside the genome.

This translocation phenomenon has been of help in research in the fields of virology, molecular biology, immunology, and cytogenetics. As demonstrated by P. Leder and C. Croce in 1982-1983, it seems to put the gene in contact with the region of the cellular genome responsible for the production of antibodies. It also became apparent that from time to time an anomaly occurred, in the form of an abnormal amplification of a gene, of which a large number of copies are found in the cell (in certain human leukemia cells, the "myc" gene is amplified 60 to 80 times!) or in the form of excess transcription.

(2) This proliferation may even survive the patient. An American, Henrietta Lacks, who was operated on for cancer of the cervix in 1955, died of her illness. Cancerous cells taken from her body and preserved in a flask continued to divide. These cells (HeLa) are still alive and being used in cancer research.

G. Roussy
Professor Gustave Roussy (1874-1948), founder of the Institut du Cancer in Villejuif, France.

But above all, as one of the essential functions of a gene is to code for proteins, research became concentrated on the proteins coded by the oncogene. In 1978, M. S. Colett and R. L. Erikson discovered that the protein coded by the oncogene of Rous's sarcoma is an enzyme that engenders a phosphorylation, that is to say, it catalyzes the fixation of a group of phosphates on other proteins. This process, already well known, was in no way abnormal. It was known at that time that all cells contain substantial quantities of these phosphoproteins, and various enzymes of the same type that phosphorylized two amino acids, serine and threonine, were also known. But what was unusual in this case was that it was another amino acid, tyrosine, that was thus phosphorylized by the enzymes (whence kinase proteins, from the Greek "kinein," to remove) coded by the oncogene, a fact simultaneously demonstrated by two research teams under Tony Hunter and Bartholomew Sefton at the Salk Institute and Owen Witte and David Baltimore at MIT. This phenomenon may occur in healthy cells, but is extremely rare (1 out of 3,000 times). In infected cells, the proportion of phosphotyrosine is five to ten times as high. But how can this change in the activity of the protein derange cellular duplication? Have not other tumorous viruses been discovered since then that do not exhibit this kinase activity, which seems to indicate that other cell-transforming mechanisms are at work?

Another recent discovery was that the chemical substances known as "growth factors" inasmuch as they send signals that stimulate or inhibit cellular division are perhaps linked to certain other activities of the protooncogenes. Thus the receptor of the epidermal growth factor (EGF) was identified at Vanderbilt University by Stanley Cohen and a team of researchers, who discovered that this protein had a kinase activity and specifically phosphorylized tyrosine. This is equally true of the receptor of PDGF (platelet-derived growth factor) and perhaps of the receptors of all growth factors. At the moment they become attached to the receptor, the signal for cellular division is transmitted from the outside to the inside of the cell and increases kinase activity. The kinase proteins of certain oncogenes may thus divert these control mechanisms to their own benefit.

In May 1983, H. W. Hunkapiller at Caltech determined the sequence for PDGF, and in July of the same year, Russell Doolittle at the University of California at San Diego and Mike Waterfield at the Imperial Cancer Fund in London showed that the protein induced by one oncogene was nothing else than a PDGF!

Could the mad proliferation of cells be caused by an oversupply of a growth factor? The proteins secreted by the protooncogenes (from which the oncogenes come) seem to play an essential role in the control and regulation of cells and the differentiation and development of embryos (now being studied particularly by Inder Verma's team at Salk). Still other teams, including Robert Holley's, also at Salk, are studying certain factors inhibiting cellular growth that seem to act more effectively on cancerous cells than on healthy cells.

Is a single oncogene enough to induce a tumor? Does not the progress in stages of carcinogenesis result from the action of several distinct oncogenes – two of them perhaps? This at least is what is sug-

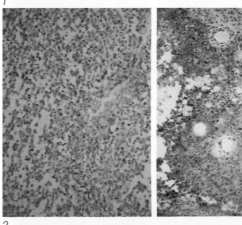
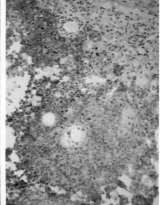
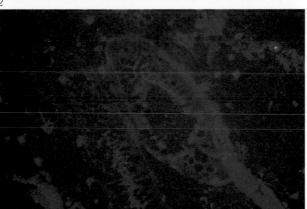
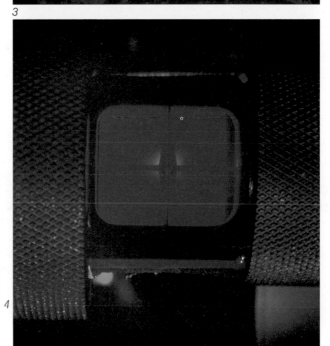

1. *D. Baltimore*
David Baltimore, Renato Dulbecco, and Howard Temin shared the 1975 Nobel Prize in Medicine "for their discoveries concerning the interaction of tumoral viruses and the genetic material of the cell."
2. *Melanoma*
Tumors of cells producing melanine (the brown pigment that gives the skin, hair, and iris of the eye their color) "labeled" by monoclonal antibodies (Biogenex test). On the left, a negative test; on the right, a positive test.
3. *"Labeling"*
The fluorescent marking at Hybritech of a cancer of the colon in a man by the monoclonal antibodies of mice.
4. *Growth factors*
The purification of growth factors from cancerous cells could lead to the development of essential products for diagnosis and treatment. Genetic Systems estimates that the market for cancer-related products would come to over $1 billion in the United States alone, and in view of this founded Oncogen in 1983 in conjunction with Syntex.

PREMIER JOUR D'ÉMISSION
Nº 716   HISTORIQUE   F.D.C.

LUTTE CONTRE LE CANCER

LUTTE CONTRE LE CANCER

1, 2. Solidarity
General mobilization to fight against modern-day plagues: racial fraternization during a meeting against AIDS held in New York in May 1983 (1); worldwide anticancer day, 7 April 1970 (2).

3, 4. AIDS
On 19 May 1983, millions of viewers of the television program 20/20 witnessed the death of Kenny Ramsauer, a handsome young homosexual (3 left) who had contracted AIDS and was celebrating his birthday the day before his death (4). A memorial service was held in a Catholic church in New York on 13 June 1983, followed by a mass meeting in Central Park.

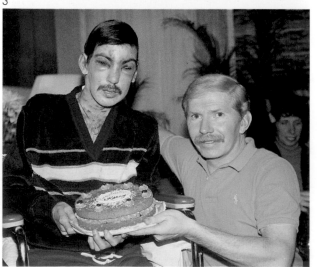

gested by recent experiments carried out by François Cuzin at the University of Nice (1981) and then by Earl Ruley at Cold Spring Harbor and R. Weinberg at MIT. Said the latter, "If we are to think of treating cancer, we will first have to discover the agents capable of recognizing precise targets – undoubtedly oncogenes or oncogenic proteins – present in cancerous cells and essential to their proliferation." We might also consider reversing the process. We could then repress the expression of the oncogene, inhibit the enzymatic activity of the oncogenic proteins, and block the receptors of the growth factors.

Reversing the process has become one of the lines of research in this field. Naturally enough, this implies an understanding of phenomena that are as yet ill defined. Another line of investigation, explored by J. F. Nicolas at the Institut Pasteur and recently crowned with success, consists of inhibiting the messenger RNA of the oncogene: the process consists of sending a signal to start copying DNA into RNA with a silent strand of DNA (the useless strand, which is not employed in the copying process) in such a way that a second strand of RNA may be obtained that attaches itself to the other and prevents it from expressing itself. It should be remembered that, during the 1960s, it was in this way that F. Jacob and J. Monod regarded the repressor, just at the time that the repressor proved to be a protein (and not RNA). The latest experiments, conducted at the Institut Pasteur as well as in the United States, seem to indicate that it may be possible to block the action of a given gene. Research is continuing on what may prove to become one of the most fabulous of all molecular-biological tools of the future.

The Japanese strategy in the field of cancer research is quite different. As we have already seen, the Japanese Institute for Cancer Research is concentrating more on developing viable molecules for industrial exploitation. Nevertheless, various researchers working in industrial laboratories, including Kyowa Hakko, have participated in research on oncogenes in the United States and have continued with this on their return to Japan. But Japan, which was one of the pioneers in the work on chemical carcinogenesis (with K. Yamagiwa and K. Ichikawa in 1915) and the synthesis of the first antitumoral product (nitramine), has specialized above all in the production of anticancer antibiotics.

The "Diseases of Civilization" and Genetic Engineering

While cancer could be considered one of the major plagues of our civilization, there are others against which genetic engineering may prove an invaluable tool.

Thus in 1980 there appeared on the scene a new and devastating disease, AIDS (Acquired Immune Deficiency Syndrome), transmitted through blood or sperm. The research teams led by Luc Montagnier at the Institut Pasteur in Paris and Robert C. Gallo at the National Cancer Institute discovered the virus causing this terrible disease and in January 1985 determined its structure. The way is now open to start medical tests and perhaps find a treatment. In July 1984, the Institut Pasteur and Genetic Systems founded the Blood Virus Diagnostic Company to sell AIDS diagnostic aids. And in

September 1984, Chiron announced that with the aid of the virologist Jay A. Levy, the company had succeeded in isolating and cloning the gene of the virus.

Other "diseases of civilization," and even more deadly ones, are those caused by an imbalance in the formation and elimination of thrombi: heart attacks, strokes, pulmonary embolisms, and thrombosis of the veins are good examples. The thrombi contain fibrin, a protein manufactured permanently in the bloodstream. To eliminate fibrin formed pathologically in small blood vessels, a process is needed that breaks down and dissolves the thrombi.

This process occurs thanks to activators that transform a dormant agent, plasminogen, into active plasmin capable of decomposing the fibrin and converting it into soluble products that can be easily eliminated. Various activators are produced by the body: urokinases or other enzymes given the generic name of PA (plasminogen activators). Urokinases or streptokinases (enzymes isolated in bacteria) are employed in the treatment of vascular diseases, but the production of urokinase is extremely costly, and there is a great risk of strong immunological responses. Furthermore, these enzymes activate the plasminogen throughout the entire circulatory system without discrimination, with a potential risk of internal hemorrhages. As a result, various laboratories (Ciba Geigy in Basle, Mitsubishi Chemical Industries along with Genentech and Mitsui Toatsu Chemicals in Tokyo) are trying, using genetic-engineering techniques or cell cultivation, to produce TPAs (tissue plasminogen activators), which will have the advantage of acting more specifically and concentrating on the thrombus.

Research is also being intensified on blood factors and the possibility of avoiding contamination during blood transfusions while lowering the cost of certain treatments, such as those of hemophiliacs (which today cost between $5,000 and $10,000 per person per year). This market would affect some 20,000 people in the United States. Of the twelve substances of which the blood is composed, one protein, factor VIII, is missing in hemophiliacs. Genentech recently announced the production of factor VIII by genetic engineering (the largest and most complex protein produced in a laboratory to date), and research is under way to determine its molecular structure. Parallel research is being carried out on substances capable of identifying factor VIII, for this would make possible the prenatal diagnosis of this illness, which is transmitted by women.

Another important blood factor, factor IX, is being produced by Transgene and the Institut Mérieux, traditional methods of fractionation having proved costly and extremely complex.

Also worth mentioning is the original initiative taken by Mitsubishi Chemical Industries in Japan, which has concentrated its medical research on perfecting cardiovascular medicines, psychotropes, and immunomodulators, rheumatism, embolisms, some cancers, and certain forms of mental illness, are presently in the clinical testing stage.

In any case, it has become a common strategy in Japan to invest in new products (above all, anticancer antibiotics) rather than in research. Such an outlook has contributed to the commercial success of several companies, though they are now becoming aware of their weakness (with respect to Europe and the United States) from the standpoint of long-term basic research aimed at determining the causes of cancer.

1

2

3

1. *Queen Victoria*
Hemophilia is transmitted through women to male children only. Queen Victoria was a "carrier" of this hereditary illness.
2, 3. **Red gold**
The by-products of human blood have become an industry with promise. The Institut Mérieux in France has built a giant factory at Marcy-l'Etoile with a series of 45,000-liter tanks used in the first phase of treating maternal blood (2). A filter rotating in a vacuum is used to eliminate the hemoglobin. Thanks to an extraordinary refrigerated transportation network connecting the Marcy factory with 5,000 maternity clinics throughout the world, hundreds of millions of liters of material blood are received by the institute to make albumin and gamma globulins.

Cancer
Is
a Very
Personal
Illness

by Shigetoshi Wakaki, Vice President and General Manager of Kyowa Hakko.

In Japan, as we have seen, research efforts are concentrated mainly on anticancer medicines. It was in 1956 that Dr. S. Wakaki, in conjunction with Dr. Tohu Hata of the famed Kitasato Institute, developed mitomycin C, whose story he tells us here. But Dr. Wakaki also considers the impact of these medicines in a broader context, that of existing anticancer therapies and of laboratory research. For while Japan may be less advanced than the United States or Europe with respect to research on oncogenes, it is the most thoroughly committed to lymphokines, as can be seen from the examples of Suntory, Ajinomoto, Takeda, and... Kyowa Hakko.

A crystal of mitomycin C The Japanese market for anticancer medicines increased tenfold between 1975 and 1980 as the result of a general mobilization against this disease, the nation's number one killer. Today, the strategy in this field is to improve antibiotics such as mitomycin and bleomycin (isolated by Umezawa in 1962) while reducing their toxicity.

"What is cancer?" We have not been able, as of yet, to give a correct answer to that question. Until now, cancer has been abstractly defined, from the standpoint of pathologists, as "abnormal, excessive, and autonomic proliferation of cells whose biological character has altered from normal tissue cells." Even today, the only reliable and acceptable method of cancer diagnosis available is microscopic examination of tissues or exfoliated cells. In the cancers encountered daily by clinicians, there is without a doubt a great variety, with differences as to natural history and immunological and biological natures. For example, even with cancers occurring in the same region, it is a well-known fact that drug sensitivity and degree of malignancy vary with each patient. In this sense, it can be said that cancer is an extremely personal disease. Although research on cancer today has brought forth vast new quantities of knowledge, the personal aspect of cancer has been the factor in preventing a single unified theory and method in the development of effective new drugs.

To discover cancer in the early stage and remove it surgically is the best method. But although it may be in a relatively early stage, there is the possibility of microscopic micrometastasis already occurring. In many cases, cancer cells dissociated from the primary tumor are transferred by lymphatics or by blood vessels in the body, forming metastases, resulting in clinical recurrence. In these advanced cases and in hematologic malignancies such as leukemia, resection and radiotherapy are both inadequate, and drug therapy (chemotherapy) plays the principal role.

Outline of Changes in Drug Therapy
Chemotherapy is regarded as the third modality after resection and radiotherapy. Since first initiated by the National Cancer Institute (NCI) in the United States in the late 1940s, vigorous screening of anticancer agents has been undertaken by numerous research organizations. Since then, a large number of improvements have been made in screening methods. However, the fundamental work – finding compounds capable of inhibiting tumor growth implanted into mice and improving the survival time of tumor-bearing mice – continues. The number of anticancer drugs used on a worldwide basis is few. Among these are the anticancer antibiotics (adriamycin, mitomycin C, and bleomycin), the alkylating agents (cyclophosphamide), and plant alkaloids (vincristine). All of these compounds kill cancer cells – for example, by impeding DNA formation, severing the DNA chain, or arresting cellular segmentation.

On the other hand, the recent advances in cancer immunology and biotechnology have made it economically feasible to supply large quantities of biological products produced by various cells. Furthermore, through the development of the concept of multidisciplinary therapy for cancer,

vigorous efforts have been made to produce future chemotherapeutic drugs with compounds possessing different (drug) action mechanisms useful in treating cancer: beginning with BCG, examples are immunotherapeutics (interferons, lymphokines, and other biological products), radiosensitizer (which is used to increase the effectiveness of radiotherapy), and compounds called biochemical modulators (which do not have anticancer activities themselves but which, in combination with existing anticancer drugs, augment their effectiveness). From among these ventures, a drug that has actually been utilized clinically has yet to be produced. However, this trend has within it the possibility of greatly changing the role of chemotherapy in the treatment of cancer in the near future.

Mitomycin C and *E. coli* L-Asparaginase
In the beginning of 1955, I was entrusted with the mass production of mitomycin A, a new anticancer antibiotic, made by fermentation of mitomycin producer, *Streptomyces caespitosus,* which was isolated by Professor Hata of the Kitasato Research Institute. My colleagues and I employed various techniques in the mass cultivation and the purification of the product. After approximately one year, we isolated a component possessing potent anticancer activity that clearly differed in nature from mitomycin A. Later it was named mitomycin C. This purple crystal demonstrated strong activity in the various types of experimental tumors implanted in laboratory animals. This compound was well received by the leading Japanese oncologists at that time with great interest. Its efficacy was confirmed in leukemia and digestive cancer clinically. In 1959, permission to market it was achieved. Mitomycin C has been the basis of cancer chemotherapy in Japan, performing the role of the standard drug in clinical research. It is not an exaggeration to say the developmental strides made in cancer chemotherapy in Japan have centered on mitomycin C.

The myelotoxicity of this compound met with problems in Europe and the United States, delaying wide clinical application. However, in recent years, due to the development of intermittent administration of mitomycin C and the lack of cross-resistance with other anticancer drugs, it has come to be highly regarded and has received great interest. The number of patients undergoing treatment with this compound has been calculated to be in excess of 300,000 yearly throughout the world.

Just when our mitomycin project was completed in 1961, Dr. J. D. Broome of Cornell University pinpointed the component in the serum of the guinea pig used to treat a mouse lymphosarcoma as an enzyme that hydrolyzed the amino acid L-asparagine into L-aspartate and ammonia. It was believed that the mechanism of action of L-asparaginase is due to the fact that for certain leukemia cells, L-asparagine is an essential amino acid, so that by administering L-asparaginase they fall into a subalimentation state and die. If this is true, it is conjectured that L-asparaginase, unlike conventional antileukemia drugs, does not inflict damage on normal marrow cells; in addition, it has the advantage of not showing any cross-resistance to other drugs. Later, it was confirmed that the L-asparaginase of the bacterium *E. coli* acted on mouse lymphosarcoma in the same way as guinea pig enzyme. On account of this, I decided to produce an antileukemia enzyme drug using our knowledge of fermentation technology and enzymology. From the fruit of my coworkers' efforts, spanning several years, L-asparaginase was isolated in crystal form from *E. coli* and proven clinically effective in some leukemia cases. This enzyme is the first pharmaceutical product of a protein of high molecular weight (141,000) to have been administered to humans. For the past ten years, since first it was marketed, L-asparaginase has been in continual use by leukemia patients both in Japan and abroad. It is unusual among anticancer drugs in utilizing the biochemical differences between the characteristics of leukemia cells and those of normal tissue cells.

Some New Approaches Related to Chemotherapeutic Drugs

Chemotherapeutic drugs used on a worldwide basis were either discovered or synthesized in the 1960s. So far, one compound sufficient for all clinical applications has not yet been obtained. But basic and clinical approaches have been carried out with the objective of achieving more effective application of existing drugs.

The first was combination treatment. As the anticancer drugs used in clinics act basically as cytotoxics, they bring on some side effects. Because cancer cells forming a tumor mass exhibit heterogeneous responses to anticancer drugs, it is impossible to destroy all of the cells in the tumor with one drug. To overcome these difficulties, various drugs with different mechanisms of action and side effects are used in combination.

The second method is regional chemotherapy. The objective of this method is to increase drug concentration in the locale of the tumor while reducing the systemic side effects as much as possible. It is a method that involves administration by injection into the spinal, peritoneal, pleural, etc., cavities or intraarterially into the artery supplying nourishment to the tumor. This method is frequently employed in clinical practice.

The third approach is the development of new systems for drug delivery. Utilizing new pharmaceutical technology, this approach increases the specificity and efficacy of drugs while suppressing the emergence of side effects or undesired responses in non-target tissues. The research in this field is focused on carrier compounds and highly affinitive vehicles (e.g., liposome) for specific tissue.

On the other hand, research on chemical modification of existing anticancer drugs is being vigorously pursued. The objectives are (1) to reduce toxicity of mother compounds (2) to improve the anticancer activity and spectrum, and (3) to yield new features (overcome resistance, change administration route, improve properties, etc.). The dose-limiting factor of adriamycin is its cardiotoxicity; because of this, a large number of derivatives are being experimentally synthesized throughout the world to mitigate this undesirable effect. Perhaps in the near future some of these will be ready for use in treatment. With the purpose of reducing the myelotoxicity of mitomycin C, we have synthesized about 700 derivatives over a ten-year period. Some of these compounds have progressed to the stage of clinical trials; however, it cannot be said that a derivative surpassing mitomycin C has yet been produced. Knowledge of the biochemistry and chemistry of mitomycin has expanded greatly through this research, and I firmly believe it will give birth to an improved mitomycin derivative.

Hormone drugs are frequently more effective than cytocidal drugs for hormone-dependent cancers (such as cancer of the breast, uterus, and prostate). As the overall population ages, the importance of research in this field will probably grow even more.

Biologicals and Biological Response Modifiers as the Fourth Modality of Cancer Therapy

In the latter half of the 1960s, great strides were made in cancer immunology. Immunological treatments demonstrated a capacity to suppress the growth of tumors in experimental animals. Numerous clinical studies were conducted in the hope of establishing immunotherapy as the fourth modality of cancer treatment. However, in the majority of these studies only marginal or negative results were achieved. In consequence, immunotherapy would not become the fourth modality of cancer treatment. However, due to the use of nonspecific immunoactivators (such as picibanil and krestin) in adjuvant treatment, interest in compounds with host-mediated antitumor activity has continued to grow in Japan.

Later, due to technological advances in genetic engineering and mass cell cultures and the discovery of hybridoma, it became possible to obtain large quantities of highly purified biological products.

Dr. Oldham of NCI has defined "biologicals" as "the products of a mammalian genome" and "biological response modifiers (BRMs)" as "agents and approaches whose mechanism of action involve the individual's own biological responses"; by these definitions, biologicals and BRMs may become the fourth modality in cancer treatment. They include, among other things, immunomodulators and/or immunostimulating agents; interferons and interferon inducers; thymosins, lymphokines, and cytokines; monoclonal antibodies; antigens (tumor-associated antigens, vaccines); and effector cells (macrophages, T cell clones, etc.).

There are many other forms of cancer treatment under study; one such is the screening of differentiation-inducing agents, which views cancer as a disorder of cell differentiation. In addition, anticarcinogens and compounds called chemopreventives will become increasingly important subjects of future study.

In tumors consisting of 10^9 cells (approximately 1 cm^3 in volume) the destruction of 99.9% of the population by chemotherapy still leaves 10^6 cells to proliferate. The remaining variants become, because of the drug treatment, even more malignant than the cells in the original tumor, and thus more resistant to treatment. Cancer is an extremely resilient living creature. Most likely it will be a long time before a drug is discovered that can completely cure it. For this reason, I believe that we must pursue research on developing drugs based on their types of mechanism of action.

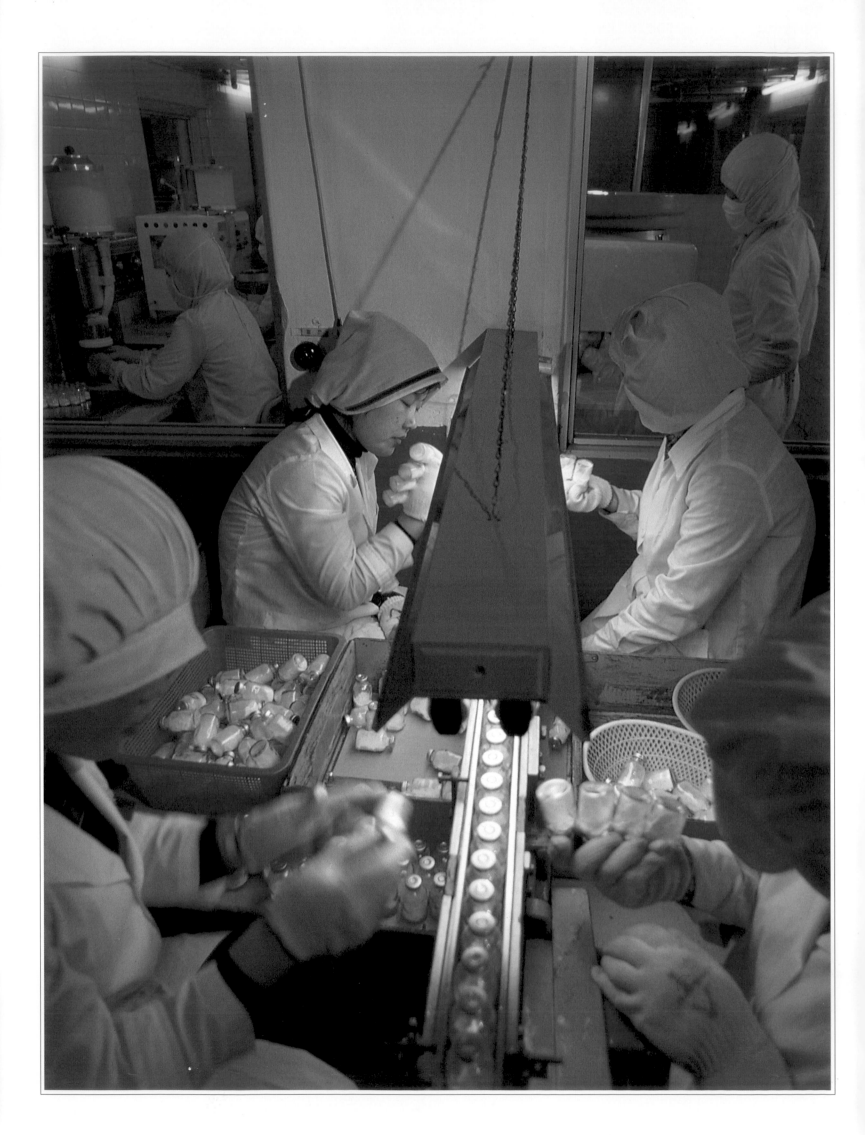

11

Pharmaceutical Economics

The cost of bringing a single major new pharmaceutical product to the market is of the order of $65 million-$100 million. This research and development (R&D) cost has to be recovered before the innovators can begin to earn profits on sales.

In the economics of drugs research there is no such thing as a lucky break – only unlucky ones, when an idea has to be abandoned no matter how much money and intellectual effort has been invested, because a flaw has been discovered that makes it too dangerous. Unlucky breaks are all too common.

It takes up to seventeen years to develop a major new drug – say, a drug to treat heart disease or malaria – from the first definition of the target. The research phase of this long R&D cycle, discovering the new therapy, will absorb about a quarter of the cost over eight to ten years. The development phase of the R&D cycle, which takes the discovery to the point of product launch, will absorb the other three-quarters of the cost, over five to seven years.

The first point to remember is that most of what you have heard about biotechnology, and will continue to hear, relates to the research phase, the cheaper portion. Biotechnology is essentially an "enabling technology," which may allow the scientist to do something that simply cannot be contemplated by other technologies at present, such as the synthesis of a compound too complex to construct by more traditional methods of synthesis. Or biotechnology may allow him to do something that he can only do now at the price of a tedious and costly purification process; or to make something purer or more "natural" than the present product. What biotechnology will not allow him to do is to circumvent the development phase of the R&D cycle, the more expensive phase, which has to demonstrate convincingly that the drug is efficacious, acceptably safe, and can be conveniently administered to the kind of patient who needs it. There is some evidence that biotechnology, far from making development easier and cheaper, may have complicated it.

Given the speed and enthusiasm with which the world has been introduced to biotechnology, it is hardly surprising that some misconceptions about what it might offer have gained wide currency. Interferon, for example, the most highly publicized of the early biotechnology targets, was a fertile medium for the culture of misconceptions. Interferon had tantalized and frustrated medical researchers for two decades after its discovery in the late 1950s as a natural substance that seemed to have a role in keeping viral infections and cancer at bay. By the late 1970s, the entrepreneur biotechnologists were advancing the view that, if interferon could be made by genetic engineering, a substance that, in the words of Lord Rothschild, the famous Cambridge scientist and London banker, was then £1 billion a gram, could be made cheaply and far purer than the interferon hitherto painstakingly extracted from blood. To give the biotechnologists credit, they fulfilled their promises to deliver pure and plentiful interferon. By the end of 1982, Biogen claimed it was giving interferon away free to medical researchers for clinical trials. A world avid for a "cure for cancer" was led to expect a miracle by the mid-1980s.

But there were complications. Interferon was proving to be not a single – albeit complex – compound but a whole family of substances of widely varying idiosyncrasies. Different biotechnology approaches were making available various versions of interferon with quite different characteristics. But one common characteristic emerged: that the interferons were not so free from side effects as was previously supposed. The first big clinical trials in the United States and Great Britain were very disappointing. Not only was the substance proving far less efficacious; it was giving the patients the symptoms of influenza – a particularly unfortunate side effect in cancer patients, whose morale already tends to be low. Questions began to be raised whether the rush to get genetically engineered interferon on to the market might not be more in the interests of the investors than of the

Medicines
A medicine production line in South Korea.

119

patients, and the regulatory authorities became more cautious, even though interferon had been publicized as a potentially life-saving drug.

From genetically engineered insulin, used to treat diabetes, comes other evidence for caution on the part of investors in biotechnology for new drugs. The target here was not a life saver, for most diabetics are well served with the standard insulin products already available, namely, porcine insulin purified to a point where it matches remarkably closely the human hormone. The goal of the biotechnologists was a perfect match with human insulin, more readily tolerated by those few diabetics allergic to the porcine product, but above all cheaper, able to capture the market on price.

Eli Lilly introduced genetically engineered human insulin in 1983. It has captured only a small percentage of the insulin market. The medical profession has not been sufficiently convinced of its advantages to advise patients to change. More significantly, in the context of this chapter, neither initially nor since has Eli Lilly tried to expand its market share by undercutting porcine insulin. The inference must be that genetically engineered insulin is expensive to make. The scientific targets of the research phase were achieved with brilliant speed (in this case, again by Genentech). But the targets of the development program must have proved costly to implement.

The present glamour products of genetic engineering – human growth hormone, tissue-type plasminogen activator, tumor necrosis factor, interleukin 2, etc. – could all suffer the fate of interferon and human insulin. They proved relatively easy targets for the "cloners," yet fell foul of other factors in the lengthy and hazardous development chain. Some, inevitably, will never satisfy the regulators that their ratio of efficacy to risk is high enough to win approval, or will remain restricted to a small market, unable to recoup the development costs. Some may be overtaken by the rapid progress of computer-aided drug design.

Diagnostics

In 1984, Lord Rothschild, as Chairman of the investment trust Biotechnology Investments Ltd. used his scientific influence with the Royal College of Physicians in London to canvass the views of eminent doctors on what the profession most needed from scientists. The results were quite different from public perception of the priorities.

The doctors had picked out six priority needs:
• simple, cheap DNA probes for diseases, including cystic fibrosis, Down's syndrome, and muscular dystrophy;
• simpler, more specific techniques for rapidly diagnosing viral, bacterial, and parasitic infections;
• infection-free blood products, such as factor VIII, factor IX, and albumin, made by genetic engineering;
• better vaccines for common diseases, such as whooping cough, measles, and hepatitis;
• a cheap, first-class supplement to replace milk at weaning;
• more effective and specific ways of aiming agents to specific cells for the diagnosis, monitoring, and therapy of cancer.

Their priorities plainly stress diagnosis rather

1. Sterile unit
2. Fleming, Waksman
1949: the visit to Merck of Sir Alexander Fleming (right), who discovered penicillin in 1928, and Selman A. Waksman (left), who discovered streptomycin. They are talking to the Director of Research and Development, Randolph T. Major (standing). Merck was the first company to save a patient's life in the United States by the administration of penicillin (in 1942); it also financed research on streptomycin.
3, 4, 5. Selection, screening, sales
Progress in data processing has provided researchers with precise analytical tools with which to observe the action of a substance on microbes and the degree of their resistance to it (3). The problem lies in the transition from the laboratory (4) to industrial production. After this, there only remains giving the medicine a last check before packaging it (5).
6. Skeleton of a mouse fetus
To better observe discrepancies during experiments.
7. Storage tanks
Storage tanks for blood derivatives.

1

2

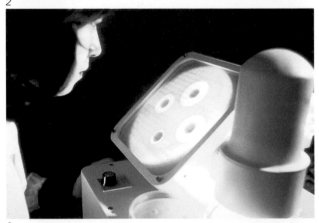

3

4

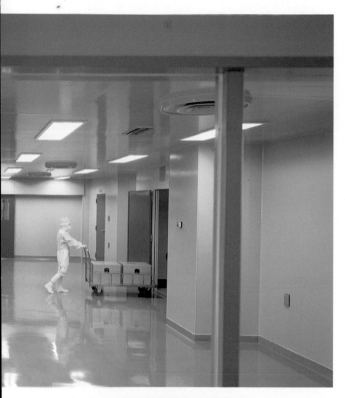

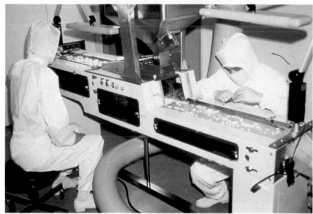

5

6

7

than miracle cures. They still urgently need ways of recognizing many serious diseases quickly and unambiguously. They want ways of following the course of an illness more closely in order to assess the efficacy of their treatment.

The relatively unsung aspect of the new technologies is the immense contribution they can make here. Except, perhaps, in the case of a disease such as AIDS (Acquired Immune Deficiency Syndrome), diagnostics will never capture public attention to the same degree as "magic bullets" for incurable diseases. But the professionals recognize their value, and the biotechnologists know that once its science is mastered, the development phase is only a year or two. The rigorous safety studies required of a therapy are obviously unnecessary for a test that may only be administered once.

Manufacture

A third outlet for innovative biotechnology in the pharmaceutical industry is in the manufacturing process. Although there is no reason to suppose that biotechnology is inherently a cheaper way of making a compound than any other – the example of Eli Lilly's Ilumulin insulin suggests it is not – biotechnology may be the best way of making some step in a complex process, some intermediate between the starting materials and the finished product. As an "enabling" technology, biotechnology may succeed in cutting the cost of manufacturing pharmaceutical products not only already on sale, thus prolonging their life for a company once patents expire and competitors offer copies, but also to go on sale in the future.

These three opportunities for biotechnology in the pharmaceutical industry therapeutics, diagnostics, and manufacture – have helped pinpoint healthcare as the industry of greatest interest for the next decade or so. The industry itself needs biotechnology – for innovation, testing, evaluation, and manufacture. It is a research-based industry, second only to electronics and information technology in the proportion of sales invested in R&D. From the start of the "biotechnology revolution" of the late 1970s, it has been the main sponsor of the new biotechnology firms. The 1984 Report of the US Congress's Office of Technology Assessment (OTA) on commercial biotechnology concludes that almost two-thirds (62%) of these NBFs (New Biotechnology Firms) in the United States were pursuing pharmaceutical uses, compared with less than one-third (28%) in pursuit of the next most popular target sector, animal agriculture.

Pharmaceutical Economics

Heirs
Australian test tube embryos, heirs to a fortune after the accidental death of their "parents," will remain frozen until legislation is passed, for there is nothing covering such an instance. A cartoon by Loup.

Life Is Asymmetrical

by Hans Leuenberger, Assistant Director of Research, Hoffmann-La Roche.

Since the work done by Pasteur (mentioned in chapter 1), we have known that life is asymmetrical. Contemporary methods in biotechnology now enable us to make use of this property to obtain only the active (L) form of amino acids (such as glutamic acid). H.G. Leuenberger explains the importance of these techniques, not only for the pharmaceutical industry (cortisone), but also for the industries processing human or animal foods (vitamins, sodium glutamate, lysine, etc.).

(1) From the Greek word "enantios," meaning "opposite.".

(2) Acidity/alkalinity factor.

The observation that some natural substances rotate the oscillation plane of polarized light led to the concept of optical activity. Each carbon atom (C) has four binding sites. If each of these sites is occupied by different atom groups, depending on the sequence in the arrangement, two distinguishable molecules are found, of which one is the mirror image of the other. We then speak of an asymmetrical carbon atom (C*).

For example, the simple amino acid serine has such an asymmetrical carbon atom (figure 1).

The mirror-image molecule types have, as a rule, the same chemical and physical properties and can be distinguished only by one physical characteristic, namely: one kind rotates the oscillation plane of polarized light passing through the substance concerned to the left; the other rotates it by the same angle to the right. Such compounds are called optically active compounds, and the two mirror-image forms are called enantiomers.[1] A uniform mixture of enantiomers in which the dextrorotatory and levorotatory effects cancel each other out is called a racemate.

It is interesting to note that, as a rule, enzymes convert only one of the enantiomers of a racemate and have no effect on the other. If a new asymmetrical center is introduced into a molecule as the result of an enzymatic reaction, only one of the two possible enantiomers is formed, and the product thus becomes optically active. For example, all cells produce only L-amino acids, such as the dextrorotatory L-serine. By contrast, if a chemist synthetizes serine in the laboratory by the usual methods, he obtains a uniform mixture of dextrorotatory and levorotatory serine molecules, i.e., a racemate, which in this case is called DL-serine.

In other words, in the synthesis and modification of natural compounds, enzymes observe strictly the spatial relations of the atoms in the molecule. For this reason, natural compounds containing one or more asymmetrical carbon atoms are optically active. It has been found that the enantiomeric forms of a compound often have quite different physiological or pharmacological effects in an organism. In the chemistry of natural products, as well as in pharmaceutical chemistry, we are therefore very interested in the production of true-to-nature optically active products, and we have for this purpose also incorporated enzymatically catalyzed reactions into synthesis schemata.

Salient Characteristics of Biotransformations

Biotransformations (enzymatic reactions) can be superior to the conversion methods used by chemists in the following respects:
- *Reaction selectivity*: Each enzyme is limited to catalyzing one single reaction, so that no side reactions occur.
- *Site specificity*: An enzyme grasps its substrate in such a manner that the chemical reaction always takes place at the same site in the molecule. This remains true even if the molecule has several functional groups that would also be eligible for a reaction.
- *Stereoselectivity*: Enzymes can distinguish between the enantiomers of a racemate in that they convert exclusively, or at least preferentially, only one of the two enantiomeric forms. If an enzymatic reaction introduces a new asymmetrical carbon atom into the molecule, then one of the possible enantiomers is formed exclusively, or at least preferentially, and the product thus becomes optically active.
- *Mild reaction conditions*: Enzymatic reactions take place in aqueous and nearly neutral solutions (pH of about 7)[2] at low temperatures (mostly under 40°C). For this reason, even sensitive compounds are converted without damage, and the biotransformation requires little energy and does not damage the environment.

Chemists and Microorganisms Work Together

Vitamin C: Vitamin C has been manufactured for over 50 years according to the classical Reichstein's synthesis. In the first step, the starting material, D-glucose, is reduced chemically to D-sorbitol. The second step, namely, the conversion of D-sorbitol to L-sorbose, is nearly impossible for the chemist. The sorbitol molecule carries a hydroxyl group (OH) at each of the 6 carbon atoms. In order to obtain L-sorbose, it is necessary to oxidize only one of these 6 hydroxyl groups in the molecule.

This task is performed with a very high degree of precision by acetic acid bacteria. This biotransformation is an impressive example of the ability of microorganisms to alter only one among several functional groups with very similar reaction properties (in this instance, the 6 hydroxyl groups in the sorbitol molecule) owing to the site specificity of the reaction. The L-sorbose produced in this reaction can then easily be converted to vitamin C in a series of chemical steps.

Cortisone: Cortisone is a steroid hormone of the adrenal cortex and plays an important role as an inflammation-inhibiting agent. It was therefore desirable to be able to produce cortisone and other structurally related steroid hormones synthetically. The starting materials were steroids of plant or animal origin. It was soon found that the inflammation-inhibiting activity depends on, among other things, an oxygen function in position 11 of the steroid skeleton, which, however, is lacking in the readily available raw materials. The introduction of this oxygen function by chemical means proved to be exceedingly difficult. Thus, in the early stages of this work, only 1 gram (g) of cortisone acetate was produced in a laborious 32-step synthesis from a starting amount of 576 kilograms (kg) of desoxycholic acid,

Fig. 1

mirror axis

L-serine   D-serine

In this mode of presentation, if the amino group (−NH₂) at C*, characteristic of amino acids, is oriented to the left, we speak of L-serine; if it is oriented to the right, we speak of D-serine. All 20 amino acids used in nature as protein building blocks belong to the L-series.

which can be obtained from animal bile. Later, a gigantic step forward was achieved by direct hydroxylation of the steroid skeleton in position 11 with the help of microorganisms. In this conversion, a compound called "substance S," which can be readily obtained by chemical degradation of vegetable diosgenin or stigmasterol, is converted directly to cortisol by 11β-hydroxylation with the fungus *Curvularia lunata,* and the cortisol is then easily converted to cortisone by chemical oxidation.

Additional microbiological and chemical reaction steps led to related compounds with even better inflammation-inhibiting action. In the course of further research work, many other microorganisms have been identified that are capable of specifically altering the steroid skeleton at almost any desired point. The microbiological hydroxylation of "substance S" in position 11 demonstrates the ability of microorganisms to functionalize a particular nonactivated position in a substrate molecule in a site-specific manner.

L-amino acids: The natural, optically active L-amino acids are used as foodstuff supplements for humans and animals as well as for pharmaceutical purposes. The main difficulty of their synthesis lies in the introduction of the optical activity. Here again microorganisms can help the chemist. One possibility is that the chemist first synthesize the racemic DL-amino acids and attach an acetyl group ($-COCH_3$) to the amino group ($-NH_2$). The acetyl-DL-amino acid produced in this manner is now brought into contact with an enzyme from the fungus *Aspergillus oryzae.* This enzyme works stereoselectively; i.e., it can distinguish between the D-form and the L-form of acetyl-DL-amino acids and deacetylates only the L-form (with uptake of a molecule of water) without affecting the D-form. As a result we obtain the desired L-amino acid, which can easily be separated from the unaltered acetyl-D-amino acid. This example illustrates the possibility of using a microbial enzyme to separate the enantiomers of a racemate.

If suitable starting compounds are available, certain selected microorganisms are capable of building up an asymmetrical center (C*) on their own. The enzyme responsible for the anabolic action directs the four different atom groups at C* in such a manner that they are always arranged in the same sequence in each molecule of the product. This results in the direct formation of an optically active product consisting of only one enantiomer. This possibility is used, for example, in the production of L-aspartic acid (figure 2).

In addition to the two examples given in figure 2, if fumaric acid is brought into contact with *Brevibacterium flavum* instead of *Escherichia coli,* then, instead of a molecule of ammonia, a molecule of water (H_2O) is attached to the double bond of fumaric

acid and the product obtained is optically active L-malic acid. These examples show that it is of the utmost importance to identify from among the innumerable microorganisms and to use as reaction mediators those that can specifically mediate the reactions desired by the chemist.

In other methods for the production of L-amino acids, the entire work of synthesis is done by microorganisms, namely, by selected strains that produce the desired L-amino acid in excess and release it into the surrounding medium. It should also be noted that much more complex compounds, such as vitamin B_{12} and many antibiotics, are produced directly by microorganisms.

Search for the Most Suitable Microorganisms

The examples presented above demonstrate that particularly critical reaction steps can sometimes be conducted with the help of microorganisms. This poses the problem of how to find the microorganism most suitable for carrying out a particular reaction. This is by no means simple. It is often necessary to test hundreds of microorganisms obtained from culture collections or isolated from soil or water samples in order to find the one that can transform the starting compound concerned in the desired manner. The chances of success can be increased by selecting for these screening tests those microorganisms of which it is already known that they mediate the desired type of reaction in a similar compound.

Once a suitable microorganism is found, it is necessary to find the conditions under which the cells multiply best and can perform their chemical work most effectively. The composition of the nutrient medium,

the temperature, and the aeration must be adapted accurately to the requirements of the mircoorganism concerned. The intrusion of foreign microorganisms that could disturb the desired biotransformation must be prevented by clean working conditions and by prior sterilization of the breeding vessels and nutrient media. It is often advantageous to separate the process of cell culturing from the process of biotransformation. With this technique, the cells are first cultured under conditions that ensure good growth, are next separated from the nutrient medium by centrifugation, and only then are introduced into the medium containing the substance to be transformed. Favorable working conditions must also be ensured for the biotransformation that now follows. Finally, methods must also be developed for isolating the reaction product from the fermentation broth.

Unfortunately, the search for efficient biotransformation processes is often unsuccessful because a suitable enzyme is simply not to be found anywhere. Difficulties also arise if the starting substance cannot penetrate into the cells or is toxic for the microorganism concerned.

Biotransformations are mediated by live cells – more specifically, by their enzymes, which may behave differently towards each particular substrate. In contrast to many chemical methods, biotransformations are not applicable generally, so that an individual solution must be found for every individual case. Many examples, especially in the chemistry of natural products, have shown how useful it can be to incorporate microbiologically catalyzed reaction steps into a synthesis sequence. The possibilities are by no means exhausted as yet.

Fig. 2

In the presence of fumaric acid in the nutrient medium, a bacterial strain of Escherichia coli *attaches ammonia (NH_3) to the double bond of fumaric acid in such a manner that enantiomerically pure L-aspartic acid is formed.*

Another bacterium, namely, Pseudomonas dacunhae, *specializes in eliminating one of the two acid groups ($-COOH$) of L-aspartic acid by cleaving off a molecule of CO_2, so that another L-amino acid, namely, L-alanine, is produced. L-alanine can thus be produced in a 2-step reaction by bringing fumaric acid into contact with both microorganisms.*

12

Green Gold

Separating a cell or a tissue from a living being and then recreating that being from it is an old fantasy of science fiction. It is quite impossible where men or animals are concerned. But not for plants. For, using a single cell, organ, or tissue, it is possible to regenerate an entire plant. "In this single flask," says Carole Meredith of the University of California at Davis, "we can study a hundred million plants. In nature, we would need thousands of acres." And dozens of years, we might add.

In France in 1938, Roger Gautheret began to exploit cultures of undifferentiated tissues. The culture of isolated cells, for its part, was only successfully accomplished in 1954 by Muir, Hildrebrandt, and Riker. Between these two dates, the culture of meristems (the terminal dome of the bud, which assures the growth of the plant, and which is spared by viruses) had made it possible to cure diseased plants. This involved, first of all, saving "Pink Dream," a superb dahlia, which was followed by the regeneration of a series of threatened species – carnations, chrysanthemums, strawberries, etc. During the 1960s, researchers studied the hormonal balance of plants and the improvement of culture media (sugar, vitamins, hormones, etc.). Varieties of disease-free plants were reproduced in thousands of examples, including potatoes, which until then had been ravaged by rust. Unilever took an interest in cloning Malayan palm trees, which were resistant to illnesses and produced 20-30% more oil than other palms.[1] Moët-Hennessy, famous throughout the world for its champagnes, started taking an interest in roses to the point of buying 34% of Delbard's capital in France and a controlling interest in America's second-ranking producer of roses, Armstrong Nurseries. For are not the rose bush and the grape vine horticultural cousins, and cannot the successful in vitro multiplication of roses find an application in the champagne business? In the course of only nine months, a thousand rose bushes can produce a million new plants (and, if pushed to the limit, up to 1 million). And all this has been accomplished away from the vagaries of climatic or seasonal factors, using only controlled light and heat.

Growing vegetable cells in culture thus makes it possible to identify interesting features, isolate them, and regenerate plants with only the desired qualities – resistance to diseases, as we have already seen, but also to herbicides. One need only introduce an herbicide into a culture to see which cells are more resistant than others, either through fusion with a more resistant strain, or spontaneously, or as the result of genetic recombination, and then regenerate them. Unfortunately, the plants whose regeneration would be the most interesting are the monocotyledons (which sprout with a single leaf), such as cereals (wheat, barbey, rice, etc.), and until now only the dicotyledons, such as petunias, tomatoes, datura, and alfalfa, have been successfully regenerated.

A still more recent technique has the advantage of simplifying research. This consists of the in vitro cultivation of pollen from anthers; these make it possible to work on haploid cells, which have but a single chromosome. This began in 1966 with the research on datura done by S. Guha and S.C. Maheshwari at the University of Delhi and in 1967 by J.P. Bourgin and J.P. Nitsch in France. These "motherless plants" made it possible to avoid the sometimes disastrous consequences of mixing characteristics while providing for the transmission of a consistent genetic heritage. These techniques seem to have produced results in the case of wheat, barley, rape, peppers, and eggplant. Tests were also made with "fatherless plants" (requiring a simpler manipulation, as it was a matter of extracting a virgin ovule to cultivate it in vitro) in the case of rice, wheat, corn, sugat beet, sunflower, and lettuce. But these "half-plants" proved sterile and could only be multiplied by means of cuttings. If an interesting mutation is achieved, the entire plant must once again be used by interspecies hybridization and doubling the number of chromosomes.

Seeds and pesticides
ICI, which chose this photograph to launch a new chemical herbicide to weed crops of dicotyledonous plants in 1985, announced its investment in the American firm Garst Seed, along with a research and development agreement between the two companies concerning cereals.

(1) The market of palm oil is appreciable: about $275 million annually.

(2) Bringing pollen from one plant to the anthers of other plants (wheat-rye, cabbage-turnip, roses, etc.).

1

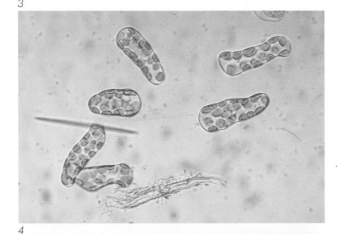

2

3

4

1. *Meristem*
Meristem of a carnation taken under aseptic conditions to carry out the regeneration of the flower at Barberet et Blanc.

2. *Palm oil*
Unilever's experimental facilities in Malaysia for the cloning of palm trees. Researchers had already been able to increase yields by 30% in the early 1980s. Multinationals are taking an increasingly great interest in the cloning of commercially valuable plants (oil palms, sugar cane, bananas).

3. *Cells*
Dissociated cells (of asparagas) obtained by simple grinding.

4. *Calluses*
Calluses are formed from dissociated cells: shown is Atlas, the latest experimental variety of corn, produced in the Ciba-Geigy laboratory.

Still another technique, perhaps even more promising, is somatic hybridization, which makes it possible to avoid some of the pitfalls of sexual hybridization[2] (the incompatibility of most plants to each other, the transfer of undesirable features, etc.). This technique above all makes it possible to manipulate not only the genes of the nucleus but also those of the mitochondria and the chloroplasts.

This is achieved by protoplasmic fusion – the protoplast being the plant cell deprived of its wall of cellulose by a process developed in 1960 by the British scientist Cocking. The first plant to be produced by regeneration after protoplasmic fusion achieved a certain ephemeral notoriety: this was the pomato (potato-tomato), produced in 1974 by D. von Wettstein at the Carlsberg Institute and by G. Melchers at the Max Planck Institute in Tübingen. Unfortunately, however, this handsome vegetable proved sterile. But this did not discourage Heinz, Campbell, and Atlantic Richfield from investing tens of millions of dollars in a protracted search for that pearl without price, the perfect tomato, simultaneously meatier and juicier and thus capable of "multiplying to infinity the number of bowls of soup made from each tomato."

In the case of coffee bushes, a somatic embryo can also be obtained directly from a primary explant without going through the intermediary stage of callus formation, which limits risks. Coffee is in fact the world's leading agricultural export product and ranks second only to petroleum in international trade. Of all the coffee drunk, 95% comes from only two species of coffee bush: Arabica (accounting for 70% of world consumption), with a delicate aroma and low caffeine content (1-1.3%); and Robusta, a variety of Canephora virtually without aroma but with a high caffeine content (2-3%). The first is grown mainly in South America and is highly susceptible to rust, a devastating disease caused by a microscopic fungus, while the latter is found principally in Africa and Asia and has genes making it rust-resistant.

The ideal solution would thus be to cross Robusta with Arabica to make the latter resistant to the disease, then back-cross it with an Arabica parent to reinforce its aroma. This could be done by traditional techniques, but a viable end product would take at least 40 years. Rust is gaining ground steadily, however (there were epidemics in Salvador in 1979 and in Guatamala in 1980). If this mysterious aroma is to be preserved (requiring a combination of 500 known components, to say nothing of the unknown ones) something will have to be done quickly. The problem is made even more complicated by the fact that the reproductive methods of each species must be exploited. This is autogamy for Arabica (autofecundation of the hermaphroditic flowers of the same plant) and allogamy for Robusta (cross-fertilization between different plants). Horticulturists are also trying to develop coffee bushes that produce caffeine-free coffee by crossing Coffea canephora with Mascaracoffeas from Madagascar, which are free of the factors biosynthesizing caffeine. But for the moment, the resulting brew is far too bitter and unappetizing.

There is a specific problem concerning genetic engineering when applied to plants. Though research has often been of fundamental importance, from Gregor Mendel to Barbara McClintock, biological and genetic knowledge of plants has fallen far behind that of other living organisms. Nevertheless, there is every likelihood that new discoveries in this field will be revolutionary, particularly if we consider the secrecy surrounding the work being done by companies in the business.

Vectors

In the case of plants, transferring genes by genetic recombination poses one essential problem, that of the vector. If the gene is transferred by a virus or bacterium, the gene risks not expressing itself in this new environment – or, if it is expressed, it risks disappearing in the following generation.

Research into two plant diseases, crown-gall and hairy root, seem to point to the start of a solution. The former is a form of cancer that affects dozens of different varieties and is communcated by the transfer of genetic information from one bacterium (*Agrobacterium tumefacens*), which was discovered in 1907 by two Americans, Erwin Smith and C. Townsend. In 1942, working on tissue cultures at the Rockefeller Institute, Armin Braun noticed that crown-gall cells resembled tumor cells in their anarchic proliferation. Later, Allen Kerr of the University of Adelaide assumed that the segment of DNA responsible for the vegetable cancer was located not on the bacterial chromosome but on a plasmid, and this was confirmed by the work of R. Hamilton and M. Fall at the University of Pennsylvania in 1971. It was localized in 1974 by Joseph Schell and Marc van Montagu of the University of Ghent, and it proved to be the Ti (Tumor inducing) plasmid, which has sites into which foreign genes could be inserted (as was discovered later).

The second illness appears as a sort of mane of roots growing at the point of inoculation. It is induced by the Ri (Root inducing) plasmid harbored by a related bacterium, *Agrobacterium rhizogenes*.

In both cases, the transfer RNA becomes integrated with the genome of the plant cell and gives it the power to synthesize proteins (opines, identified in 1960 by Georges Morel and his team at INRA, Institut National de la Recherche Agronomique, in France). Initially, it became possible to inactivate the pathogenic portions of the Ti and Ri plasmids without altering the information required to synthesize opines. In this case, it was wondered, why not transmit through these plasmids the genes responsible for synthesizing other proteins? Almost simultaneously, two teams at Monsanto and the University of Ghent succeeded in doing this, and they were soon followed by one at Agrigenetics.

At Monsanto, the gene of resistance to an antibiotic (kanamycin) is isolated and hooked onto the bits of DNA that control the signals to produce proteins; this chimera gene is then inserted into a petunia cell. The plant, once regenerated, subsequently grows in a kanamycin-rich medium, which normally would kill it. It has thus been proved that the plant can not only retain a foreign gene but can pass it on to the next generation.

Other possible vectors have also been developed, like Calgene's microinjection system. At first it was thought that as Agrobacteria only attacked dicotyledons, they could not be used as a vector for monocotyledons, but two recent experiments seem to have disproved this. The potential of the

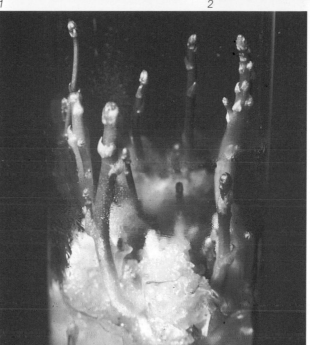

1, 2, 3, 4. **Plants in a tube**
Regeneration and clones of roses at Delbard, tomatoes at the CNRS, asparagus in Japan, and oil palms in Malaysia.
5. **China**
Professor Hiroshi Harada of the University of Tsukuba, in Japan, in company with a young Chinese agronomist, visits an experimental farm operated by the Agronomic Research Institutes of Shangai, where a new variety of rice is being grown that was created by the culture of anthers.
6. **Anthers**
A culture of tobacco anthers: the formation of haploid plantlets from immature pollen grains.

1
2
3
4
5
6

cabbage mosaic virus and other (not very promising) vectors is also being studied.

Nitrogen Fixing

Quite another research sector that may prove interesting is that of nitrogen-fixing plants, for this might eliminate the need for high-cost chemical fertilizers ($1 billion for corn alone in the United States) as well as the fuel needed to manufacture, transport, and spread ammonia (with the attendant risk of pollution, intensified by the relative brevity of its effectiveness).

During the nineteenth century, the French chemist Jean-Baptiste Boussingault, the German Baron Justus von Liebig, and the Dutchman G. Muller were the first to demonstrate the indispensable role played by nitrogen in the life of plants and to describe the nitrogen cycle. Odd as it may seem, while plants cannot get along without nitrogen, they are incapable of using the natural reservoir of this gas in the atmosphere (80%), and can only assimilate it in the form of nitrates or ammonia. Yet one bacterium (*Rhizobium,* identified in 1888 by Hermann Hellriegel and H. Wilferth) can naturally provide the plant on which it is a parasite with the required nitrogen in the form of assimilable ammonia. This phenomenon had been exploited since Roman times, though its origins remained a mystery. It had been noted in the case of leguminous plants (the only ones on which this bacterium lives as a parasite), such as beans, peanuts, alfalfa, soybeans, peas, clover, and lupin. This symbiosis betwen rhizobia

and legumes is responsible for nearly half of the nitrogen fixed each year on the earth's surface. If we eventually succeed in understanding this phenomenon, which is still a mystery, we may be able to apply it to nonleguminous plants. We already know that nif (nitrogen-fixing) genes fix the nitrogen, that nod genes are the source of the nodules on the roots of leguminous plants, and that an enzyme complex, nitrogenase, catalyzes the nitrogen and converts it into ammonia. But exactly how does this symbiosis function? How do the plant and the bacteria coordinate their nodules to avoid the intrusion of oxygen? So far, we have only been able to determine that the position of the nif genes, which are to be found on a very large plasmid of over 500 genes, is the determining factor in their recognition by the plant.

Klebsiella, another nitrogen-fixing bacterium, is now being studied by Winston Brill, University of Wisconsin at Madison, F. Cannon and R. Dixon, University of Brighton, A. Pühler at Bielefeld, C. Elmerich at the Institut Pasteur, and Frederick Ausubel at Massachusetts General Hospital. The latter has identified a group of 17 genes, of which 2 or 3 either engender or halt the nitrogen-fixing process, while 3 code for nitrogenase. These genes have been cloned and inserted into *E. coli* and have succeeded in fixing nitrogen.

Other organisms are also capable of fixing nitrogen, including blue river algea (cyanobacteria), *Azospirillum brasiliensis* and *Rhodopseudomonas capsulata.* Quite recently, it came to light that alders and about a hundred other trees and shrubs of the

*1, 2, 3. **Protoplasts***
The digestive juice taken from a snail (1) contain at last 30 different enzymes, some of which digest the cell walls of the plant. One may obtain protoplasts (2) of various kinds from a dilution of 10 to 1 of this juice, which can be fused (3): the protoplasts of corn are shown in green, that of sugar cane in white.
*4, 5. **The pomato***
Potato-tomato hybrids were achieved by Professor Georg Melchers and his team at the Max Planck Institute in Tübingen, and it was in the same laboratory that Tony Holder and Carsten Poulsen demonstrated that these hybrids contained potato as well as tomato genes. The potato flowers (5, above) are white, those of the tomato (5, center) yellow. The hybrids from different flowers depend on the protoplasmic fusion from which they issued.

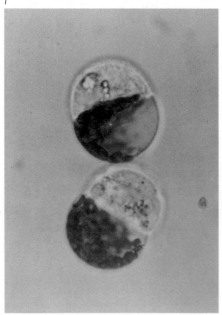

1

2

3

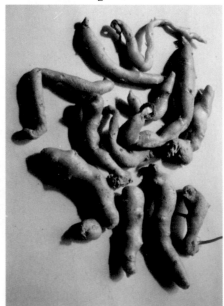

4

5

same family (Angiospermae) have leaves that are rich in nitrogen compounds and that their humus is rich in organic nitrogen. This phenomenon is due to microorganisms whose study began in 1974 and that have subsequently been cultivated as fermenters. The Laval University in Quebec has announced that it has perfected an inoculation process that should permit the reforestation of James Bay.

Applications

One of the very first commercial applications of this research will, as we have already said, be the "vaccination" of plants against diseases and insects. In 1981, North America was invaded by a devastating sort of mite, the gypsy moth. The following year, several of the trees attacked in 1981 had developed a system of natural protection by secreting a quantity of tannin that made their leaves completely inedible to the insects. One solution thus might be to cross a domestic species with a savage species of the same kind that was resistant to insects or viruses. Attempts could also be made to inoculate plants, as animals are with vaccines, with attenuated forms of viruses or bacteria, but little is as yet known about the immunity defenses of plants. Today, the tendency is rather to reinforce the natural microorganisms attacking weeds or insect pests, the biopesticides. One bacterium full of promise, Bacillus thuringiensis, was used to paralyze the digestive systems of insects during the 1981 epidemic. Isolated in Japan by Ishiwata in 1905,

studied since the 1950s, and sold commercially by Solvay-Biochem in Belgium and by Sandoz and Abbot in the United States, this bacterium has the advantage of being far more selective than chemical insecticides. But there are two sides to this coin: Bacillus thuringiensis may prove to be too specific. The solution might be to find an other bacterium with a similar effect and to combine in a single organism a group of genes each of which is capable of selectively fighting a single type of disaster.

Biopesticides could be considered a factor in the new awareness of the damage to men and the environment that can result from chemical pesticides, which was so vividly illustrated during the 1950s by Rachel Carlson's book The Silent Spring. Paradoxically enough, the major chemical companies have been the first to take an interest in the research, which would enable them to sell both their chemical herbicides and the seeds of plants that resist this herbicide and are capable of protecting surrounding plantings. Recently, Calgene succeeded in isolating a gene resistant to the glyphosate contained in Monsanto's Round-up herbicide, by producing a mutation (using chemical methods and radiation) in a bacterium vulnerable to Round up. The Calgene team then isolated the gene in the resistant mutant providing this resistance.

After that it was a matter of finding the appropriate vector to introduce it into tobacco, soybeans, and cotton – the crops that risk being involuntarily attacked by Round-up. This ecological research should make it possible to preserve, if not expand, the world market

1, 2, 3, 4, 5, 6. **Shikonin**
The plant Lithospermum erythrorhizon, *which flowers in summer and from whose roots (shikon) medicines as well as coloring agents are derived (1). Japan, from which it has almost disappeared, imports them from China and Korea. Today, biotechnology and tissue cultures have made it possible to produce massive quantities of shikonin (2). Mitsui Petrochemical produces it in two forms, as a powder (3, left) and in stick form (3, right), and sells it in a cream for wounds, burns, and hemorrhoids (4), as a cosmetic (5), and as a vegetable dye (6).*

1

2

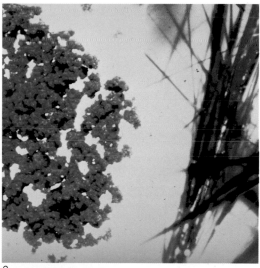

3

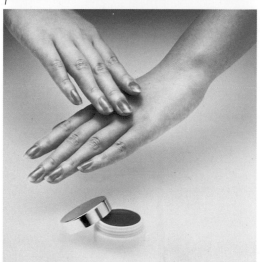

4

5

6

for pesticides, which is presently worth about $12 billion. It should be possible to sell pesticides and herbacides along with seeds for neighboring plantings.

Again it was the chemical industry that was responsible for the initial sales in Japan in February 1984 of the first chemical compound produced by the cultivation of plant tissues. This was Shikonin, a secondary metabolite extracted from the roots of a plant found in China and Korea, supplies of which were running increasingly short. Shikonin is used as a coloring agent, as a remedy for wounds and burns, and as an additive in cosmetics. Mitsui Petrochemical Industries, which perfected the process of production by tissue cultivation, along with M. Tabata of the University of Kyoto, sells its product for $4,000 per kilogram.

Biomimetics are also on the distant horizon. But natural mechanisms must first be understood before they can be imitated. With this in view, Du Pont de Nemours, for instance, is studying the mechanism of photorespiration: there is an enzyme in plants that fixes carbon dioxide; being relatively unspecific, however, it reacts with the oxygen in the air and thus makes a product that the plant, which does not know what to do with it, transmits from organelle to organelle before releasing it into the air in the form of carbon dioxide, uselessly consuming energy in the process. Only three cereals (corn, sorghum, and sugar cane) develop a sort of natural pump, which enables them to escape this phenomenon. If this phenomenon could be imitated, it would be possible to double the production of soybeans, as has been shown by artificially replacing the oxygen with carbon dioxide.

"When you observe a plant, you see it grow, turn green, wither, and then die. Before this aging process begins, however, an enzyme is expressed which destroys the enzymes fixing carbon dioxide. If we perfect a system of commutation, we will be able to delay this process by identifying the molecules which, at a certain stage in life, irrevocably initiate aging. Here is one of the possibilities offered by research on plants."

So states Ralph Hardy, President of BioTechnica International, and he is not the only one to consider the longterm consequences of research on plants in the field of the "life sciences." This is true too of Hoechst, which over a ten-year period invested $50 million in a contract with the Massachusetts General Hospital. Howard M. Goodman, who is in charge of this research, says that this research "can give us a new perspective on the normal and abnormal processes which take place in animal (and therefore human) cells" (*MGH News,* April 1983, p. 1).

The International Stake

Yet in the medium term, agriculture and the plants on which we are the parasites (to use an expression of Carl Sagan's) represent a fantastic economic opportunity for the biotechnologies. The world market for seed is estimated to come to about $50-$60 billion. According to the L. William Teweles Consultant Firm, this figure could almost triple by 1990 for genetically modified seeds. Plants that are resistant to pesticides, herbicides, diseases, and biopesticides could easily carve a sizable slice from the $10 billion market for agricultural chemical products.

Seed companies have not been among the pioneers of biotechnological research. They must give way to the leaders in the pharmaceutical and chemical industries – and first of all to the kings of agrochemistry: Bayer, Ciga-Geigy, Monsanto, and Shell. But there are also Du Pont de Nemours, Union Carbide, Sandoz, Upjohn, Hoffmann-La Roche, Lubrizol, Cardo, BASF, Elf Aquitaine, Total, Rhône-Poulenc, and Arco, to mention only a few. Small research companies have been proliferating with greater or lesser success: Calgene, Sungene, Plant Genetics, Phytogen, IPRI, Ecogen, Native Plants, Plant Breeding Institute, Agricultural Genetics, Plantech Research Institute (founded in Japan by Mitsubishi Corp. and Mitsubishi Chemical), Agracetus (resulting from a merger of Cetus with Grace), etc. The weak point of these small companies is the patience and heavy investments required for such research, as interesting initial results cannot be expected for five, ten, or even fifteen years. Agrigenetics had to agree to be bought up by Lubrizol to produce oil from plants. IPRI was making no headway when Lafarge-Coppée invested in its research. Cetus had to get rid of its agricultural division in Wisconsin. In Japan, a few companies are conducting research on improving seeds and resistance to disease: Mitsui Toatsu, Sumitomo Chemical, Mitsubishi Chemical, and Takeda (which has a botanical garden in Kyoto and a pilot farm where the quality of rice from around the world is tested). But, as Hiromu Sugama, Director of the Bioindustry Department at Mitsui Toatsu, says, "The market for seeds is a captive market in Japan, locked up by the State, which holds a monopoly on the sale of rice, the only important cereal in this country." At Asahi Chemical, Yasushi Ichikawa recalls that, during the 1970s, the government refused to authorize use of a hybrid rice developed by a team at the University of Okinawa. And he adds, "Japan is an island, or a string of islands covering a vast-north-south distance. Another problem is that we must take variations in climate and species into account." Yet recently the State seems to have unlocked the door; research is being intensified, and Japanese companies will soon have to join the fray.

It should also be remembered that people are far more ready to invest in medical research, for, as William Rains of the University of California at Davis says, "We can easily understand foundations for cancer research, but who's going to will money for the study of plants?" An article in *Newsweek* that appeared on 12 November 1984 estimates, in fact, that of the $3 billion invested in biotechnology, less than a third is devoted to the latter sector. Furthermore, according to Marc Van Montagu, who along with J. Schell directs the famed plant genetics laboratory at the University of Ghent, there are but few researchers in this field, perhaps a thousand throughout the world who specialize in the molecular biology of plants (a ratio of 1 to 50 with respect to researchers in general molecular biology). Various strategies have been adopted in the chemical industry, and Monsanto's may serve as an example. This firm threw itself into biotechnology in 1979 when it bought up the wheat department of an important American seed concern, De Kalb AG Research, as well as another Arkansas seed firm, Jacob Hartz Seed Co., which were merged into a subsidiary, Hybritech International Inc. This was not exactly an original move, as Shell International had already taken over Plant Breeder Ass., Orsan, Wilson Hybrids Inc. (Orsan had in addition become associated with Claeys-Luck), Lubrizol, and Agrigenetics, while Rhône-Poulenc had acquired Seedtec International, and five of America's leading seed companies were taken over by the "chemicals" in 1984. But Monsanto also invested some $170 million in an international biological research center, which opened its doors near St. Louis, Missouri, in October 1984 complete with microclimate simulation rooms (from the Brazilian Rain Forest to the Saudi Arabian Desert). The company has also opened agrochemical laboratories in Japan, Brazil, and Belgium. It has signed contracts with various universities (Rockefeller, Cornell, Harvard, and Washington in Missouri) and agreements with such genetic engineering companies as Biogen and Genentech. It has even ventured into risk capital through Innoven (established in 1972) and Advent-Eurofund (1982). One can readily see the power exerted by these large companies and the variety of their approaches to a field where it seems appropriate to fight a battle on several fronts.

While the pharmaceutical industry may be equally powerful, its interests are not quite the same. Research on plants may make it possible to understand certain physiological and biochemical mechanisms. In addition, if we consider that in the United States alone one-fourth of all medicines are derived from plants, it may well be possible to use cultures of plant cells to produce natural substances of a new kind with interesting biological properties, or already known plant-derived substances (like scopolamine) at far lower cost.

Between 1930 and 1975, the improvement in varieties of corn, wheat, and rice, the development of irrigation, and the use of fertilizers, pesticides, and herbicides have made it possible to increase the world's food production substantially, but progress has slowed down considerably over the past ten years due to a boomerang effect (the secondary effects of pesticides and herbicides, pollution, and the cost of fertilizers). The companies that today have taken up the challenge of bioindustry are concerned, above all, with being in the forefront of the battle to feed the world's hungry ten to twenty years from now. And if, on top of this, plant cells prove an asset to medical research and the market for pharmaceuticals, in fact this will be but a return to our earliest beginnings, for the oldest form of medicine in the world was the medicine from plants.

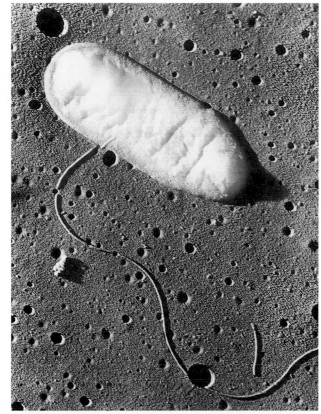

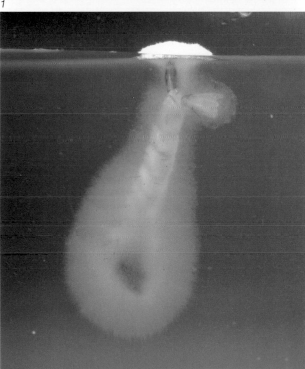
1

2

1. *Rhizobium*
The study of plant mechanisms for assimilating nitrogen has already led to an understanding of how they absorb water and phosphorus.
2, 3. **Biopesticides**
Yellow fever mosquito larvae killed by a fungic pathogenic agent (2), and an insect killed by a microfungus (3). Mycogen in the United States is devoted to this research.

3

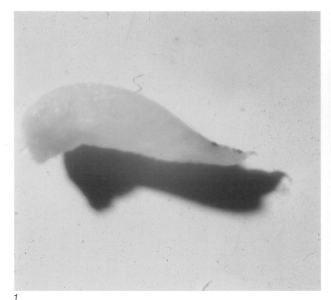

1

2

1. *Monocotyledon*
2. *Dicotyledon*
3, 4, 5. *Microinjection*
Microinjection techniques have been transposed from animal to vegetable cells. They had to be changed because of the presence in the latter of a pectocellulose cell wall. Treatment with an enzyme makes it possible to eliminate this, but the protoplasts obtained are fragile. They are immobilized on a glass slide (3), or by suction using a holding pipette. Dr. Ann Crossway at Calgene completes the delicate preparation for the immobilization of the protoplast.

6. *Vector*
The Ti plasmid having revealed itself an efficient genetic vector for the dicotyledons, teams of researchers have sought to extend the method to the monocotyledons. J.P. Hernalsteens in Ghent has thus shown that it is possible to induce tumors in asparagus: the transformed cells continue to biosynthesize opines after several months. Here research is being pursued at the Max Planck Institute.

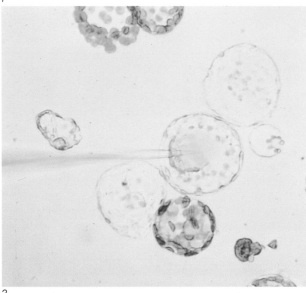

3

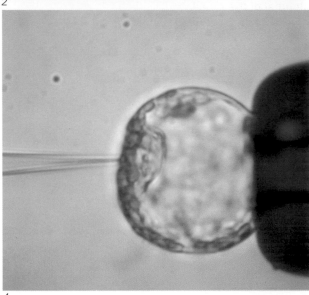

4

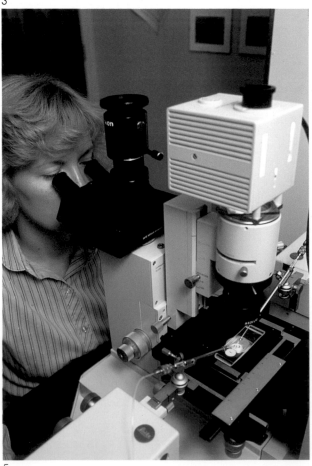

5

6

Plants, Seeds, and Biotechnologies

by Ernest G. Jaworski, Director of Biological Sciences in Monsanto Company's Corporate Research and Development Staff.

Dr. Jaworski holds degrees in chemistry and biochemistry from the University of Minnesota and Oregon State University. He is a member of the Editorial Board for the Journal of the American Society of Plant Physiologists *and a member of the Board of Trustees of the Gordon Research Conferences Inc.*

Plants, seeds, and vegetable matter: a gigantic future market, but one that still faces uncertainties. It is always difficult to distinguish between hopes and reality.

Introduction

Biotechnology is defined as the application of biological systems and organisms to technical and industrial processes. This technology not only encompasses the classical genetic selection and/or breeding methods for improving yeast strains for beverage manufacture, improved microbial strains for antibiotic and amino acid production, new plant introductions for agriculture and vaccine development for animals, but also includes the more recent techniques for in vitro modification of genetic materials (recombinant DNA) and other novel techniques for modifying the genetics of living organisms (cell fusion, hybridoma technology, somatic cell culture). The chemical industry is actively participating in the development and application of these modern techniques to problems related to plant agriculture, animal production, food processing, human health care, specialty chemicals, and waste disposal. More specifically, the agricultural chemical industry has developed interdisciplinary scientific teams to utilize these new techniques in both basic studies of various important plant processes (photosynthesis, stress, translocation of nutrients, nitrogen fixation, etc.) and the development of genetically transformed plants and microbes. Some of these techniques and molecular tools will be described.

Plant Tissue Culture

Plant cell and tissue culture, including micropropagation, while dating back to the beginning of the twentieth century, was exploited minimally until the late 1950s, when regeneration of plants from tissue culture was achieved. In the 1960s, this technology (micropropagation) was applied to the commercial multiplication of the orchid meristem (tip culture). This involved the propagation of the growing points of plants normally found at the tips of the stem, root, or branches. This method of clonal propagation allows rapid multiplication of the species and has been widely commercialized for a number of horticultural plants. One of the major problems still remaining to be resolved for the broad application of micropropagation resides in the difficulty of regenerating certain plants from single cells, especially important crop plants, such as the cereals and soybeans. Since plant cells can be grown on solid and in liquid media (cell suspension culture), the mass culture of cells provides a useful tool for screening at the cellular level for traits of potential use to the plant breeder. Since as many as 1-10 million cell aggregates can be cultured in a single 250-milliliter flask (representing a space requirement of perhaps 10-100 acres if individual plants were to be put into the field for screening for the same desired characteristics), it is apparent that this tool could be very powerful not only in increasing the numbers of propagated materials following regeneration from cells but in reducing the time required to select for genetically interesting traits.

It has been noted by many researchers that during propagation of tissue cultures and cell suspension cultures, genetic variants can be generated. These variants have demonstrated differences in plant size, flower morphology, pigmentation, growth habit, disease resistance, and other characteristics. While not all traits are desirable, such variants arising from somatic cell culture (somaclonal variation) can find many applications in modern plant selection for breeding purposes. While somaclonal variation is not well understood and, in fact, is somewhat controversial, it is leading to studies in such crops as potato, tomato, sugar cane, corn, tobacco, wheat, oats, rice, and rape. The cell culture technology allows large populations of potential genetic variants to be screened simultaneously and also permits the application of selection schemes analogous to those used with bacteria. The latter approach would be an impossibility with intact plants because of the land mass required to screen such large populations of material. Furthermore, mutagens can be used to treat cells and enhance the level of variation. Thus, the time needed for selection for certain characteristics, such as herbicide tolerance or disease resistance, could be significantly reduced.

A key requirement for certain uses of the above-mentioned system is the capability for regeneration of whole plants from individual cells that have a true genotypic change, that is, one that will be expressed when whole plants are regenerated and is stably transmitted through sexual propagation. The limitation of the system is that only traits expressed at the cellular level will be identified. Traits such as improved nutrient transport or yield or those involving interactions between differentiated cells (e.g., oil production) probably cannot be selected using this system.

Plant cell and tissue culture can also be utilized to produce protoplasts, that is, cells with the cell walls digested away by enzymatic means. When the cell walls are removed (a technology developed in the late 1960s), protoplasts from different types of plants can be fused by using techniques that cause the cell membranes to melt into one another. While this technology has been a very exciting one, it has not yet led to significant advances in the development of hybrid plant materials. One of the limitations has been the poor formation of cells from protoplasts following fusion, a characteristic of many crop plant species. Furthermore, in many plant types where plant cells can be induced to regenerate into whole plants, the regeneration characteristic is lost when protoplasts are used. Since this technology permits unique combinations of nuclear and cytoplasmic materials, much emphasis is still being placed on the development of regenerable, fertile hybrid fusion products. The technology also provides a means for directly introducing genes (naked DNA) from unconventional sources, albeit without the precision of directed transformation.

Plant cell culture and cell protoplast fusion has caught the attention of plant geneticists and breeders throughout the world. Since somatic cell genetics could provide an alternative means for creating modified plant systems with enhanced disease tolerance, insect tolerance, stress tolerance, and potential growth enhancement, much research effort is being devoted to the study of the potential of applying somatic cell genetics to agricultural problems. While these programs are ongoing and evolving, no major impact has yet been noted using these technologies applied to major field crops. However, in ornamentals and certain vegetable crops, plant cell culture has provided a useful means for mass asexual propagation of horticulturally important species, some vegetable crops, and as a means for generating virus-free plant materials. Much work remains to be done in this area, and significant research activities are proceeding in both industrial and university laboratories.

Two key issues require more extensive exploration. These involve the ability effectively to regenerate different crops, such as soybean, maize, and wheat, from either protoplast or cell cultures and the definition of appropriate selection techniques and strategies for the isolation of cells with properties desired by the plant breeder.

Genetic Engineering

In agriculture, the potential for modifying whole plants using genetic-engineering techniques (transformation) is estimated to be one of the most revolutionary applications of gene splicing. While the genetic modification of plant cells has not yet reached the stage of development that microbial modifications have, the combination of traditional plant breeding, plant tissue culture, and genetic engineering is likely to have an enormous impact on plant agriculture and silvaculture.

Genetic changes can be engineered into the cells using tailored vectors containing appropriate DNAs for the introduction of the desired traits. Traits that have been and are being considered include general ones, such as development of male sterility, fertility restoration, chemical transformation (for example, nitrogen conversion to ammonia), improved protein quality of seed grains, enhanced protein levels in forage crops, plant disease resistance, insect tolerance, drought and flooding tolerance, salt tolerance, metal tolerance (Al^{+3}, and Pb^{+2}), herbicide and pesticide tolerance, enhanced photosynthetic carbon fixation, improved response to fertilizers, and enhanced heat and cold tolerance, to mention just a few. While many of these traits are complex (multigenic) and will require considerable basic research to define their genetic loci, others are believed to be due to single genes and may be achieved in the near future. Thus, significant efforts are being devoted to the isolation of herbicide tolerance genes for insertion into plants in order to create plants less sensitive to nonselective herbicides.

Although single-gene modification, such as resistance to herbicides, is foreseeable, it is likely to be a decade or more before multi-gene transformation becomes technically feasible, and even longer before it becomes a commercial reality. The task, however, is of great social and economic significances and is being pursued with vigor by both industries and universities alike. To realize these goals, more effort must be mobilized to develop improved vector systems for the transformation of plants.

To introduce genes into plants, initial efforts have involved the use of *Agrobacterium tumefaciens*, a naturally occurring soil-borne bacterium that can introduce genetic information stably into plant cells, creating tumors called crown-gall tumors. The Ti plasmid, a circular piece of DNA normally carried in the bacterium, has now been engineered to disarm its capabilities for producing tumors but still maintain its capacity for the insertion of appropriate genes into plant cells (e.g., genes that confer antibiotic resistance).

A number of research groups throughout the world have now demonstrated that a foreign gene can be inserted stably into plant cells and be expressed at the cellular level, the whole plant level, and, in subsequent progeny, seed developed by self-fertilization of the plants. In order to obtain expression of genes in plants using the Ti plasmid system, appropriate regulatory sequences that allow the turning on of genes (promoters), signal sequences for the appropriate transport of the gene product, and necessary termination signals at the end of the gene must be present. The sequences utilized in the initial studies with the Ti system included those regulatory sequences derived from the nopaline and octapine synthase regions of the Ti plasmid, which were known to be recognized by the plants'own molecular mechanisms. Scientists at Monsanto Company, Washington University, and the University of Ghent were successful in obtaining expression of the kanamycin resistance gene (neomycin phosphotransferase II) in cultured plant cells. Further studies indicated that the kanamycin resistance gene was stably inherited by petunia and tobacco in a Mendelian fashion and was expressed by the recombinant plants seed progeny.

Scientists at the University of Wisconsin and Agrigenetics have reported the successful expression of bean storage protein genes in tobacco plants. Continuing efforts have also demonstrated the expression of the gene coding for the small subunit of ribulose bisphosphate carboxylase derived from peas in petunia cell cultures and the expression of human alpha-chorionic gonadotrophin hormone and mouse dihydrofolate reductase in petunia cells.

Ti vectors carried by *A. tumefaciens* have been quite useful in the transformation of dicotyledonous plants, such as petunia, tobacco, tomato, potato, and sunflower, but crown-gall disease normally does not infect monocots such as the cereal grains. While methods are being studied for using Ti systems to transform monocots, additional efforts are being devoted to the study of transposable elements first recognized by Dr. Barbara McClintock at the Cold Spring Harbor Laboratory. Transposable elements are thought to be segments of DNA that have the ability to move around the genome and control or alter many different genes. Since these elements are capable of inserting DNA into new locations in a plant, they are of interest for application to the insertion of foreign genes, especially into corn chromosomes.

In addition, plant viruses are being studied as potential gene transfer agents. While initially much of the viral work was focused on the use of double-stranded circular DNA coding for the cauliflower mosaic virus (CaMV), to date the use of CaMV has been primarily in providing strong promoters for transformation vectors involving the Ti plasmid system. More recently the Gemini viruses have received increasing attention, since they infect a wide variety of plant species, including cereals and legumes.

Finally, the use of microinjection technology for directly introducing DNA (genes) into plant cells (or organelles) is receiving considerable emphasis, since this approach may require less engineering of the gene and permit the possible bypassing of some of the restrictions mentioned earlier in protoplast and cell culture technologies. This procedure has been successfully applied to mammalian cells, but there have been few published reports to date describing positive results in plant genetic transformation.

While there is little doubt that the application of genetic engineering to the transformation of plants has advanced rapidly in the past two years and holds great potential for useful agricultural modifications in the future, it may be prudent to point out that a number of significant issues and problems remain to be resolved. On the positive side, it can be predicted that the approach to these issues and problems will produce basic molecular data that will have general value in advancing our understanding of gene organization, regulation, and expression. Issues currently being addressed that face molecular biologists attempting to transform plants include the basic mechanisms of gene transfer, gene transcription, and translation, the molecular nature of promoters, ribosomal interaction sequences and terminating regions, the nature of leader and transit sequences that regulate the movement of gene products

from one organelle to another, and the understanding of regulatory sequences involved in the tissue-specific, temporal, and/or developmental regulation of gene expression. Finally, it will be essential to develop a better basic understanding of the biochemistry of plant systems if one is, in fact, going to identify those genes that are capable of improving crop plants.

Microbial Pesticides

A second aspect of biotechnology potentially important to agriculture is the development of genetically modified soil bacteria to enhance crop productivity. Such genetically modified bacteria would be added to the soil or seed. Many plants form biological partnerships with soil microorganisms in which an intimate association between the root system of the plant and the microbe results in the transfer of nutrients from the soil to the plants. Other soil microbes may produce chemicals that protect the roots from attack by pathogenic bacteria and fungi; or, in the most well-documented case, the organism (namely, *Rhizobium* species) forms a symbiotic relationship with the plant, fixing nitrogen for the benefit of the legume and deriving from the plant its source of energy and nutrition. The potential for improving the efficiency of nitrogen-fixing bacteria by genetic-engineering technologies is receiving attention and could have an enormous impact on crop productivity. Although the notion of introducing the nitrogen-fixing complex directly into plants has attracted a great deal of publicity, it is improvement of nitrogen-fixing bacteria that is most likely to be successful in the near term.

Another role for genetic engineering involves the biotic control of pests, which dates back to man's recognition that insects are subject to diseases. The first real application of biotic control was the discovery of milky disease, a disease that involves spores of a bacterium in the genus *Bacillus* that contain a toxin lethal to the Japanese beetle. More recent advances have resulted in the commercialization of *Bacillus thuringiensis* spores containing a crystalline endotoxin that is the active ingredient for the control of *Lepidoptera* larvae. Thus the introduction and use of bacteria for agricultural crops is not especially novel or new, but the new molecular tools will provide the means for more specific improvements in these microbes. Some 1,500 naturally occurring microorganisms or microbial by-products have been potentially identified as useful insecticidal agents. Some of these are already being marketed, such as the *Bacillus thuringiensis* spores mentioned earlier. Toxins made by the *Bacillus thuringiensis* subspecies *Israliensis* also kill mosquitoes and black flies.

In addition, soils in many areas of the world appear to have the ability to suppress the development of certain plant diseases and are suspected to contain factors that can be manipulated to control soil pests such

as fungi by the engineering of root-colonizing bacteria. The experimental uses of such root-colonizing bacteria applied to either seeds or roots of plants have already resulted in significant increases in plant growth and production. Since microbes lend themselves to more rapid genetic engineering than plants, it is expected that the transformation of useful root-colonizing microbes with appropriate genes could lead to major benefits in crop productivity. Examples would be the introduction of the *Bacillus thuringiensis* toxin gene into root-colonizing microbes.

It is conceivable that specific rhizosphere microbes could be engineered to secrete allelopathic chemicals that would ward off bacterial and fungal pathogens and even secrete herbicides that would be selectively toxic to weeds. Some of these microbes produce iron-binding siderospores that have been associated with increased plant growth and vigor through improvements in iron availability.

Summary

The major biotechnological approaches cited above, including plant cell and tissue culture selection, plant breeding (both traditional and newer approaches, such as haploid production, embryo rescue, and pollen culture), genetic engineering, and transformation of plants and the genetic engineering of rhizobacteria, are likely to have a major impact on the production of new plant varieties and on the productivity of current plant varieties. While the targets for crop improvement are essentially the same as those of traditional breeding approaches, that is, increased yields, improved qualitative traits, reduced labor costs, and reduced production costs, the newer technologies offer the potential to accelerate the rate and type of improvements required to facilitate the work of plant breeders, agriculturists, and growers. Of the technological methods being used for crop improvements, plant genetic engineering and genetic modifications of rhizosphere microbes can be expected to have a major impact on crop improvement, with the likelihood that significant increases in agricultural productivity will be achieved by the year 2000.

13

Aroma, Flavor, and Other Benefits

From Flavor to Aroma

What would food and drink be without flavor, aroma, or color? Throughout history, men have always sought to make their food more appetizing, first by using spices and herbs, and then by the "spirits" of fruits or aromatic plants drawn out by alcohol or by essential oils in which water replaces the alcohol. More recently, chemically synthesized aromas have made their appearance. Biotechnology continues this tradition, pursuing research on food additives, aromas, and coloring agents (adding a carotenoid extract to salmon, for instance, to give it a pink glow, in contrast to the whiteness resulting from intensive fish-farming methods).

One of the new stars among these biofoods is sodium glutamate.

A food additive that reinforces salty or bitter flavors, sodium glutamate was the foodstuff whose discovery by S. Suzuki in 1908 led to the establishment of one of Japan's leading food-processing companies, Ajinomoto ("the essence of taste"). Originally extracted from such natural sources as gluten, caseine, or the residue liquors from distillation (stillage), it began to be produced by fermentation in 1956 using a bacterium whose enzyme system had been deranged in such a way that, in a fermentation medium, glutamic acid would accumulate. Two Japanese companies immediately dominated the market (Ajinomoto and Kyowa Hakko); they were joined two years later by Asahi Chemical – a company specializing in chemical products and synthetic fibers – for whom this was the first step toward biotechnological diversification, working in conjunction with Toyo Jozo, the saki specialists. However futile the production of sodium glutamate may seem from the standpoint of meeting mankind's basic needs, it does, as we have already seen, represent an important step in the history of fermentation techniques and opens the way to the production by fermentation of complex molecules, the chemical synthesis of which is both complicated and costly on an industrial scale. In addition, in contrast to chemical methods, this system makes asymmetrical synthesis possible (in the form of the L acid alone, which eliminates the stages of separating the L and D forms). In this field, we must also mention Takeda, which, in addition to sodium glutamate, produces other amino or nucleic acids used to enhance the flavor of dishes – an example is ribotide, a mixture of sodium inosinate and sodium guanylate – or to strengthen the nutritional content of rice, like shingen (with the addition of vitamins and mineral salts).

It is often difficult to determine whether the use made of certain amino acids is nutritional or pharmaceutical: thus Ajinomoto's argine 2 could be considered the former (as it is a diet drink based on arginine), while elental, based on amino acids and vitamins, is closer to the latter.

Aspartic acid, glutamine (antiulcer), and arginine are Kyowa Hakko's three leading products, along with threonine, histidine, ornithine, isoleucine, phenylalanine, and tryptophane, which Kyowa Hakko produces by fermentation and which could compete with Ajinomoto's tryptophane, which is produced by chemical means, or the tryptophane produced enzymatically by Mitsui Toatsu. (The basic raw material used by the latter to produce tryptophane is another amino acid, serine – used in medicine.) On the other hand, Ajinomoto is now trying to produce phenylalanine by fermentation.

The principal derivatives of nucleic acids – inosine and adenosine – are essentially aromatizers that have a synergistic effect on sodium glutamate. But some derivatives of nucleic acids are also used as medicines (like adenosine, which activates blood flow) to correct metabolic disturbances or serve as analytical reagents. This is true of orotic acid, NAD, FAD, and ATP (in which Kyowa Hakko is the world leader). Some companies, such as Asahi Chemical, have started to produce amino acids by the hydrolysis of yeast and turned to fermentation techniques. In addition to inosine, Asahi produces

Mycoprotein
Steaks, meatloaf, vegetables? No, microorganisms on your plate instead. Or more accurately mycoproteins produced by Marlow Foods in Great Britain, which meet dietetic and medical requirements in both the United States and Europe.

adenosine and ATP and compounds derived from nucleic acids (like surabine, an anticancer drug that acts by inhibiting the enzymatic system). Yamasa, a traditional producer of soy sauce, has turned the biotechnological corner with an aromatic additive based on a nucleotide perfected by Akira Kunika in 1957 and produced by the enzymatic breadown of a yeast that reinforces the sodium glutamate. Ajinomoto initiated the production of amino acids by genetic engineering in 1981.

The sensitivity of our taste buds is undoubtedly less highly developed than our sense of smell. In any case, smell and taste combine in what is known as flavor: hops flavor beer, plums or pears some spirits, juniper berries gin, coca or quinine soft drinks, lemon lemonade – the natural flavors in use for so many years are mainly the secondary metabolites of plants.

Research is now being carried out in several countries to improve the quality of flavors on an

1. *Glutamic acid*
2. *Arginine*
3. *Tryptophane*

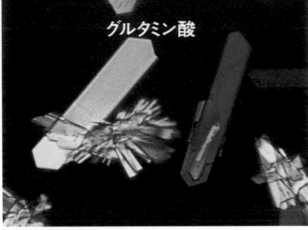

industrial scale. In 1974, Pernod-Ricard opened an agronomy laboratory for the cultivation of aromatic plants in vitro – black currant, liquorice, vanilla, mint, quinine, fennel – all capable of replacing in the famous brandname drink the star anise, which comes from China (at great expense) or North Vietnam (inaccessible). Work is being done under the direction of Professor Roger Gautheret in Montpellier, where as early as 1938 he perfected an in vitro method of cultivating plant cells and noticed that certain fruit juices (principally orange and grape) stimulated cell mutliplication. Three systems are being investigated: the micropropagation of plants with exceptional aromatic characteristics, selection in vitro of plantlets having acquired new genetic characteristics or having been changed by mutagenesis, and a study of the biosynthesis of aromas in fermenters. An attempt is also being made to draw a "computer portrait" of the aromas one wants to create using a variety of analytical methods (flavor profiles, mass spectrometry combined with a chromatograph and the results read by a computer, etc.). But even more research time is being devoted to the study of the traditional fermentation processes during which various microorganisms produce the aromas, and to the determination of the number of molecules behind an aroma (1 in the case of pear, 344 for raspberries, for instance).

The sweet and buttery aroma released by diacetyl in the case of milk, the bitterness produced by butyric acid in the case of hard cheeses, perfumed alcohols and esters for beer and wine, the unctuous compounds in the peach produced by lactones: all of this stems from the hydrolysis of proteins and the autolysis of yeasts, but our knowledge of the metabolism that brings forth one aroma or another or the part played by the strains remains limited. Several research teams have now begun a study of yeast strains. An attempt is being made to perfect immobilized cell reactors, and bioconversion is also being tried, using chemical means, to convert a precursor compound to the aromatic state, the microorganism in this case serving to initiate the controlled enzymatic reactions (methylations, oxidations, etc.). One of the procedures for aromatization during the course of production was tested by Bongrain in France for cheeses: a culture medium of *Penicillium roqueforti* strains is mixed with cream, and rennet is added to this. This produces soft cheeses with the aroma of blue cheeses. Ripening is quicker, but the aroma is still too volatile to make commercialization possible. Other processes are being studied at Sanofi, which has taken over the Granday Laboratories and works with Dairyland Food (Milwaukee) on cheese aroma. These techniques of producing aromas could supply answers to some of the problems posed by traditional techniques: expensive and sometimes polluting methods of extraction, losses in organoleptic qualities in the course of drying and dissolution, oxidation and conservation problems, etc. Furthermore, they correspond to changing trends in tastes and eating customs and the irruption of fast-food techniques.

Where acidulents are concerned, acetic acid remains the most important of the industrial organic acids, having many more applications than vinegar (pharmaceuticals, coloring agents, insecticides, and plastics). Lactic acid, for its part, became the

1, 2, 3. **Fennel**
As part of the Pernod-Ricard group's research program on aromas, a fennel plantation was started in 1974 in an effort to find another source of anise than the badiane essence imported from China. Over a period of seven years, improvements in cultivation techniques have led to a fourfold increase in yields per hectare (1), in vitro culture (2), and the creation of a pilot plant. A rectification process for fennel essence was also developed, which made it possible to build an industrial rectification (3) column in Herault.
4, 5, 6. **Carotenes**
Canthaxanthine (4) is one of the brilliantly colored carotenes (5) that gives the flamingo its delicately pink feathers (6), just as another carotene, astaxanthine, produces the pink in salmon. The flamingo absorbs it through the little shrimp it eats or, in the zoo, through substitutes added to its meals.

first organic acid to be produced industrially by fermentation. As for citric acid, it is made by a yeast that has been put on a diet: when its ration of minerals, such as iron or manganese, is limited, a nutritional imbalance occurs and the mold produces citric acid.

Sweeteners: Nibbling Away at the Market for Sugar

Obesity, diabetes, hyperglycemia, dental caries – the manifold evils to which lovers of sugars and sweets are prone. To say nothing of soaring prices, like those in 1973-1974. Yet for many years now (since 1879), we have known of an artificial sweetener, saccharin, which has far fewer calories than saccharose (as it is not metabolized by man), and which has been used in soft (carbonated and nonalcoholic) drinks in America since 1968. But because of its disagreeable metallic aftertaste, it was mixed with other artificial sweeteners, the cyclamates. In 1969, however, cyclamates were banned in the United States, to the great distress of their leading producer, Abbott Laboratories: they were considered dangerous as the result of tests on animals showing that they had a carcinogenic potential. Canada banned saccharin in 1977 for the same reasons. Doubt had thus been sown: the market is wide open.

Two substitutes for saccharose (table sugar) have proved quite interesting commercially: aspartame for diet beverages and isoglucose, which has an equal caloric content. Aspartame was discovered by J.M. Schlatter working at Searle in 1965. It is a dipeptide with a sweetening power 200 times as great as saccharose – compared with 300 for saccharin – but above all it has no disagreeable aftertaste and is completely harmless to man. Its price is still very high ($200 per kilo, compared with $9 for saccharin) because it is produced from phenylalanine ($55 per kilo), of which the leading producers are Japanese (Tanabe, Ajinomoto, Kyowa Hakko). Genex has developed a bioreactor that makes it possible to produce phenylalanine from cinnamic acid using a strain of yeast that remains secret. An initial contract for the enzymatic production of this amino acid was signed with the Italian firm Pierrel, in April 1983, and a second four months later with Searle to supply several hundred tons of this precious amino acid. Searle already made use of quantities of phenylalanine synthesized chemically by Hoechst.

And what about synthesizing it enzymatically (as it is a dipetide)? This had already been started in January 1983 by a Japanese firm, Toyo Soda, which in September isolated three microorganisms capable of fixing aspartic acid on the phenylalanine methyl ester. The production of aspartame by cloning a gene of polyaspartame, for its part, was successfully achieved by Searle in Great Britain at its High Wycombe research unit. For the moment, the two systems (chemical and biotechnological) are competing.

Aspartame has been used in soft drinks in Canada since 1981, and subsequently in the United States (Coca-Cola, Royal Crown Cola, Squirt, Pepsi-Cola, Seven-Up, etc.). Products sweetened with aspartame have been launched by numerous companies – General Foods, Quaker Oats, Heinz, Procter and Gamble, etc. Hundreds of millions of dollars are at stake, particularly in the market for "diet" soft drinks. In 1984, Searle and Ajinomoto established a joint subsidiary, Nutrasweet, and aspartame has been given official approval in Brazil, Great Britain, Ireland, and the Scandinavian countries. Sugar's monopoly is slowly being nibbled away, but it is still protected by legislation in various countries, notably in Europe.

On the market for sweeteners a new product, isoglucose, has begun to compete with aspartame... and opened a new breach in sugar's ramparts. It can already claim 33% of the American market for the sugars and sweeteners used in beverages. The first patents, applied for during the 1960s, were filed by Japanese (Samatsu Kogyo) or American companies (Miles Laboratories, Corn Product, Standard Brands). But the initial production processes were extremely costly. In 1972, as we have seen, the appearance of fixed enzymes made it possible to lower prices. Other companies joined the rush: Anheuser Busch, Baxter Laboratories, Amylum, and those important enzyme producers Novo and Gist Brocades. It was in 1976 that the HFCS (high-fructose corn syrups) first appeared, with a fructose content of from 55% to 90%. Enzyme techniques (invertases, Novo's novenzyme, glucose isomerase, Genex's fructosyl, transferase, etc.) and media for fixing them are being constantly improved. Considerable use is being made now of the hard portion (lignocellulose) of corn husks, an abundant and low-cost material that makes continuous production a possibility.

The industrial development of isoglucose has been hampered by legislation in various countries, and been further slowed down by the fact that is has not yet been crystallized and is only available in the form of syrup.

Other sweeteners may also join in the competition. There is acesulfame K, an artificial sweetener discovered in 1973 by two Hoechst researchers, which could be used in combination with aspartame, and above all thaumatine, a natural sweetener derived from a Sudanese fruit whose basic proteins had been isolated in 1973 by Van der Wel at Unilever (since cloned by Unilever and expressed by Plant Genetic Systems in Ghent). It would be interesting to express these in citrus or other fruits that have no flavor but are of high nutrional value. And there are the L glucids (semisynthetic sweeteners), whose synthesis was patented in 1981 by the American firm Boise Cascade Corporation.

The thaumatines, for there are several forms of these, have a sweetening power 5,000 times that of saccharose. They may also be used as exhausters of aromas. Starting in 1978, Tate and Lyle, Unilever, and Kent University studied the possibility of cloning their genes, and Unilever finally succeded in this in 1983. This is interesting inasmuch as the price of thaumatine is prohibitive: $16,500 per kilo.

Price is still an obstacle to the production of the L glucids to be found in algae and plantains (L fructose), sugar beets (L arabinose), and linseed and red algae (L galactose).

Increasingly serious thought is being given to the possibility of mixing various sweeteners so as to obtain a better combination from the standpoint of taste and cost. Biotechnology should also make it possible to derive some value from the by-products. But such new competition will undoubtedly lead to a new international agreement on sugar.

Texturing Agents

The manufacturers of sauces and ice creams, as well as pastry cooks, frequently need what are known as texturing agents to make their sweets, soups, and instant foods more appetizing. Gelifiers, such as gelatines and alginates derived from seaweed (Sanofi), may be used for this purpose, as may gums.

One of these gums is a biopolymer, produced by fermentation, which seems to have many advantages: the gum xanthane, secreted by a bacterium *(Xanthomonas campestris),* which increases the viscosity of water enormously and can tolerate a high degree of salinity (qualities which, as we shall see, are of use in another field: drilling for oil). But in the food industry, xanthane is above all a thickening and stabilizing agent of exceptional quality: "Thickening in itself is not an original quality," says Alain Sedent of Rhône-Poulenc, "but thickening in small quantities with great stability with respect to variations in temperature and acidity and in a saline medium is a privilege reserved for xanthane alone."

This polysaccharide (composed of long chains of sugar molecules) was discovered in the later 1950s by Department of Agriculture researchers in Pretoria, United States, who isolated it from rotten cabbage leaves. In the mid-1960s it was produced by Kelco in San Diego and granted authorization as a food additive by the redoubtable Food and Drug Administration (FDA). Kelco, a subsidiary of Merck, was the only firm on the market for almost ten years, but today other companies, such as Rhône-Poulenc (and soon Sanofi), are selling the miracle gum.

Vitamins

So much for the civilization of the man in a hurry – the overfed man. But three-quarters of the world's population is still living in the civilization of the underfed man, and it is here that biotechnology can also play a part. Scurvy, beriberi, and rickets are some of the diseases directly attributable to a lack of vitamins. Yet most of the vitamins needed by the body of an animal, and above all a man, are to be found in the tissues, though the body is often incapable of synthesizing them: in general, they are supplied by food.

The discovery of vitamins dates back to the work done by Funk in 1911, when he isolated vitamin B_1. This was followed by McCollum with vitamin A, Mellanby with vitamin D, and then Albert von Szent-Györgyi with vitamin C. It is interesting to note that vitamin C is produced at one stage of his synthesis by fermentation using a wild bacterium discovered by Tadeus Reichstein.

Today, biological methods are particularly useful in producing two vitamins in particular, vitamin B_{12} (or cobalamine) and vitamin B_2 (riboflavin). Vitamin B_{12} was isolated for the first time in 1948 by two teams, one in Britain (E.L. Smith) and one in the United States (K. Folkers and R. Rickes) from liver extracts and the must from the fermentation of antibiotics. Within the body, it is produced by the microflora in the digestive system and is supplied mainly by meat products, with the result that vege-

1. *Separation of fructose*
2. *Production of aspartame at Ajinomoto*
3, 4, 5. *Xanthane*
Trout aspic, quiche, condiments, jelly beans....

1

2

3

4

5

tarians often suffer from a vitamin B_{12} deficiency, and this could lead to pernicious anemia. This vitamin also plays an important part in the formation of blood and in growth. The production of this vitamin has resulted in a protracted industrial battle in which Rhône-Poulenc has emerged the winner (60-65% of world production). In 1955, Dorothy Hodgkin revealed the structure of its molecule and discouraged any thought of chemical synthesis. It would require no less than seventy separate and distinct chemical reactions!

Today, vitamin B_{12} is produced directly during the course of fermentation by propionibacteria (bacteria that synthesize propionic acid) and certain strains of *Pseudomonas*. To obtain more efficient strains of *Pseudomonas denitrificans* (isolated for the first time at Merck during the 1950s) mutagenesis is obviously used, but genetic recombination in vitro or in vivo is also being considered. But efforts are also being devoted to the fermentation media, where during the course of biosynthesis, four compounds play an essential role: saccharose, betaine, cobalt, and 5.6DBI – a substance synthesized from riboflavin. A particular effort is being made to try to replace saccharose by far less costly molasses, with the additional economic advantage that beet molasses, for instance, also contains betaine. All of this research may prove exceptionally fruitful on an industrial scale, particularly if we remember that, since vitamin B_{12} was first produced, yields have been increased 200-500 times.

Lesions of the skin, mucous membrane, or eyes are among the ills that can result from a lack of vitamin B_2. This vitamin, which is found essentially in cereals, vegetables, and brewer's yeast, can even, it would seem, stave off cancer of the liver. Produced for the first time in 1938, it is still another example of what can be done by selecting and screening strains: in just over thirty years, it has become possible to obtain 100-300 times more vitamin B_2 than with the original wild strain of mold (*Ashbya gossypii*).

Attempts are now being made to use biosynthesis to produce other vitamins whose production requires many stages, including vitamin A, which is essential to growth and is sold mainly for mixing into compound animal feeds, or vitamin E, whose extraction from palm oil by molecular distillation is now being attempted in Malaysia.

The Hoffmann-La Roche Company is taking a special interest in the production of beta-carotene, not only a source of vitamin A but also a natural coloring agent for foodstuffs, with a range of shades from yellow to red. Studied in particular by Paul Karrer of the University of Zurich, who determined its structure in 1931, it has been synthesized since the 1950s. Recent discoveries in molecular biology have made it possible to study in greater detail its metabolism, biosynthesis, and functions in the body.

Microorganisms:
The Food of the Future?

The hors d'œuvre tastes and chews like breast of chicken fried in butter and breadcrumbed. Pulled apart with the fingers, the scampi-sized "bites" have the appearance and fibours texture of poultry. Yet the protein is fresh from a fermenter. It is pure mycoprotein, a fungus harvested from a seething

1, 2. **Vitamin C**
3, 4. **Vitamin B_{12}**
5. **Beta carotene**

1

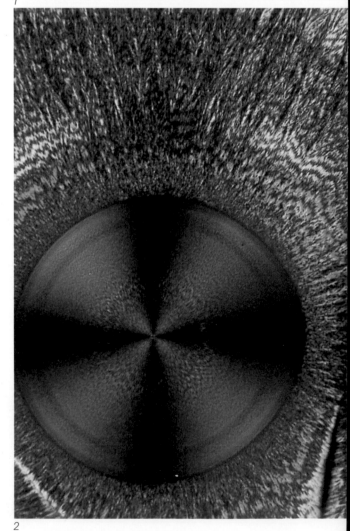

2

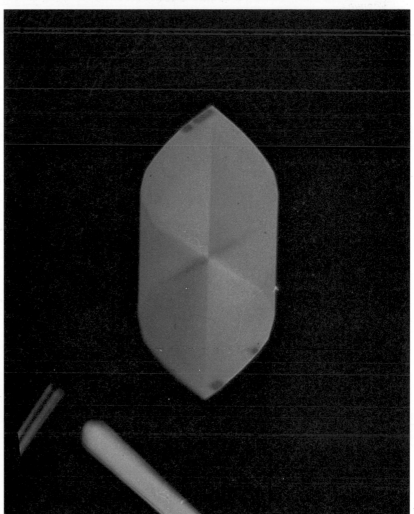

broth in a stainless-steel vessel closely watched by a computer. The biotechnologists at the Lord Rank Research Centre of Ranks Hovis McDougall are confident that their mycroprotein will be the world's first entirely new kind of human food to get government blessing and public acceptance. This British firm specializing in bakery products uses as raw material the starchy portion of plants from which it bakes some of its present products. This "microfungus" is in fact a plant and comes from the same family as the truffle and the source of flavor in some kinds of blue cheese. Other food-processing groups, including Unilever, are also working hard on the challenge of trying to simulate familiar foods (Unilever even constructed a whole "chicken"), but also of creating entirely novel foods. The common factor is that, although all are high in protein and fiber, none contains flesh, fish, or fowl.

Ranks grows its mycoprotein on a syrup made from food-grade starch or sugar. Ammonia salts are added to supply nitrogen for conversion to protein, plus the trace elements found in a nutritious food. This is run through a fermenter under the watchful eye of a computer controlling, above all else, the acidity, and out pours mycoprotein – as much as a ton a week from the pilot plant at High Wycombe. This project was initiated in 1973 by Professor Arnold Spicer, and it took almost twelve years for it to reach the industrial stage. Obviously enough, to concoct this porridge of the year 2000, synthetic colors and aromas must be employed. Nutrionally, mycoprotein is the equivalent of milk protein low in fat and sodium, high in fiber, and with no cholesterol – "all the factors the modern nutrition establishment considers important," says Jack Edelman, who succeeded Professor Spicer, and who sees mycoprotein as the ideal way to convert any surplus of the indigenous carbohydrate crop in any country (wheat in Britain; potatoes in Ireland; rice, sugar, or cassava in hotter climates) into food products of far greater nutritional value.

In the summer of 1984, RHM and ICI announced a joint venture (Marlow Foods) to produce this "Peter Pan of biotechnology" and raise production from one to twenty tons a week. To succeed, these proteins will have to prove less expensive than animal proteins, and consumers will have to learn not to wrinkle their noses or grit their teeth at the idea of having microorganisms on their dinner plates. In any case, there can be no doubt that they will provide a solution to the problem of hunger in underdeveloped countries and halt the ravages of dietary deficiencies.

Without microbes, as we have seen, there can be no bread, no wine, no soy sauce or saki – in short, none of the basic foodstuffs on which East and West depend. Why, then, not skip the intermediate steps and eat the microbes? The idea is not a new one, inasmuch as sixteenth-century Mexicans ate little cakes made of algae from their lakes, Africans in Chad feasted on *Spirulina platensis,* Germans during both world wars sprinkled brewers' yeast or *Candida* on their soup, and the British use this same brewers' yeast in their diets. Edelman estimates that by 1984 about fifty tons of mycoprotein had already been eaten, much of it by RHM employees, but some by intrepid consumers in limited test marketing, and all without apparent ill effects. So perhaps tomorrow....

Food Today and Tomorrow

by Toshinao Tsunoda, Vice President and General Manager of Ajinomoto.

Born 21 January 1921, T. Tsunoda joined Ajinomoto in 1948. A Doctor of Agronomy, he has worked in the private sector (member of the Board of Morishita Seiyaku) as well as in the public sector: he is presently Vice President of the Fermentation Techniques Association. The production of glutamic acid was the basis upon which was built the Ajinomoto Company, which since its origins has specialized in foodstuffs, but which has now diversified into numerous other fields, including pharmaceuticals. Japanese cuisine is among the most refined in the world, and this article brings up the meaning of the name Ajinomoto – the essence of taste. Our readers will learn that the basic materials of life (amino acids, nucleic acids) can also accentuate the tastes of our sauces.

Background

L-glutamic acid, an animo acid instrumental in the composition of proteins, was discovered in 1866 by H.L. Ritthausen. But it was not until 1908 that a Japanese scientist, K. Ikeda, discovered its usefulness as a condiment when he showed that L-glutamic acid was an essential element in the taste of konbu,[1] one of the ingredients of the Japanese national dish known as dashi.[2]

K. Ikeda used the word "umami" to describe the flavor of L-glutamates, particularly sodium glutamate, and advanced the thesis that it was a fifth basic flavor along with sweetness, acidity, saltiness, and bitterness. Later research into the psychology and physiology of taste, as well as into the different branches of food chemistry, confirmed his hypothesis.

Once its specific flavoring properties had been demonstrated, sodium L-glutamate (MSG) was officially recognized in Japan and throughout the world, and is now an essential condiment.

5'-Inosinic acid (IMP), a substance similar to nucleic acids, was discovered in meat extracts in 1847 by J.F. Licbig. In 1913, S. Kodama showed the essential role that IMP plays in the taste of dried tuna, an ingredient in traditional Japanese bouillons. Likewise, 5'-guanylic acid (GMP), another substance similar to nucleic acids, was isolated in the pancreas in 1893 by I. Bang, although its flavoring capacity ("umami" in Japanese), analogous to that of IMP, was not discovered until 1960 by A. Kuninaka.

Like MSG, substances related to nucleic acids are currently used as condiments in the form of sodium salts. A. Kuninaka and others have also shown that combining these susbtances with MSG reinforces their "umami."

Today, total world production of MSG is 340,000 tons, with 80,000 tons produced in Japan alone.

MSG is usually associated with small quantities of nucleic acid sodium salts, and is a popular condiment used both by Japanese housewives and food-processing companies in the preparation of meat, seafood, and farm produce, to which it gives increased flavor and taste.

Production

Up until thirty years ago, sodium glutamate and nucleic acids were obtained by decomposition of foodstuffs rich in these elements. It was only in 1956, the year of the first biosynthesis of L-glutamic acid in Japan, that production techniques involving fermentation and exploitation of microorganisms came into widespread use and were applied also to production of 5'-inosinic and 5'-guanylic acids.

Under normal conditions of fermentation, microorganisms strictly regulate their production and make only the amount of substance necessary. The mechanism controlling this phenomenon is called metabolic regulation. Production techniques based on fermentation must therefore induce the microorganism to manufacture and excrete large quantities of the desired product by somehow bypassing the mechanism of metabolic regulation.

Techniques for fermentation of amino acids – in particular L-glutamic acid – have been considerably improved. Peripheral techniques have undergone intensive improvements, as have production techniques (production of the useful substance by enzymatic methods, which involve harnessing the enzymes in a wide range of different microorganisms, continous production of useful substance in bioreactors, and immobilizing microorganisms or enzymes synthesized by these microorganisms). These are all examples of industrial application of fermentation techniques.

Today 5'-inosinic acid and 5'-guanylic acid are generally produced by biosynthesis, using the same principle as for L-glutamic acid and other amino acids, i.e., by partly bypassing the metabolic regulation mechanisms in microorganisms. However, the system of biosynthesis and the metabolic regulation mechanisms involved are more complex than in amino acids.

The most advanced techniques (DNA recombination and cell fusion) applied to the biosynthesis of amino acids – in particular, L-glutamic acid and substances similar to nucleic acids – are now being examined, the aim being to produce enhanced microorganisms. This research has recently begun bo bear fruit.

Properties

The usefulness of these components as condiments resides in their flavoring capacities. Western scientists have long thought that taste may be divided into four basic categories: sugary, acid, salty, and bitter. The most recent psychological and physiological tests have shown that what the Japanese refer to as "umami" is a separate property, which cannot be classified in any of these four categories, and there is a great deal of scientific evidence that this property should in fact be considered a fifth basic flavor.

It has thus been demonstrated that amino acids (L-glutamic acid, etc.) and the 5'-nucleotides (5'-inosinic acid, etc.) act as indicators of edible substances via the chemical receptors of almost all living organisms, from the most simple microorganisms to the higher land animals. In the course of research into taste-receptor nerve endings, Kurihara's team has recently demonstrated the existence of proteins acting as specific receptors for the "umami" substances mentioned above.

Likewise, Y. Kawamura's team, which stud-

Dr. Kikunae Ikeda
The "father" of "umami" flavors.

ied the transmission of impulses by gustatory nerves from the gustatory receivers (taste buds) to the brain, has revealed the existence of nerve endings that have a specific response to "umami" substances.

Finally, quantitative psychological research carried out by S. Yamaguchi's team with human subjects had defined a multidimensional model locating the four main flavors, their different combinations, and MSG on the basis of their degree of similarity; they have shown that MSG cannot be classified in any of the other four categories, and that it must in fact be represented by an additional dimension.

The flavor threshold of "umami" substances for human gustatory nerve endings is between $4 \times 10\text{-}4 \times 10^6$ and $4 \times 10\text{-}3 \times 10^6$ for MSG, $1.9 \times 10\text{-}4 \times 10^6$ and $4.7 \times 10\text{-}4 \times 10^6$ for IMP-Na2, and $6.6 \times 10\text{-}5 \times 10^6$ and $3.8 \times 10\text{-}4 \times 10^6$ for GMP-Na2.

The existence of "umami" substances does not modify the flavor thresholds of the other four basic flavors (sweet, acid, salty, bitter). In other words, "umami" substances do not increase sensitivity to the other four flavors but actually constitute an independent flavor. If we take a watery solution containing only MSG, the intensity of the "umami" flavor obeys the Weber-Fechner law and follows a linear progression according to the logarithm of the concentration of MSG in the solution. In the case of nucleic acids, this intensity hardly increases with concentration. However, by mixing MSG with certain nucleic acids, we can greatly increase "umami" intensity. This property, due to a combination of effects discovered in 1960 by A. Kuninaka's team, is found only in "umami" substances, and no equivalent is known for the other four basic flavors.

Electrophysiological and biochemical methods have established that the combined effect of MSG in certain amino acids is registered at the nerve endings of taste buds.

Research teams led by Y. Kawamura and M. Sato have studied this effect in cats, and in rats and hamsters, respectively, using electrophysiological methods to register the response of the tympanic membrane when a watery solution containing "umami" substances is placed on the tongue. This work has shown that a combination of MSG and 5'-inosinic acid sodium salt elicits a response greater than the sum of responses elicited by these substances taken individually, and that the amplification of the response coincides with known data on the functional response of taste in man.

K. Torii and R.H. Cagan have removed the epithelium from the tongue of a cow and prepared homogenized fractions, some containing taste buds and others not, and have measured the presence of bonded MSG in each. They have thereby demon-

strated significant bonding in the first case and an almost complete absence of bonding in the second, along with a measurable increase in the quantity of bonded MSG in the presence of 5'-guanylic acid sodium salt.

"Umami" substances can be found in most natural foods, although in very variable proportions, and constitute an essential gustatory principle. L-glutamic acid is found in many foddstuffs, such as meat, fish, seafood, milk, vegetables, and marine plants. Its content is 2,240 mg% in the Japanese dish konbu, 1,200 mg% in cheese, 280 mg% in sardines, and 140 mg% in tomatoes. 5'-inosonic acid is present in large quantities in meat and fish: 285 mg% in tuna, 122 mg% in pork, 107 mg% in beef. 5'-guanylic acid is found in large quantities in mushrooms (20-60 mg%), and in meat, fish, and seafood.

In order to examine the contribution made by "umami" substances to the flavor of natural foodstuffs, researchers have set up elimination tests. The first stage involves an exhaustive analysis of the composition of extracts in a specific foodstuff; the second stage involves reconstitution and comparison using identical solutions minus one or several components.

The team led by S. Konosu carried out similar tests to examine the relationship between the composition of food extracts and their flavor by focusing on types of seafood selected for their distinctive flavors. They noted that in certain crabs and types of abalone with a very distinctive taste, the elimination of "umami" substances, such as L-glutamic acid, 5'-inosinic acid, and 5'-guanylic acid, resulted in a significant lowering of intensity of the flavor.

Effectiveness
A primary factor in the "pleasing taste" of a foodstuff is of course its particular flavor as perceived by the five senses: taste, for the gustatory quality and intensity, smell for aroma, touch for texture, sight for preparation (shape, color), and hearing for the noise of mastication. Other factors also intervene, although in a more complex way: outside environment (ambiance, temperature, humidity at the time of the meal, etc.), nutritional traditions (dietary habits and culture, religion, etc.), and the human factor (state of the diner's physical or psychic health, etc.).

"Umami" substances such as MSG and IMP-Na2 are added to foodstuffs to bring out their flavor, which constitutes the main component of the "pleasing taste" mentioned above. However, these susbtances are not themselves "pleasing" when tasted in isolation.

If we administer to a representative sample of the population individual solutions of cane sugar, cooking salt, vinegar, and caffeine, and test for the degree of gustatory pleasure elicited by each solution, we can see that only sugar (cane sugar) is considered pleasing, and that acidic (vinegar) and bitter (caffeine) substances are systematically perceived as unpleasant when the gustatory intensity exceeds the threshold of perception, while "umami" and salty substances (cooking salt) are either perceived as being not pleasing or unpleasant at medium intensities, and always unpleasant at higher intensities.

This shows that gustatory pleasure is the result of interactions between certain characteristics of the foodstuff, such as its aroma or flavor. In practice, it is known that adding MSG or salt to a watery solution giving off an aroma of beef or soy sauce, for example, considerably increases gustatory pleasure without changing the original aroma (S. Yamaguchi).

MSG increases the gustatory pleasure elicited by most foodstuffs (soups, meat, vegetables, etc.), whether cooked in the home or prepared industrially. However, this is true only up to a certain limit of concentration. MSG demonstrates the same properties of built-in limitation as salt, i.e, the sensation of pleasure or its opposite limits the amount used.

Gustatory tests have shown that "umami" substances at an appropriate concentration reinforce not only the "umami" taste of the food in question, but also increase gustatory pleasure by adding certain qualities: body, fullness, texture, smoothness. Flavoring tests practiced in the American laboratory of Arthur D. Little have proved that the gustatory effect of MSG, IMP-Na2, and GMP-Na2 provides a sense of richness, volume, and harmonious blending of tastes.

This ability to give body, fullness, and smoothness is not the sole property of "umami" substances; it can also be observed with sugar and salt. This means that "umami" substances are added to foodstuffs for the same reasons as sugar and salt, and have much the same effect. We might also recall at this stage that their effect may be increased by combination. This is why tiny proportions of MSG, IMP-Na2, or GMP-Na2 suffice to bring out and accentuate the "umami" flavor of foodstuffs in which this savor is latent, i.e., beneath the threshold of perception.

Bouillons and consommés used traditionally as a basis for soup stocks today serve to bring out "umami" qualities by a combination of effects, insofar as they associate ingredients of animal origin (meat, fish, seafood) containing 5'-inosinic acid and ingredients of plant origin (mainly vegetables) containing L-glutamic acid.

Future Prospects
MSG, IMP-Na2, and GMP-Na2 play a primary role in nutrition and the human metabolism. L-glutamic amino acid is a fundamental component of proteins present in the composition of the human body, while 5'-inosinic acid and 5'-guanylic acid are similar to nucleic acids and play an important regulating function in the biosynthesis of proteins. This is why "umami" substances have considerable physiological importance, just as sugars are important sources of energy.

The fact that "umami" substances, which are indispensable components of gustatory pleasure, also have a key role to play in human physiology is a quite remarkable piece of "natural" engineering.

Researchers will now undoubtedly begin to explore the links between "umami" and gustatory pleasure, particularly in terms of aroma and flavor, and to pursue and widen the scope of scientific studies of gustatory pleasure and the mechanisms involved, so as to encourage healthy eating patterns in man.

Moreover, it would be useful to develop new and more effective techniques for production of "umami" substances (MSG, IMP-Na2, GMP-Na2, etc.) as well as other related substances (amino acids, peptides, etc.) using the new biotechnologies now available – in particular, DNA recombination, cell fusion, and bioreactors.

(1) Foodstuff manufactured in Japan.
(2) Traditional Japanese stock.

147

14

Down on the Farm

Animals have fewer scruples or, rather, no choice. So it was only natural that they were kept in mind when the great single-cell protein adventure began. At the time it seemed like the ideal answer to the eternal question of where to find greater quantities of more nutritional but cheaper animal feeds. Unfortunatly, however, it was not all smooth sailing for the businessmen concerned.

Nevertheless, things seemed to be off to a good start when Bel established its first lactoserum unit in 1956. At that time, two economic factors had pushed it in this direction: the relatively low cost of feedstock derived from petroleum and the relatively high cost of soybean cake (the prime source of protein for animal feeds). In any case, it was a major oil company, British Petroleum (BP), that was the first to enter the field when it decided to manufacture a protein, Toprina, produced by a yeast that fed on residues from crude oil (the viscous paraffins present in diesel fuel or pure paraffin). In 1971, BP joined with the Italian company ANIC to found Italprotein in Sardinia to mass-produce this famous protein. The situation was completely reversed, however, by the world oil crisis and a series of new agreements to lower the price of soybeans. After losing $100 million, BP called it quits.

But ICI continued the struggle in which it had become involved in 1968. ICI's experience with single-cell protein, built upon a basis of experience in antibiotics, contains important lessons for the biotechnology business. It serves as a valuable model in many respects. In any case, it remains the largest investment made by a single firm in biotechnological research. Dr. Peter Senior, of ICI's Agriculture Division at Billingham, has summed up the investment in these words: "From my experience, the single-cell protein processes that have been developed have stretched the imagination and innovative skills of all those involved in this development to a degree that the conventional chemical industry has not experienced before."

The research teams developing both single-cell protein and mycoprotein had reached the conclusion early in their projects that they could be competitive only if the fermentation ran continuously. This was a major and difficult departure from batch fermentation, where the fermenter could be readily sterilized with steam between batches. With continuous runs lasting for weeks there was a real danger of an infection by the wrong sort of microbe, contaminating or even outgrowing the product.

Initially the company wanted to use methane and signed a long-term contract for supplies with the British Gas Corporation. In the early 1970s; however, ICI changed course, abandoning methane, partly because of the high capital cost of protecting a fermentation process feeding on methane-oxygen mixtures from the risk of explosion. It turned, instead, to methanol (methyl alcohol), which has the advantage of needing less oxygen, thus further reducing capital cost. Pruteen (as the product was called) uses two of the most abundant substances on earth – carbon (from methane) and nitrogen (from the air) – and marries them by means of a microbe to make a food. It is the factory farming of microbes – whole microbes, not just one constituent of a microbe – secure from the changing seasons and the vagaries of weather.

This factory farming takes place in a single 600-ton pressure vessel – the fermenter – over 60 meters high and 1,500 cubic meters in capacity. Its purpose is to keep a microbe, which ICI has christened *Methylophilus methylotropus,* feeding on a balanced diet of methanol, ammonia, and air, together with traces of such nutrients as phosphoric acid and magnesium sulfate. At a temperature of 35°-40°C, on this diet the microbe divides every 150 minutes. In other words, each microbe produces about 1,000 new ones in a single day. In practice, the productivity is lower on so large a scale, not least because of the immense problems of maintaining ideal conditions for cosseting the microbe throughout such a large volume. ICI has been obliged, for instance, to install tens of thousands of

The Bio-Mini system
In this process designed by Hoechst, microorganisms enclosed in spheres of natrium alginate make it possible to convert raw materials and residues into proteins.

149

(1) In the United States by surgery, in France transvaginally.

Pruteen in Billingham

injection nozzles for methanol and pressure-pad sensors to follow the fermentation, to the point where one engineer describes the internals of the fermenter as resembling a watch movement in its intricacy.

ICI has tried to attack the feedstock cost by genetic engineering. Early in 1980 it announced the development of a new microbe, capable of making Pruteen with 7% less methanol than the natural organism, and it has been demonstrated successfully at the Billingham pilot plant. But ICI has not made use of the new genetically engineered microbe, as it appears that the fermenter's entire internal design would have to be changed.

If we have chosen to describe ICI's experience in detail, it is because it provides one of the best examples of the problems that must be constantly overcome to switch from laboratory to industrial scale. Furthermore, Robert Margetts, director of research of ICI's Agricultural Division, sees the big fermenter as an immense intellectual investment through which the company has demonstrated that, by doing enough science at bench scale (5-liter fermenters), it can confidently take immense leaps in scale.

John Brown Engineers and Constructors designed and installed the continuous heat sterilizing system used by the Pruteen process. The plant is now being used to produce RHM's famous mycoprotein. It took ICI twelve years and tens of millions of pounds to go from its initial experiments to industrial production. But this daring experiment opened the way for less costly pilot plants and the mastery of the essential techniques required for the bioindustry of the future.

Other single-cell proteins include Hoechst's Probion (600-800 tons per year from a pilot plant, compared with 150,000 tons for Pruteen), and other feestocks than methanol are being tested: cane molasses (Speichim), sulfite liquors from pulp factories (in Finland), lactoserum (Bel), effluents from candy factories (Tate and Lyle), corn residue, and cellulose residues.

ICI is keeping a close watch on the profitability of its system. In view of the fact that the cost of Pruteen remains high (twice that of soybeans), it may perhaps prove wiser to turn to well-tested methods and produce two avidly sought amino acids: lysine and methionine.

Methionine and lysine, which are essential to animals, are all the more valuable in that they are not found in cereals and must thus be added to forage. While, despite recent research, methionine continues to be made by chemical methods, 80% of the lysine sold is produced by fermentation using a bacterium, *Cornybacterium glutamicum,* which we already encountered in the production of sodium glutamate. The market is of considerable size, and estimated to be worth some $200 million. It is one of the eight amino acids essential to the body, which cannot synthesize them itself.

Cornybacteria produce this for their own use, along with methionine and threonine. But they do not need any surplus, with the result that, when the required amount has been produced, the combined action of threonine and methionine brings into play an enzyme (aspartase kinase), which inhibits the production of lysine. Ajinomoto has developed a mutant capable of excreting 70 grams per liter of lysine, which is now being used by Eurolysine (founded by the Japanese concern and the French firm Lafarge-Coppée). It is thus a matter of short-circuiting the natural switch. If the enzyme (homoserine dehydrogenase), required for the production of threonine, but useless for lysine, is withdrawn, the bacteria continue to produce lysine in quantity. Nevertheless, as the bacteria need a dose of threonine to live, small amounts are provided that

are insufficient to stop the production of lysine. The Eurolysine factory in Amiens is now the largest in the world and produces more than 21,000 tons of L-lysine per year.

Research is being intensified in an effort to improve the quality of meat and increase dairy production by means of growth hormones, proteins, additives of all sorts, vaccines, and antibiotics mixed with forage to prevent diarrhea and other epidemics affecting livestock, but embryo transplant is also being tried. Attempted experimentally on a small scale in the United States and in France[1] as early as 1972, the technique has been improved considerably since then, above all by deep freezing, which makes it possible to preserve 7-day-old fetuses, and by embryonic division (so as to make homozygote twins).

Thanks to sperm banks, bulls can now father a thousand calves. And with embryo transplants it is now possible to choose both a pedigreed bull and a high-production cow to produce fleshy beef cattle and super milch cows. From 4 to 10 embryos can be collected at the same time, and the operation can be repeated up to 4 times per year; the embryo is then transplanted into a "carrier" cow, to use the modern parlance. Such transfers can also be carried out with sheep, goats, and other liverstock.

In vitro fecundation is for the moment less successful with calves (despite the famous Virgil conceived in this fashion at the University of Pennsylvania) than it is with mice, hamsters, guinea pigs... or people.

In its advertising, one American firm even goes as far as asserting, "You can fly with 2,000 cows in your handbag," but the ambient temperature would have to be pretty low for this! These portable embryo banks will be of no use whatsoever without veterinarians capable of collecting and replacing the embryos... highly paid skills indeed.

What then is the next step? Determining the sex of embryos, no doubt, making it possible to choose either a good dairy cow or a hunk of beef on the hoof. To do this, it will be necessary to separate X and Y chromosomes, and various methods are now being tested, ranging from antibodies to electrophoresis. One apparatus designed by the Nomura Research Laboratory even makes it possible to separate spermatozoa, in a gravity-free atmosphere... in other words, in space. Such work may make it possible to eliminate certain illnesses connected with sexual chromosomes, as well as certain metabolic deficiencies or a lack of immunity defense. All of these manipulations will undoubtedly permit the exploration of biological processes that still remain a mystery, such as cellular differentiation.

Artificial insemination, embryo transplants, and sex selection will undoubtedly become the three mainstays of the livestock industry of the future. But what is good for animals may not necessarily be good for people, and cries of alarm have consequently been raised – for example, by Roberta Steinbacher of Cleveland State University, who cites most parents' "preference for the male,"[2] and on the basis of American reports, foresees "a rise in the crime rate, more wars, the accentuation of sexual stereotypes, polyandry, a rarefaction of women's merchandise," and a top of all this, "still greater accentuation of the inferior social status imposed on women today if big brother-little sister

becomes institutionalized due to most parents wanting the eldest child to be a boy." One may tend to smile at this dark, male-dominated vision of the future, but we need only recall the multitude of legal and emotional tangles that have already resulted from artificial insemination, test-tube babies, and surrogate mothers. What's going on down on the farm may in the long run be more important than we think, and affect more than just our taste buds.

(2) *Biofutur*, July-August 1983, p. 27.

1

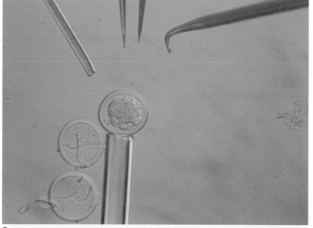

2

3

4

1. *Embryo flakes*
Embryos, taken 7 days after fecundation, already consist of from 100 to 200 different cells. They are placed in a plasma solution and introducted by suction into a plastic flake. The flake is cooled to $-35°C$, then stored in a container of liquid nitrogen (at $-196°C$). To thaw the embryo, the flake is immersed in a bath at $37°C$. The sensitivity of embryos to thawing varies and often depends on the species.
2. *Splitting an embryo*
When the embryo has arrived at a stage of initial differentiation, the membrane is opened and is cut in half by microsurgery (the embryo is generally immobilized by suction). This couple of embryos is then transplanted into the uterus of the foster mother.
3. *Superovulated ovary*
Superovulation make it possible to recover a greater number of embryos.
4. *Twin lambs*
The technique of dividing and transferring embryos makes it possible to accelerate the rate at which prize livestock can be reproduced and produce genetic twins. These in turn make possible comparative tests, in which one twin is treated with a certain substance and the other serves as a control.

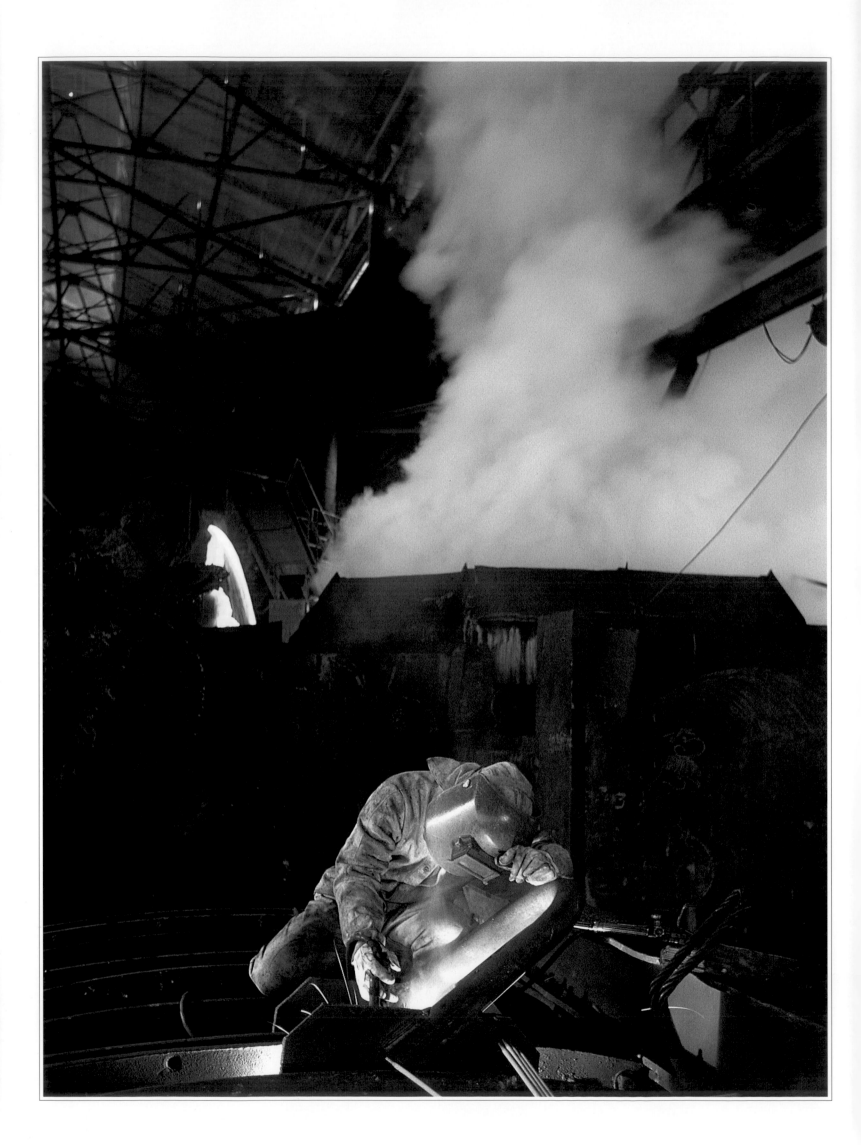

15

Microbes
as Metal Traps

In the year 1,000 B.C., the people of the Mediter ranean recovered copper in solution by draining the water from mines. And in eighteenth-century Spain, this practice was still commonplace in the Rio Tinto mines. But it was only in 1957 that the role played by bacteria in the recovery of metals in solution (leaching) was discovered.

The best known of these microbes is *Thiobacillus ferrooxidans*, which lives in ore deposits. While most metallic ions kill other bacteria, it would seem that a strain of *Thiobacillus* has evolved in such a way that it can feed on this poison and even flourish on it: the enzymes of the bacteria (aerobic) directly attack the atomic structure of the metal, detaching electrons from iron and sulfur, whose atoms, in their turn, steal electrons, from their neighbors. The metals, deprived of their electrons, then become highly soluble in sulfuric acid formed by oxidation of the sulfur, and the metals thus liquefied can be recovered. The *Thiobacillus* also acts as a trap, accumulating many metals in its cell in concentrations greater than those in the surroundings in which they grow (copper, nickel, cadmium, zinc, lead, but also cobalt, strontium and rubidium, arsenic, and antimony).

During the late 1970s, there was growing interest in *Thiobacillus* for extracting metals from sulfur ores. The stakes were considerable: metal-extracting techniques are expensive, and the bottom had just fallen out of the markets for copper and uranium. Furthermore, some veins of copper contain gold and silver, and the use of these somewhat special microminers might have made it possible to recover the precious metals at far lower cost.

Two methods have been used traditionally to extract metals: floatation and subsequent smelting in the case of metal-rich ores, and chemical treatment fro the poorer veins. Both methods are not only costly but create heavy pollution (sulfuric acid, which falls as acid rain). Today, 10% of the copper produced in the United States is obtained using bacteria – sprinkling surface deposits (dump leaching), sprinking chunks of crushed ore (heap leaching), or pulverizing the ore in a vat (vat leaching).

For almost forty years, BC Research, the Canadian company with headquarters in Vancouver, has pursued its work with bacterial extracting methods, and its teams have developed a leaching process capable of competing with smelting systems for veins rich in copper (and perhaps in nickel and zinc). A. Bruynesteyn, who is in charge of research, has chosen to use a silver catalyst, which stops the oxidation reaction initiated by the *Thiobacillus* halfway between sulfur and sulfuric acid, all of which makes it possible to produce sulfur as a by-product (of interest to chemists) and to reduce costs by 28%. International mining concerns have advanced $500,000 to build a commercial-scale reactor. The companies in question include BP Minerals, Wright Engineers, Vancouver Mining Firms, Placer Development, and Somito Mining.

For ores with a low metal content, Charles Burton's team in Dublin (Institute for Industrial Research and Standards) is experimenting with spraying *Thiobacillus* directly into the mine: when the water is pumped out, the copper is recovered. This process, which seems obvious enough, is in fact quite complex, for secondary products are formed in the course of the reaction that, depending on ambient parameters of a hydrological, geological, and physical nature, can inhibit it. Yields are still quite low, but the in situ method of recuperation seems promising. Yet there are few companies pursuing research in this field, for it will probably take years and require heavy investments without necessarily resulting in a system that will revolutionize the mining industry. Among these companies we might mention Noranda in Quebec, Chevron, and Inco (in conjunction with Biogen).

The precious-metals industry is already giving greater thought to the advantages of microbes, and the Engelhard Corp. has developed two systems for recovering metals using fungus organisms in

A copper mine in Chile
For mine ores with a low metal content, some copper mines are trying bacteria.

153

waste water. At Kodak, Robert Belly and G.C. Kydd have isolated a *Pseudomonas* bacterium from photographic emulsions that can extract silver from silver sulfide solutions.

In contrast to the situation in other sectors of biotechnology, there is but a single firm specializing in this field: Advanced Mineral Technologies (AMT) of Socorro, New Mexico, which is managed by Corale L. Brierley. She has faith in the future: she assumes that there will be a potential market of $5 billion in the year 2000.

Unfortunately, *Thiobacillus ferrooxidans* has proved desperately slow. It is slowed down still more by traces of arsenic or lead present in certain veins or by too high a concentration of copper or zinc. Thoughts naturally turned to genetic engineering. At General Electric, David S. Holmes has isolated three different plasmids of a strain that is resistant to uranium: cloned and expressed in *E. coli,* the genes seem to code for substances that become attached to metals and destroy their toxicity. At McGill University, in Montreal, researchers have achieved the same results for iron. In South Africa, a microbiologist at the University of Cape Town has isolated three plasmids of a strain that tolerates arsenic – and many veins of gold are rich in arsenic. A *Thiobacillus* that tolerates arsenic could lead to the elimination of expensive methods of getting rid or arsenic, and above all the fearsome production of sulfurous gases using traditional methods, while at the same time raising yields to 86-90%. A South African company, Gencor Group Laboratories, is now studying these strains of bacteria, which the General Mining Union Corporation will test in a pilot plant. Yet there are many problems in cultivating bacteria: it is difficult to do on a solid feedstock (it requires cultivation in suspension), which is indispensable in enabling researchers to isolate a mutant and clone it. Furthermore, it cannot cope with an environment as acid as that of the human stomach – inhibiting the action of most of the antibiotics used in the selection process.

Thiobacillus also "works" in uranium mines, attacking iron disulfide (pyrite) and oxidizing it into sulfate, which makes the metal soluble.

Biodepollution

Recovering toxic metals from industrial waste has the double advantage of fighting against pollution at the same time. At the University of Liège, Jean Remacle, using a mixture of bacteria (including *Thiobacillus*), has been able to recover lead, copper, zinc, cadmium, nickel, and cobalt in this way. The trap works even better if the microorganisms are cultivated in media: using a "fluidized microbe field," 90% of the metals in industrial effluents can be removed. To eliminate metal salts in solution, use has traditionally been made of ultrafiltration or ion-exchange resins: these are inert matrices containing mobile ions that, passing through any given solution, change place; as microorganisms can grow and multiply in waste water, cleansing and regenerating the resins (with a solution that rids the resins of the ions that have become attached) can be dispensed with and uninterrupted operations made possible. Their principal defect remains their poor selectivity. Bacteria could also be used to convert sufurous fumes into iron sulfate, which could be easily eliminated with carbon, as has been demonstrated by T.F. Huber of the University of Delft. It will also become necessary – and this is no minor problem – to train interdisciplinary teams, as few microbiologists have any acquaintance with metals or mining engineers with microbes.

Perhaps one day it will become possible to fabricate immobilized bacteria whose proteins will attach themselves to metals, like the "metallotioneins" (MTs) found in certain eucaryote organisms, which show a distinct affinity for heavy metals (gold, silver, and lead). Researchers at the University of California at Berkeley and at Genex are considering cloning and expressing MT genes in *E. coli*. But it will first be necessary to know in detail the identity and functioning of the microorganisms that entrap metals.

For *Thiobacillus ferrooxidans* is not the only microbe of this type. The family also includes a close cousin, *Thiobacillus thiooxidans,* and a more distant cousin, *Thermothrix thiopara,* a thermophilic microbe that can withstand far higher temperatures (60°-75°C): this comfort-loving creature nests in hot springs and volcanic fissures. Research is also continuing on TH bacteria, which behave similarly to *Thiobacillus.* The TH strains develop on ferrous or mineral feedstocks (pyrite, chalcopyrite, covellite, and pentlandite), as has been shown by James A. Brierley at New Mexico's Institute of Mining and Technology and Norman W. Le Roux of Britain's Spring Warren Laboratory. The TH bacteria can be distinguished from the *Thiobacillus* by their tolerance to heat (up to 50°C) and their different metabolism (above all, their inability to convert carbon dioxide by themselves). The most rugged of all these organisms is undoubtedly *Sulfobolus* (which can stand temperatures of 60°C and even more), which is found in volcanic cracks and sulfurous waters. This bacterium seems able to survive even near-boiling temperatures. Another advantage: though aerobic, it can behave like an anaerobic bacterium in the absence of oxygen; it can also attack metals that are indigestible for other microbes, including chalcopyrite and molybdenite.

Organisms as simple as brewer's yeast or the fungus *Rhizopus arrhizus* could even absorb uranium from waste water. *Pseudomonas aeruginese* seems capable of accumulating in its cell a high concentration of metals (100 milligrams of uranium per liter of solution in 10 seconds). This capacity to absorb metals, which has been noted without being understood, may prove extremely valuable in trying to eliminate the pollution in water. Still other microbes have been discovered that eihter accumulate metals on their surface or help to precipitate them.

So why not a cocktail of microbes? Donovan P. Kelly's team at the University of Warwick has in fact confirmed that the combined action of different strains increases the efficiency of the procedure and accomplishes synergistically what a single organism would have been incapable of.

The primary mission and first use of these "metal traps" will undoubtedly be to purify waste water. But perhaps Corale Brierley's optimism will in the end be justified, and microbes one day become miners.

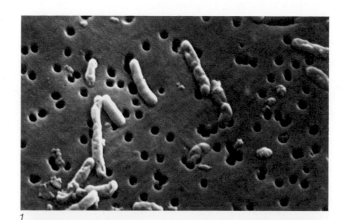

1. **Bacteria TH3**
Moderately thermophilic
bacteria known to act as
catalysts in extracting the
metals from certain ores at a
temperature of 50°C. Seen
under an electron scanning
microscope.

2. **Ion exchangers**
Ion exchangers, used for the
retreatment of the water
from power stations, and as
catalysts in chemical
reactions. They are
particularly useful in
eliminating traces of heavy
metals from industrial waste
water.
3

3. **Bacterial lixiviation**
The Santa Rita mine in New
Mexico (background) and
lixiviation basins
(foreground right). Bacteria
(often Thiobacillus) oxidize
ferrous iron into ferric iron.
The soluble copper sulfate
obtained at the end of the
process drains into recovery
basins.
2

16

Energy from the Source:
The Biomass

Energy, etymologically, means "force in action." Biomass is the name given everything that lives and breathes and exchanges energy in an endlessly repeated cycle from vegetable to animal, to man, to the earth, and to water. Why not, man finally asked, tame and domesticate this force in action in the biomass, this energy in which we are bathed?

Biogas

Disneyworld, the realm of Mickey Mouse and Sleeping Beauty, is also the kingdom of the water hyacinth, which purifies its ponds. Water hyacinths in tropical countries, water chestnuts in Upper New York State's lakes, algae in the Venetian lagoon, aquatic vegetation without root or bark, all bathed in light, all champions of photosynthesis, whose age-old reserves of energy we are now beginning to tap. Why not simply grow them, harvest them, recover the fuels or chemical products resulting from their decomposition (by anaerobic digestion), employing them upstream to purify water and produce biogas and downstream to provide a protein-rich source of animal feed (by processing the residues)?

Written almost 3,000 years ago, the Chinese book of divination, the *I Ching,* mentioned the swamp gas used by people in the province of Szechuan. This gas, produced by rotting organic matter, is one of the essential components of the natural gas that is today known as methane. For millions of years, microbes have been decomposing the dead cells of plants and animals and producing methane. Bacteria of the same type are found in the stomachs of livestock, enabling them to digest grass (by breaking down the cellulose and liberating the methane).

Biogas (which contains 50-80% methane) is today produced by fermentation, using domestic garbage and agricultural waste, thus simultaneously reducing pollution and producing energy. It was back in 1900 that for the first time gas from manure was used in a leprosarium near Bombay. At the present time, there are tens of thousands of digesters in India and almost five million in China. Installations in the industrialized world are often still experimental, though Great Britain installed its first digesters in sewage purification stations in 1911, and starting in 1935, English purification centers were provided with gas-fired boilers. But as things stood, the principal goal was the elimination of pollution and not the production of energy. In the United States, Getty Synthetic Fuel is now engaged in thoroughgoing research. The United State's energy potential (from garbage) is estimated at 5.6 billion cubic meters of methane – the equivalent of 5 million tons of petroleum.

But innumerable problems must be overcome to produce methane industrially. The bacteria concerned have a low growth rate and are extremely sensitive to heat, acidity, and a host of other factors that vary constantly in nature. Anaerobic, they produce little heat and need energy from outside. As the ingredients or organic matter are extremely varied and disparate, as can be imagined, a veritable armada of microorganisms must interact in methane generators. And we are barely beginning to determine the nature and function of most of them.

Fuels and Solvents

Today, nine-tenths of the cars sold in Brazil operate with sugar cane in their engines, or at least on alcoholic fuel based on sugar cane. Alcohol-based fuels (ethanol, methanol, acetone-butanol) would seem to have a far brighter future than biogas where industrial use is concerned. In the United States, a few cars are already operating on gasohol (90% alcohol, 10% gasoline). In Zambia, the Philippines, Zimbabwe, Nicaragua, and Paraguay, there are factories producing ethanol. And in Japan, Kyowa Hakko has built a pilot plant at Hofu for the continuous fermentation of alcohol using immobilized yeast cells.

Ethanol
The Piracicaba factory in the state of Sao Paulo, Brazil, where sugar cane is converted into ethanol. Over half of the Brazilian harvest (200 million tons annually) is converted into ethanol for use in automobiles.

There is an old saying that goes, "The best soups come from the oldest pots," and, however folksy that may sound, it is essentially true, for we have returned to the fermentation of sugar obtained from animals (lactoserum), vegetables (lignocellulose: wood, straw, corn cobs), or glucides (the effluents from sugar factories, molasses, alcoholic plants). And the raw materials are local: sugar in Brazil; corn in the United States; sweet potatoes in Japan; cassava, trees, or grass elsewhere. The microorganism fulfills its ancestral duty as a converter, breaking down the glucose-fructose chain of the sucrose in sugar cane, for instance, producing carbon dioxide and water. But if the source of oxygen is cut off, the series of chemical breakdowns stops at a certain stage and the sugars form ethanol molecules. To achieve industrial yields at moderate cost, one must select, by screening, strains that are the most resistant to alcohol (a normal microbe dies in a medium whose alcohol content exceeds 12-15%).

The ideal microbe would simultaneously be able to tolerate alcohol and possess amylases capable of breaking down starch (the energy reserves of plants), of which millions of tons are produced each year by cereals and tubers. When plants need energy, these enzymes intervene, breaking down the starch molecules and freeing the sugars. They are extremely efficient and capable of playing a part in over a thousand different chemical reactions in a second. The gene that controls the production of amylase in bacteria has now been identified, and by genetic-engineering techniques could be transferred to a yeast.

Another potential raw material could be cellulose, the essential component of plants and the most widely distributed organic matter on earth (the term lignocellulose biomass covers wood, the straw from cereals and rice, corn stalks and husks, sorghum or manioc stalks, beet pulp, bagasse from sugar cane, etc.). Yet there are enzymes secreted by microorganisms that attack the cellulose wall and make it possible to produce sugar (cellulases). In 1950, microbiologists working for the US Army discovered the creature responsible for the deterioration of its wooden equipment, which they had attributed to sabotage. It proved to be a common filament microfungus, *Trichoderma virida,* which, cultivated on paper, textiles, or wood shavings, reveals its cellulytic activity. The gene of cellulase could perhaps be transferred from the fungi to yeast. One of the important handicaps encountered is the presence of lignine (highly resistant) in close association with cellulose, but microfungi that attack lignine have also been discovered. To date, only a "syrup of wood sugar" has ben obtained, in the form of a very dilute molasses with à 5% sugar (glucose) content. The USSR seems to be continuing along these lines for the production of animal feeds (SCP). In 1984, a pilot plant for the production of acetone-butanol-ethanol (solvents) was built by IFP at Soustans in France for the enzymatic hydrolysis of lignocellulose raw materials. In the United States, the Milbrew Facility is the only factory producing ethanol from lactoserum. Actually, however, the greater part of the ethanol used as a raw material by the chemical industry is synthesized from ethylene (a petroleum derivative).

The microbes concerned are generally yeasts for ethanol, and bacteria for acetone-butanol-ethanol (ABE). For cellulose, tests are being carried out with the bacterium *Clostridium thermocellum,* and for starch *Thermoanaerobacter ethanolicus,* both of which, as their names imply, are heat-resistant. This is a nonnegligible advantage, inasmuch as heat increases the speed of the reaction (and consequently the yield) and distillation is much easier when the fermented liquid is hot.

Hydrogen and Photosynthesis

But speaking of energy and the biomass, do we not have at our disposal the most fantastic factory in the universe for producing hydrogen? Do we not have at our disposal the prime energy source of them all (the sun), a raw material that covers three-fifths of the earth's surface (water), and a natural solar and chemical power station (the plant cell)? Of the 10,000 watts per square meter (W/m^2) of energy emitted by the sun, do we not receive 145 W/m^2 – already a fabulous figure and an average of 375 times the equivalent in petroleum used by the richest of countries?

Dispersed and intermittent, this energy is unfortunately not easy to capture. Photoelectric cells and thermodynamic devices do this on a small scale. But have we forgotten that there are natural captors with varied and enormous storage capacities: plant cells?

1

At the birth of the universe some five billion years ago, the atmosphere was made up of hydrogen, not oxygen. Two billion years later, photosynthesis appeared on the scene with photosynthetic bacteria with colored organites (chromatophores), which did not yet feed on water but on sulfur. Still another billion years after this, it was the turn of the cyanobacteria (including the blue algae) to make their appearance: more highly evolved, they carried out a photolysis of water. Gradually, oxygen accumulated in the atmosphere. But it took another three hundred thousand years for the organite responsible for breathing to appear. This was the mitochondrion.

Thus it was only (if we can use the expression) 800 million years before our own time that the protective layer of ozone was formed that made it possible for plants to develop and form chloroplasts – organites found only in plants, and the seat of photosynthesis.

Photosynthesis, a process still incompletely understood, was first brought to light when a Flemish chemist, Johann Van Helmont, planted a willow weighing 2.5 kilos in 90 kilos of earth. At that time, it was assumed that plants drew their energy from the earth. Van Hermont was to prove that it was from the air (and from water) that the plant derived its energy, thanks to the carbon in the atmosphere and sunlight. Five years after it was planted, the willow weighed 75 kilos and the mass of earth had only lost 60 grams. Today, we know a little bit more about the process of photosynthesis. When molecules of chlorophyll are struck by light, their "collecting antennae" (proteins and pigments) are excited by the photons. Through a system of "molecular photocells," they transfer the energy absorbed to the interior of the cell and convert it into a form that may be assimilated by the plant (ATP), then conclude by expelling the oxygen into the atmosphere. This oxygen is in its turn "burned up" by animals and men, who exhale carbon dioxide, which is absorbed by the plant – and on the endless cycle continues.

But would it not be possible to force a plant to produce hydrogen by providing its chloroplast with an enzyme (hydrogenase) capable of seizing two hydrogen ions and attaching them to two electrons to form a molecule of gas? A good idea theoretically, but with no practical outcome. In that case, would it not be possible to provide a bacterium that already has this hydrogenase enzyme (such as *Clostridium butyricum*) with the sugars needed to produce hydrogen? Unfortunately, the bacterium stops working after a few hours, though in Japan they have been able to prolong the experiment by immobilizing the *Clostridium* cells and feeding them with waste waters rich in sugars. The latest

1. *Water hyacinths*
2. *Supply through bacterial filters*
These filters are used for the fermentation of methane on a fixed bed of the waste from sugar factories. The bacteria that cause the methanization are fixed on rings, as at the Thumeries sugar factory in France built by SGN. This makes it possible to replace aeration and purification basins with an installation for the accelerated treatment of waste water using anaerobic bacteria, which require up to six times less energy than traditional biological processes.
3. *Toledo*
Plant for treating liquid animal waste (from pigs) built by SGN in Spain in 1983. Methanic fermentation on a fixed film is employed here. Its capacity is 120,000 pigs per year.

1. *Biological treatment of effluents*
The first large-scale plant for the biotreatment of effluents was built by Hoechst. Tanks containing the bacteria receive the organic matter of an effluent from vast settling tanks. This matter is subsequently oxidized or converted by a bacterium into proteins.
2. *Photosynthesis*
3. *Spiruline algae*
Cultivation at an IFP pilot facility.

1

2

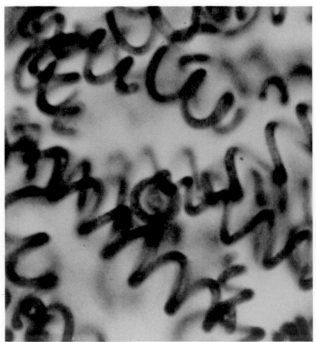

3

approach under consideration is to obtain hydrogen directly from water. When exposed to light, the alga *Chlorella pyrenoidosa* is capable of assembling two molecules of water and two molecules of carbon dioxide to form a glucolate, which, hydrolized by an enzyme (glycolic oxydase), furnishes a feedstock for a bacterium that produces carbon dioxide and hydrogen.

All this mixing and simmering of plants is a complex matter, of course, but given the need for energy – particularly among our more disinherited brethren – continuing research is a necessity.

Photosynthesis may be an interesting way of obtaining lubricants, fuels, and other replacements for petrochemical products. Was not the biblical "burning bush" simply *Dictamnus fraxinella,* whose flowers contained terpenes? And are there not plants rich in hydrocarbons, such as laurel, elder, parthenium, goldenrod, Philippine nut, and even carrots, fennel, and cloves? Back in 1968, a British geochemist discovered that a microalga *(Botryococcus braunii)* was responsible for a large part of the world's offshore oil deposits. This alga can, in fact, produce and store hydrocarbons (paraffin or terpenes) in an extracellular pocket. Research is being intensified on the energy potential of this alga in France (Ecole Nationale Supérieure de Chimie de Paris), the United States (Queen's College of the University of New York, Martin Marietta), as well as in Japan and Portugal. We should be able, in Olivier Roncin's words, "to tap this reserve without causing any damage or hampering the pace of subsequent production, just as we take honey from a hive without bothering the bees" (*Science et Avenir,* 1984, 50:62).

Here, as in every other aspect of biotechnology, everything depends upon costs. Among these many processes for capturing the energy of sunlight or that contained in the biomass, some could be interesting on a craft scale or locally, but none is competitive at the industrial level. Here again we are looking forward to the future and placing our bets, bets that it will obviously be imperative to win in developing countries.

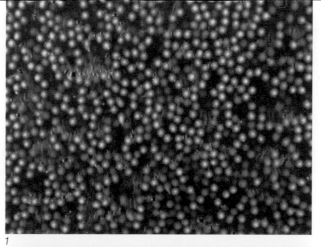

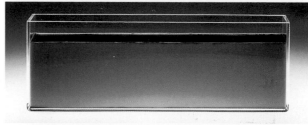

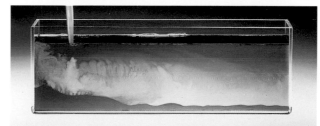

2

1. *Immobilized cells*
In its pilot plant producing ethanol, Kyowa Hakko uses fixed yeast cells. While the idea of growing microorganisms on insoluble supports dates back to Pasteur, the development of cell immobilization techniques derived from those developed for enzymes. This makes it possible to stabilize biological activities with respect to outside disturbances and to improve productivity by maintaining large concentrations of cells, as well as to use continuous processes.
2. *Treatment of effluents*
Into the effluent (a) is poured a mixture containing the bacteria (b), whose activity can be observed in the aeration basin (c). At the end of the process, the water has once again become clear, and a sediment of biomass is deposited on the bottom (d). This is the Hoechst process.
3. *The Burning Bush*
"Behold the bush was burned by fire but was not consumed."

3

Biotechnology and Water Treatment

by Jean Bébin, Assistant Manager of the Lyonnaise des Eaux-Degrémont Research Center.

Degrémont, a subsidiary of Lyonnaise des Eaux (France's second-ranking private water company) is a world leader in the treatment of water. But what can the biotechnologies bring to the elimination of pollution in this field? Very little actually, but great hopes, replies Agronomic Engineer Jean Bébin, an expert with the Agence Française pour la Maîtrise de l'Energie (AFME) and the Ministry of the Environment.

1. Coating of anaerobic methagenic bacteria on a granular surface support.
2. Fixed cultures of aerobic nitrification bacteria on a granular surface support.

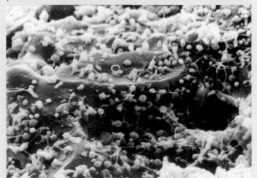

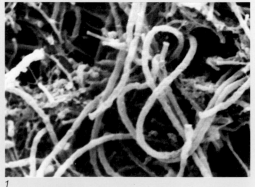

Water treatment is of two main types:
• treatment of surface water (rivers, lakes, etc.) and/or deep water to produce potable water or clean water for industry;
• cleansing urban and/or industrial waste water before evacuation or recycling.

The first type is for water that is generally only lightly polluted. It is usually not necessary to have recourse to biological processes, except for nitrification (i.e., transforming ammonia into nitrates) and denitrification (i.e., breaking down nitrates in the soil) and for some aspects of physical processes (slow filtration, absorption over activated charcoal).

The second type, however, is for very heavily polluted water contaminated by dissolved organic matter, whether colloidal or in particle form. Treatments of this type do remove pollutants, although not completely, and use a series of physical processes (removing oil, grease, decanting, or flotation). Aerobic and/or anaerobic treatments may also be used.

These biological processes are carried out in reactors with free or fixed media and generally employ complex bacterial populations developed spontaneously from natural sources: soil bacteria, human or animal wastes, industrial ferments, etc. These biological processes cannot be carried out in the usual conditions of industrial fermentation characterized by sterile environments, defined and controlled substrates, selected strains, etc. Water treatment deals with pollution in its most complex and variable state. Thus, while the industrial sector has made a heavy investment in fermentation techniques, it is difficult, a priori, to see how modern biotechnologies (genetic engineering and enzyme engineering in particular) could be applied with some hope of success.

However, this preliminary, and rather negative, approach must be tempered, and we cannot exclude the possibility that within several years, biotechnologies may well play a crucial role in water purification. A certain number of recent discoveries do seem to indicate slow but irresistible progress in relevant techniques. There are a number of different aspects: identification and measurement of pollutants, use of highly developed separation systems, development of intensive biological reactors, and so on. The following remarks are intended to open discussion of some future aspects of the use of biotechnologies in the field of water treatment, although they are not intended to be exhaustive.

Identification of Pollutants

Since water is an extremely effective solvent, rivers and underground water harbor enormous populations of very varied organic molecules. The global measuring methods generally used are no more than indications allowing us to judge the extent of pollution, but today they are only marginally useful in guiding the designer or operator of a water-purification plant anxious to eliminate all traces of pollution.

If we wish to identify the exact composition of this mass of organic pollutants, we can use sophisticated techniques, such as gaseous chromatography or high-pressure liquid chromatography, downstream of complex concentration and extraction processes. Here, the latest advance in applied techniques in the best pollution-control laboratories is a technique combining gaseous chromatography and mass spectrometry. Biological techniques are now taking their place alongside these complex and costly analytical techniques. They include enzyme electrodes and, in particular, immunoanalysis, using very specific antigen-antibody reactions, which should, over the next few years, provide us with rapid and very precise means of identifying and quantifying certain families of compounds (pesticides, various chemical products of industrial or natural origin) in special lookout stations. These innovations will quite definitely enhance the quality of the end product in water treatment.

This type of research is carried out in France by Lyonnaise des Eaux associated with Immunotech, a French company specializing in the production and application of monoclonal antibodies.

Similar techniques may also be envisaged for identification and quantification of indicator (coliforms, fecal streptococci) or pathogenic organisms (bacteria, viruses), thereby replacing traditional bacteriological methods, which demand an incubation period of several hours or even days. In France, the Institut Pasteur, Immunotech, the Compagnie Générale des Eaux, and Lyonnaise des Eaux are collaborating in this field, with the assistance of the Ministry of Health.

Rapid and specific identification of the organic compounds and microbial agents responsible for pollution will not only increase researchers' knowledge, but will in the long term contribute to the discovery of more specific processes for high-level antipollution treatments in which biotechnologies have a definite role to play.

Although research into antipollution treatments involving fixed enzymes is still in the preliminary state in a few leading laboratories, it is highly probable that industrial applications will be discovered within the next decade.

Specific Elimination of Certain Pollutants

Whereas the biological systems responsible for cleansing organic substances tend to be genuine, complex ecosystems, in which a large number of bacterial strains, fungi, and yeasts coexist along with protozoans playing the important role of predators, and whereas these ecosystems are permanently developing and adapting to qualitative and quantitative changes in their environment and variations in control parameters, in other cases the cleansing microflora used is much simpler and more stable. This is true of nitrification processes (transformation of ammonia into much less toxic nitrates), which employ autotrophic baterial strains (*Nitrosomonas* and *Nitrobacter*).

Thus, in France, researchers from the Lyonnaise des Eaux Research Center and the Université Claude Bernard in Lyon have shown that we are dealing with only a limited number of species and even of serotypes (subgroups). Under these conditions, it is quite possible to envisage cultivating selected bacterial species and conserving them for rapid seeding of bioreactors (or reseeding in the event of accidental destruction of the cleansing flora used in nitrification processes). In the long term, we could even envisage modifying the genetic makeup of these strains to enhance productivity or modify their resistance to sometimes unfavorable environmental conditions.

Researchers have also found cases of bacteria adapted to very specific situations (water evacuated from mines, for example), and work on these strains, through conventional methods of selection or mutation or by more complex genetic-engineering techniques, could pay off. American and Japanese researchers have also announced, and patented, the use of bacterial strains specific to certain industrial substrates, but widespread application of such cocultures is far from simple. It is rare that exactly the same pollutants will be found in two plants; not only may they manufacture different products, but each factory may manufacture different products at different times of the year, and conditions favorable to continued existence of a given strain are often difficult to identify, and above all to stabilize.

Some research teams, in particular Professor J. G. Zeijus's team at the University of Wisconsin at Madison, have attempted to develop anaerobic cocultures adapted to this type of substrate (lactoserum, collagen, etc.) with the aim of developing cleansing systems based on methane fermentation. If this type of biological system proves capable of working in a stable manner and overcoming problems of competition in a complex and variable environment, biotechnologists could make a considerable contribution, but this technique is still in its infancy.

The Japanese have also launched a program called BIOFOCUS, designed to develop novel microorganisms by cell fusion to eliminate organic pollution, nitrogen, and phosphorus.

Development of Membrane Separation Techniques

Inverse osmosis or hyperfiltration membranes have been successfully used in desalination of briny water and even seawater, in conjunction with thermal processes; ultrafiltration and tangential microfiltration through mineral or organic membranes are now being applied in the agricultural and food-processing and bioindustrial sectors. Application of these techniques to water treatment might, in the long term, supplant certain physicochemical processes (flocculation-filtration) and reduce levels of disinfectant, but could also modify the design of biological reactors.

Membrane reactors could separate out organic molecules from pollutant substrates upstream of the reactor, or subproducts of bacterial metabolism downstream, and could thus modify the conditions of variability characteristic of biological reactors. This would in turn allow the use of more specific bacterial cultures, since it would remove the objections noted previously to the use of modified strains.

This is the reasoning behind the Japanese Aquarenaissance 90 Project, an enormous research effort bringing together public and private research centers with the aim of developing membrane bioreactors within five years for treatment and recycling of urban and industrial waste water. A number of different possibilities will be examined – in particular, anaerobic treatments.

Although these projects may strike European or North American specialists as futuristic, Japanese competence in biotechnology and the sums invested (13 billion yen in five years) should give them cause to reflect, and this theme definitely merits closer examination.

More Intensive Biological Procedures

At the present time, membrane bioreactors for water treatment exist only on paper. However, the application of the most effective technologies, inspired by fermentation engineering, have begun to bear industrial fruit.

French specialists (Companie Générale des Eaux, Air Liquide, SGN, Lyonnaise des Eaux-Degrémont) have progressed in several years from extensive use of primitive fermenters using free cultures to reactors with cultures fixed on fine plastic or granular materials, and are now bringing out improved processes and equipment. These advances may allow reduction in size of installations and simplification of procedures for separating the liquids processed and the cleansing microorganisms, with success rates that would have seemed impossible just a few years ago. Construction has already begun on the first industrial-scale installations for high-level cleansing (tertiary treatments), removal of nitrogen (nitrification and denitrification), and anaerobic cleansing, and the impact of these new reactors is expected to grow. The basic research such novel techniques involve (in particular, research into phenomena of fixation in bacteria and the metabolic modifications they introduce) cannot fail to have an effect on our approach to the biological phenomena that they highlight.

Thus, throughout the world, researchers and technicians in the water treatment industry are reexamining basic concepts and knowledge in the light of the latest biotechnological discoveries. The impact of these discoveries is still slight, but the potential is immense, and in the next ten to twenty years, we can expect to see many changes in treatment plants, quality of water produced, and the state of our rivers.

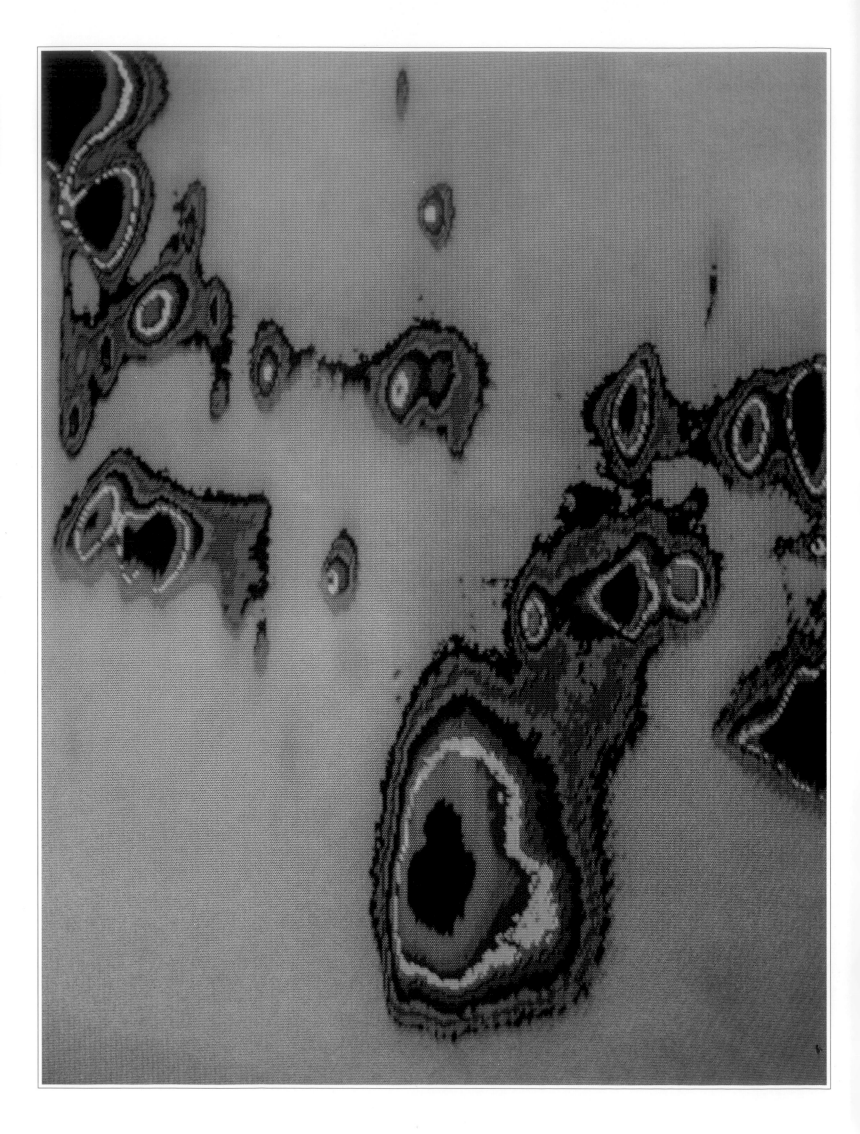

17

Biomaterials

Contemporary biotechnologists are challenging nature with their attempt to design and shape molecules with entirely new characteristics, molecules that are simultaneously more efficient, more stable, and better able to withstand the high heat and acidity imposed by industrial production.

Various laboratories in the United States (Genentech, Cetus, Genex, Amgen, Du Pont, SmithKline) and Britain (ICI, Celltech, Tate, Lyle, etc.), as well as in France (University of Strasbourg) and Japan (University of Kyoto), are concentrating on protein engineering. The initial idea is simple enough: instead of transferring a foreign gene to a microorganism to make it produce an unaccustomed substance, why not tinker with the gene before the transfer so as to produce (via the microorganism) a protein completely unknown in nature, as this may prove both a more precise and a less costly way of proceeding?

To do this requires "site specific mutagenesis" and an understanding of the extremely complex structure (above all three-dimensionally) of the molecule. Exactly what part of the molecule (in other words, what corresponding bases of the gene) should be changed? At present we work mainly with approximations, bombarding the crystallized protein with x rays from every angle and, with the aid of computer-assisted mathematical calculations, determining its structure so as to fix the atomic groups responsible for the protein's activities. Present methods of computer simulation make it possible to picture the protein in three dimensions, consider it from various angles, and observe what happens when two molecules meet (a procedure until now used by the pharmaceutical industry to observe the effect of certain substances on human metabolism). It is then up to the molecular biologist to synthesize the fragment of DNA thought to modify the active area. Genentech is continuing research of this kind on an enzyme capable of breaking down a cell wall (the lysozyme T4), Cetus on an interleukin 2 (to make it more resistant),

Amgen on an interferon capable of being used in tests, and Iwao Tabushi of the University of Kyoto on another enzyme used by bacteria to transport electrons (cytochrome 3). As Iwao Tabushi says, "It would be easy enough for us to make artificial cells that had no genes. One of our purposes in this would be to inject them with synthesized genes capable of producing this type of new protein, for this would above all make it possible to improve considerably our fermentation systems. Obvious medical applications can be seen on the horizon, but it is a matter first of all of knowing how this type of system interacts with a living system... of which we know nothing whatsoever right now". If, he continues, "we at the University of Kyoto are trying to fabricate effective artificial enzymes and implant them in artificial cells, the goal is primarily to reduce the size of bioreactors: for a similar activity we would be able to switch from cells 40 Angstroms (Å) in diameter[1] to those of about 5 Å."

But this research is apt to last some time yet, as most proteins are very large molecules indeed, with many active areas. If the wager is won, the study of nature will have paradoxically resulted in artificial creations that make it possible to use not only the L form of acids but also other forms that it does not use. Tabushi adds, "Nature is an excellent model to start with, to set off in other directions." Molecular biology has thus made it possible for chemistry to go from the stage of faithfully imitating the structures of certain compounds, so as to reproduce their natural functions, to that of copying the *concept* (with a totally different molecular structure). The limits to this form of imitation (biomiming?) obviously depend on the extent of the knowledge acquired by researchers. One of the present enigmas in this field is the nature of the interaction that occurs between human genes and proteins and how, under the circumstances, to proceed with the site specific to mutagenesis.

Before it became possible to invent new proteins, we already knew how to immobilize proteins

"Fingerprints"
A real file of protein "fingerprints" can be established by observing on a computer screen the changes in the composition of the proteins in human plasma (for example). It then becomes possible to develop health alarm systems. Similar systems to this – developed by Du Pont de Nemours – have made it possible to create sensors or biological substitutes for certain materials.

(1) 1 Angstrom (Å) $= 10^{-10}$ meters (m).

like enzymes, and thus to fabricate membranes with fixed enzymes to make biosensors capable of studying and measuring various processes – glucose content, etc. This work was conducted mainly at the Tokyo Institute of Technology by a team led by Isao Karube. It concentrated on certain basic problems: the development of enzymatic mini-electrodes and research on stabilizing artificial proteins integrated into a system subject to electrical fields. On the horizon can be seen a far more ambitious program still in the realm of science fiction: the biomicrochip and the biocomputer, in which molecules replace semiconductors. "We are already thinking," says Karube, "of small systems of microbiosensors, which we could implant in the body and enable each person to monitor his own physical condition and the chemical composition of his blood. This personal biosensor could be connected to an Information Network System, or INS, from which the signal emitted from the biosensor would be transmitted to a computer and from there to a doctor in a hospital who could make a diagnosis. It could also immediately test a medicine's effectiveness." The problems to be solved include the interaction that might occur in such a tiny "bug," the reliability of the signal, the design of the enzymatic membrane that could hold 20,000 molecules on an infinitely small area, and that of the organic membrane that would cover it to prevent the enzymes from being inhibited by factors in the blood and aqueous substances in the body.

Such sensors could also play an important part in the design of new robots capable of "touching, seeing, and hearing." Alongside this biorobot, this cyborg of the future, we might be able to invent new systems of molecular memories.

"I do not think that proteins are capable of doing such sophisticated things," says Tabushi. "Molecules are too sensitive to voltages and other forces such as cosmic rays, and the bit (information unit) could suddenly switch from 0 to 1 with disastrous consequences to the computer."

This opinion is not shared by James McAlear, who, along with John Wehrung, established the Gentronix Company in Maryland to "prepare computers of the sixth generation and the advent of three-dimensional integrated circuits."

It was in 1978 that McAlear, a biophysicist, and Wehrung, an electronics wizard, filed a theoretical patent. What was their idea? As chips they would use monoclonal antibodies capable of attaching themselves to the corresponding antigens under the effect of the electrical charge of the atoms, which would solve the problem of assembling the components. For the connections two solutions would be possible according to McAlear and Wehrung: infinitesimally small metal conductors could be made using highly specific enzymes to inscribe fine deposits of metal on and within the scaffolding of monoclonal antibodies; use might also be made of certain waves, postulated in a mathematical theorem, that are detected by proteins (these waves would modify the dimensions of the alpha helix). The so-called solitary waves, or solitons (originally proposed for biochips by Forrest L. Carter of the Naval Research Laboratory), could conduct current and relay information while avoiding the problems of the miniaturization of semiconductors;

in particular, as the wave is a thousand times slower than an electron – though it should be remembered that a biochip is a thousand times smaller than a traditional microchip – the problems caused by the dissipation of energy and the creation of heat can be avoided. As to the problems of the keyboard, supplied with electricity, and which must be connected to the biocomputer, only a laser whose semiconductor's wavelength is greater than the space between the components may be used; why not use photons, those luminous particles capable of creating and detecting solitons?

All of this research can only produce results (if they produce results) in the distant future, and this has tended to cool off investors. IBM has abandoned the work in this field carried out at its Yorktown Heights center since 1974 by Arieh Aviram and Philip Seiden, who were working on a system in which an organic molecule could exhibit two electronically stable states according to the distribution of the charge in the molecule (to obtain binary states 0 and 1). In fact, all of the systems mentioned above attempt to reproduce microchips using semiconductors by reducing the distance between the components; today a minimum distance of 1.5 microns (millionths of a meter) has been achieved – it will be impossible to go beyond 0.2 microns (below that point, the electronic system deteriorates). But we must also cope with the "bugs in bugs," that is to say, programming errors induced by the extreme sensitivity of organic molecules (compared with inorganic semiconductors). Furthermore, it seems highly unlikely that molecular systems will prove competetive with high-speed integration systems using semiconductors.

Another idea would be to fill an unexplored technological "niche" and consider systems of a totally new type, capable of solving some of the problems posed by artificial intelligence – pattern recognition, precise control of various elements in a robot, complex simultations, etc. This would involve exploring the possibilities of analog instead of numerical computers, as they are less general but better suited to a specific task, working on the basis of the extraordinarily varied three-dimensional configurations that enzymes are capable of assuming in space. This idea was proposed by Tim Poston, a mathematician working at the Crump Institute for Medical Engineering in Los Angeles. For does not an enzyme have an active area showing a degree of "memory" in that it recognizes the shape of a given feedstock? Are not its configuration and function modulated by certain variables allowing an entire range of analogical responses? An enzymatic system of this type could be included in a larger digital machine and operate sporadically under its direction. In the case of a robot, the binary computer would play a part somewhat similar to that of a brain, leaving the role of the sensory organs to molecular analogical systems.

Are such dreams to be permanently dismissed, or do they in fact foreshadow the techniques of the Third Millenium? No one is in a position to answer such a question today, not even the intrepid engineers who continue the good fight. But as William of Orange was reported to have said, "It is not necessary to hope to make a start, nor to be successful to continue."

James McAlear
The "father" of the biochip.

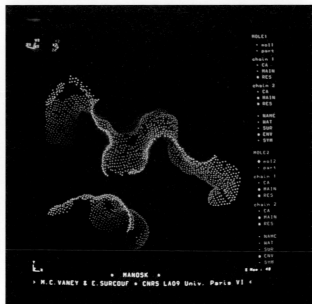

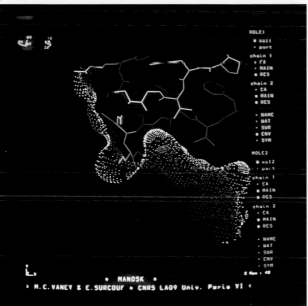

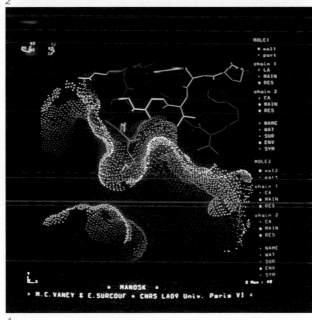

1

2

3

4

Thanks to the spectacular progress made in data processing, it is henceforth "easy" to model, simulate, and manipulate complex assemblies of molecules and macromolecules on a screen, to examine these entities in real time and in every direction, and to visualize their surfaces and the fields of force associated with them. In other words, we can apprehend an increasingly large number of the fabulous interlockings of the molecular "Mechano Set," the basis of all biochemical and biological phenomena. Laboratories everywhere in the world, like Suntory's (1), now customarily work with CAC (Computer Assisted Conception). Thus, a group at France's CNRS (E. Surcouf, M.-C. Vaney, I. Morize, J. Cherfils, and J.-P. Mornon) have developed MANOSK, France's first program for interactive macromolecular engineering (with an Evans and Sutherland screen), with the financial aid of Sanofi-Clin Midy. They then studied the molecule thought to carry the sex hormone progesterone.

The existence on this molecule of an internal cavity adapted to receive progesterone (lower left of photograph 2) was known, but they recently discovered the presence of a heretofore unknown surface site. Its shape showed apparent analogies to that of a well-known site, that of an enzyme. Consequently, a comparision was made (4) between the site of the uteroglobin (2) with that of the trypsin (3) (in yellow); the atomic skeleton (in mauve on the same photograph) is that of a part of the substance inhibiting the enzyme; a lysine occupies the pocket of the trypsin. The surface site of the uteroglobin could thus receive an elongated residue (such as a lysine). Work of this kind could, in the same way, make it possible to study the sensors of steroids – of which we still know virtually nothing – starting from structurally well-known models and using various primary sequences. The idea is thus to make a "very close" comparison between a relatively unkown molecule and a per-

fectly known molecule and work by analogy. In other cases (as on the photograph on the back cover), it is a matter of observing the impact of various new substances on the organism. Programs such as MANOSK will henceforth make it possible to extend our knowledge of the specific modes of interaction between the biochemical partners, and are particularly suitable for molecular biology, for the conception of new medicines and new macromolecules.

Biopolymers

To replace chemical polymers by substances secreted by microbes is no longer a dream but a reality. Xanthane, which, as we have already seen, is being used in the food-processing industry, could, in quite another field, compete with the polyacrylamides presently used to recover oil from solidified deposits and thin out drilling slurries. The first biopolymer to be discovered, xanthane is now being produced on an industrial scale.

Emulsan, its younger brother by twenty years, has been produced by the Petroferm Company of Florida. It is secreted by the bacterium *Acinobacter calcoaceticus*. In the early 1970s, two American researchers working at the University of Tel Aviv, David Gutnik and Eugene Rosenberg, were trying to find a bacterium capable of biologically breaking down the petroleum on a polluted beach, isolating it, cultivating it, and using it directly in the dirty holds of tankers. But the bacterium died. They then decided to isolate the substances that gave *Acinobacter* its effectiveness, and it was thus that they isolated emulsan. In this polysaccharide, the oily compounds (fats) hang from a sugared spinal column. As these lipids have an affinity for oils (such as petroleum), and as the saccharides are soluble in water, in an oil-water mixture emulsan acts as an emulsifier and stabilizer, and lowers the viscosity of the mixture to one-thousandth that of petroleum. It is presently being tested to wash tankers and the holds of barges and to lower the viscosity of petroleum being carried in a pipeline. It is also being considered for the recovery of oil deposits solidified in the rock, for in contrast to xanthane, which acts as a viscous piston, compressing the oil and forcing its way up from the depths, emulsan liquefies it so it can be pumped. We might mention that Elf has perfected a product to break down hydrocarbons at sea called Inipol. This is not a biopolymer but a nutrient based on nitrogen and phosphorus that acts a little like Popeye's spinach, stimulating the growth of the ambient flora. This "magic potion" makes it possible to maintain these soluble nutrients in the interface between the oil and the water, for otherwise they would become diffused in the column of water and be lost.

Biopolymers may also prove of value to the genetic-engineering industry in isolating substances from the "fermented soup." Starting in the mid-1960s, it became possible to obtain a polysaccharide, agarose, from agar gum by drying and washing a red alga. The primary interest of agarose is its lack of marked positive or negative charge (as the charge could interfere in a separation process). And agarose entraps certain specific molecules, allowing the others to pass by. According to tests made by Damon Biotechnology, this biopolymer can also be used as a sort of "parcel post" to deliver quantities of medicine within the body slowly (microencapsulation). Another biopolymer, dextrane, is derived from blood plasma and used to replace complete plasma (Rhône-Poulenc).

Biopolymers may also one day replace plastics in a broad spectrum of applications ranging from surgical sutures to soda bottles. This is another of ICI's wagers, as it founded Malborough Biopolymers and invested $200 million to develop Biopol (biopolymer polydroxybutyrate, or PHB), which has the advantage of being biodegradable in the body or in the environment. PHB is obtained from a bacterium that feeds on starches and sugars and that is deprived of nitrogen. In contrast to emulsan and other gums, it is not secreted but stored by the microbe. One of the essential problems thus lies in recovering the product. Another is its high cost: the least expensive glucoses on which the bacterium can be cultivated still cost three times as much as naphtha (the basic raw material of the plastics industry). Even if lactoserum (whey), molasses, or other by-products are used, PHB still remains more expensive that its closest chemical rival, polypropylene. Its greatest advantage is, obviosuly enough, that it is biodegradable, and can thus be used as a coating for fertilizers, which can be released into the ground gradually. Researchers are currently attempting an overall approach that would make it possible to extract lipids, proteins, and nucleic acids from the bacterium, along with the biopolymer, and thus lower costs. Where plastics are concerned, we might also mention that Saul Neidleman at Cetus has combined three enzymes to create a synthesis of alcene oxide, which may prove more profitable than the chemical synthesis in the production of such derivatives as gluconic acid (added to detergents) and fructose.

Among the other potentially interesting biomaterials, we must also mention a protein that is not a polysaccharide, but the principal component in connective tissue: collagen. This may be used in dermatology, surgery, or for solid supports. The Collagen Corporation, founded in the United States in 1975 with Monsanto holding 30% of its stock, is already selling a collagen-based substance for use in plastic surgery.

But we must also pay our respects to the pioneer

Ananda Chakrabarty
The "father" of oil-eating bacteria.

in the field of the biological depollution of petroleum products, the Bengali scientist Ananda Chakrabarty, who, while at General Electric, was the target of the first salvos in the battle against "filing a patent on life." He had, in fact, discovered that the genes coding for a substance that gave certain bacteria the power to break down hydrocarbons were to be found in their plasmids. Into a bacterium that loved oils, he then transplanted the plasmids of three others that shared the same tastes to create a particularly greedy chimera. The United States Supreme Court was willing to sanction the patent on the process, but not on the new organism thus created. Even today, Chakrabarty still sees red: "Pasteur certainly patented life with his yeast! And my bacterium is the pure product of human invention." The suit was only won on appeal in 1980. But by then Chakrabarty had left General Electric, which shrank the investments required to develop the bacterial system. At the University of Illinois, using a microbe, he is now attacking Agent Orange, the devastating herbicide used in Vietnam by which many veterans claim to have been poisoned. Furthermore, in certain parts of the United States, children have been stillborn after the area had been sprayed with one of the two components of this herbicide. But he is also continuing his research on greedy chimera, and in 1980 founded Petrogen to develop a biopolymer that, like emulsan, would be capable of drawing oil out of shale or asphalt sands. "A considerable market," he muses, "if you think that drillers have to leave behind 70% of the oil because they can't pump it." These voracious microbes might also be used against the black tide from oil spills, and unlike chemical agents, they are biodegradable. But two major obstacles now stand in their way: the ecologists, who refuse to hear a word about monstrous microbes being let loose in nature, and above all the necessity of employing in symbiosis an entire panoply of supermicrobes specialized to ingest specific poisons.

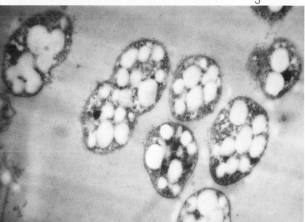

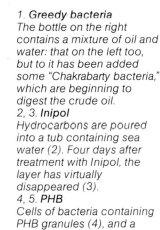

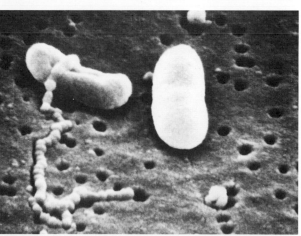

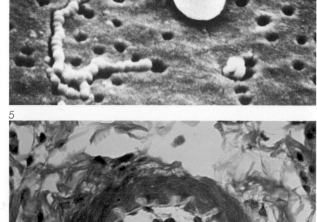

1. Greedy bacteria
The bottle on the right contains a mixture of oil and water: that on the left too, but to it has been added some "Chakrabarty bacteria," which are beginning to digest the crude oil.

2, 3. Inipol
Hydrocarbons are poured into a tub containing sea water (2). Four days after treatment with Inipol, the layer has virtually disappeared (3).

4, 5. PHB
Cells of bacteria containing PHB granules (4), and a filament of PHB that is starting to escape (5).

6. Collagen
On this section of a blood vessel, the blood vessel (red) may be distinguished from the collagen (blue). The collagen becomes integrated with the body and forms a sort of bridge across which nerves and blood vessels develop – a property that could be interesting in, for instance, reforming a breast after a mastectomy. The Collagen Company, founded in 1975, uses this particularity of conjunctive tissues and of the "material" constituing the skin, bones, tendons, and cartilages. It will henceforth be possible, among other things, to repair broken or splintered bones without resorting to surgery.

169

18

From the Laboratory to the Production "Pot"

This fantastic exploration of microorganisms and the new substances they seem able to secrete, this concocting and creating of heretofore unheard of molecules, has so captured men's imaginations that in many instances they have ignored or forgotten that essential ingredient known as technique. It has become of crucial importance today to develop the engineering technology required to mass-produce such products, and businessmen now put greater stress on the word "engineering" than they do on "genetic" or "enzymatic."

If an interesting substance such as a hormone, lymphokine, amino acid, etc., is obtained in a flask containing a few hundred milliliters, and if one wants to proceed to the next stage and produce it in a 100,000-liter fermenter, a change in scale of some proportion is obviously necessary – 500,000 times as much energy must be provided, and this cannot be done by simple replacing a low-horsepower engine with a giant motor powered by a nuclear power station; the heat produced would be such as to volatilize any microorganism and make it disappear in a cloud of steam. Identical problems arise when it is a matter of cooling such an enormous quantity of "soup," or of sterilizing or filtering it, or carrying out drying, distilling, purifying, or separating procedures. Entirely new systems have had to be invented to feed the bacteria, yeast, or mold with oxygen without at the same time cooking it. And all of these procedures require the strictest control of the environment (acidity, heat, oxygen, etc.) and increasingly precise microbiological screening.

Electronics plays an essential part in all of this: without the electron microscope, neither molecular biology nor genetic engineering would be what it is today. Without data processing and automation, regulation and control would be impossible. Without the graphic representation by computers of three-dimensional molecules and their interaction, there would be no talk of biosimulation, and investigation would be greatly retarded. Without programs, researchers would not have available those banks of living matter (like sperm or embryo banks) consisting of gene, clone, or vector banks or banks of monoclonal antibodies. In 1981, four Stanford University researchers, artificial intelligence specialists Ed Feigenbaum and Peter Friedland and biology and biochemical specialists Brutled and Kades, founded IntelliGenetics, which concentrates on sales of expert systems for molecular biologists. Says Paul Armstrong of IntelliGenetics, "Our systems have to be very flexible; not everyone uses the same vectors, but the more specific research becomes, the better able we are to help define the problems." It is the computer, for instance, that helps to cut apart and glue together the puzzle of life, that is to say, the thousands of bases of a gene. And it is the computer that helps to sort out the information provided by such measuring tools as the mass spectrometers. "We have the analytical tools," adds Armstrong, "now we must place them at the disposal of a real system of knowledge."

That is just the rub, it appears. Kevin Ulmer, Director of Research at Genex, even insists on this: "The possibility of making theoretical predictions is quite limited right now. The structure of a protein is far more complicated than that of the most sophisticated polymer. Theoreticians are unable to find one valid experiment to systematize the theory of the shrinkage in space and of the sequence of amino acids in three dimensions. We do not even have a computer program capable of pruning away the dead wood, of eliminating what doesn't work. The only thing we can do is change one or two amino acids and see what happens. It's all a matter of intuition. All this gadgetry is fine... if it weren't for the huge vacuum of the theoretical approach."

Fermenters and bioreactors are followed by a battery of machines for purifying, separating, analyzing, and sorting – the "downstream" processes that extract the "gold" of biotechnology from the souplike contents of the fermenter. The oldest of these processes is the centrifuge, offshoot of the

Liquid chromatography
A star apparatus of the biotechnologies, sold by Pharmacia, a company famed for its separation and purification equipment.

171

1. Centrifugation
To produce antibodies from lymphocyte Bs, for instance, as is being done here at Genetic Systems, one must start by separating the lymphocyte Bs of the blood by centrifugation.

2, 6. Chromatography
Liquid chromatography (2) and gas chromatography (6) are essential instruments in separation.

3, 5. Electrophoresis
Another separation process: by the difference in electrical charge, size, or shape.

4. Mass spectrometry
This apparatus makes it possible to analyze the substances liberated by a microorganism in a fermenter. In this way, an analysis of the quantity of carbon dioxide produced can be a valuable indication in providing the microbe with food during the course of fermentation.

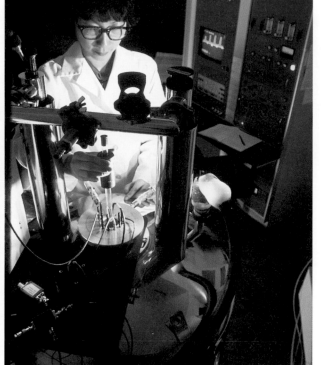

1. Ultrafiltration
In biotechnology, this
apparatus permits
continuous purification.
2. NMR
Spectroscopy by nuclear
magnetic resonance is a
powerfull analytical tool that
makes it possible to
determine the structure of
complex molecules by
observing their behavior
when subjected to very-
high-frequency vibrations
within a very strong
magnetic field.
3. Laser for cell sorting
A laser system for analyzing
and sorting biological cells
in flux designed by the CEA
(Commissariat à l'Energie
Atomique). Flux cytometry
makes possible cell-by-cell
analysis, sorting according
to several parameters
simultaneously, the
subsequent reutilization of
the cells, the selection of
rare cells (up to one cell out
of million), and great speed
in analysis and sorting.
4. Cytofluorometer
Detection by fluorescence
with the aid of a
cytofluorometer makes it
possible to study a wide
range of cellular properties.
Here, a Merck researcher is
separating and identifying
lymphocytes.

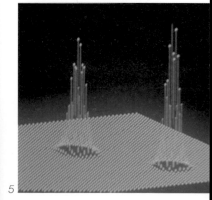

5. Measurements
The cytofluorometer also
makes it possible to make
precise quantitative
measurements of the
reactions of antibodies on
an individual cell at the rate
of 3,000 cells per second.

173

old barrel churns used by farmers. In 1878, G. de Laval, who founded Alfa-Laval, invented the continuous evacuation centrifuge to separate and concentrate cream, purify and homogenize milk, and separate such by-products as curds and whey. Apart from their value to the oil and nuclear industries, centrifuges are used in biotechnology for antibiotics, refining insulin, clarifying citric and glutamic acids, separating blood proteins from plasma, etc.

Ultracentrifugation, for its part, brings into play forces that can be hundreds or even thousands of times as powerful as gravity and applies them to cells, tissues, and molecules. This technique was developed by the Swedish scientist T. Svedberg between 1920 and 1940. Marketed immediately after the war and fitted with optical systems, apparatus of this sort makes it possible to follow the process of sedimentation and to measure molecular masses, densities, and the speed of sedimentation. Used today by all research laboratories as well as in the pharmaceutical industry, it is employed at various stages in the cloning of a gene in a bacterium (purification of the mRNA, the cDNA, and the plasmid). The vaccine against hepatitis B produced by genetic engineering is purified by ultracentrifugation. It is above all used to process large quantities of material – larger than those that can be handled by other simpler, less costly, more sensitive, and finer resolving processes, such as electrophoresis on gels or HPLC (High Performance Liquid Chromatography), both of which were invented during the 1970s.

Proteins are separated according to "isoelectric point" (focalization) and then according to molecular weight (migration) by electrophoresis. Subunits of proteins are determined by staining at the end of the process, and automation of the procedure makes it possible to save time and facilitate the analysis. But the favorite tool or researchers is chromatography, without which it would have been difficult indeed to obtain interferon, the growth hormone, or albumin. To produce albumin, the Institut Mérieux, for example, developed "columns" of incompressible balls with a positive electrical charge. The albumin (with its negative charge) clings to this while the impurities and residual hemoglobin pass through it. During a second stage (in a column of hydrophobic balls) the pigments are eliminated, and then the lipoproteins are fixed in a third column. Still another process developed by Pharmacia introduces a filtration on gel stage at the end so as to obtain a manometric albumin. The heart of the chromatograph remains the column and its filling, which depends upon the goal in view: filtration on gel for the separation of molecules according to size, ion exchange for separation according to charge, affinity for intermolecular reactions, and division chromatography for adsorption or polarity. The basic principle is that of blotting paper – when it absorbs an ink spot, a halo forms: the solvent rises higher on the blotting paper than the coloring agent. In this way, substances of different molecular weights or chemical affinities can be separated. Using HPLC, it is possible to survey the formation of substances in the process of formation as this occurs.

As the result of a seminar attended by biologists, chemists, physicists, and doctors, a new approach to cellular sorting was attempted. The end result of this joint effort was the cytofluorometer, a cell selecting and analyzing device, which studies several parameters simultaneously.

To identify specific cells and isolate them, why not make use of a known property? Certain structures on the surface of a cell bond easily with fluorescent substances, so why wouldn't it be possible to detect the marked cells with a fluorescent microscope? It was by using this principle that Leonard Hertzenberger's team at Stanford University designed the prototype of the Fluorescence Activated Cell Sorter (FACS), subsequently developed by William Bonner and Russel Hulett.

Three American companies (Becton-Dickinson, Ortho, and Coulter) are selling cytofluorometers of this type, and France's Atomic Energy Commission is developing a new version.

How does a cytofluorometer work? A cellular flux (saline solution containing millions of cells per milliliter) is projected under pressure into a micropipe with an orifice only a few microns in diameter. This injector is then subjected to high-frequency vibrations, and with each shake a droplet containing a cell falls into the field of one or more laser beams, whose rays excite the fluorescence markers fixed by the cells. This fractioning of the droplets was developed by the French physicist Félix Savart in 1833 and followed in particular by C. Boys, who had published a photograph of a stream of water modulated acoustically: a musical note, coupled to the stream, made its diameter vary regularly, and these variations – amplified by the forces of surface tension – in the end divided it into droplets. By deflecting the droplets by means of a an electrical field, Boys had shown that they could take different paths.

But to return to what happens within the apparatus. Detectors receive the light, which is refracted (according to the form and position of the particle or the internal structure of the cell), as well as the fluorescence that has been emitted. They convert this into electrical impulses (sorting impulses), which make it possible to give the stream an electrical charge at the moment at which the drop is formed containing the very cell that caused the electrical signals. At the very moment that this charged droplet becomes separated, it is deflected in a magnetic field and falls into a recipient. Thanks to data-processing techniques, graphic histograms with their parameters are obtained at the same time, and in this way, in a single second, 10,000 cells may be analyzed and 2,000 of them sorted!

These cell analyzers and sorters are now at work detecting and studying cancers, the results of chemical treatment on cells, the isolation of abnormal chromosomes, the prenatal detection of diseases, the detection of antigens through marked antigens, and the sorting of dead and living cells. This permits accurate and rapid diagnosis and medical follow-up in such diseases as leukemia, where examination by means of a cytofluorometer requires but a blood sample (instead of the usual sampling of bone marrow). Some of the consequences of the use of this device have been unexpected. In 1983, for instance, Edgard Engelman's team at Stanford University tested the blood of 100 blood donors per day to determine those who had been infected with AIDS. One of the characteristics of AIDS is a change that occurs in white blood corpuscles, which can be detected with a sorter and marked antibodies. The astounding fact that "blood can be dangerous" made the front page of the *Wall Street Journal.* "The Blood Banks," says Edelman, "fearing to see their validity questioned, launched a massive counter-offensive. Yet it has become clear that from now on the techniques of molecular biology will be making their mark on many traditional aspects of medicine. The analytical laboratories will be the first to feel the shock as the result of personal tests using mono-clonal antibodies. In this respect, biotechnology will lead to a real revolution in the traditional struc-ture of medicine."

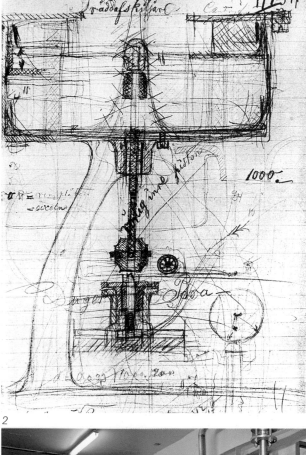

1. Gustave de Laval
Inventor of the first cream separator.
2. Separation
Drawing of the first milk skimmer, ancestor of the separator, in 1877, the invention that lay behind the Alfa-Laval Company.

3. Nostalgia
David Fishlock's still in the English coutryside.
4. Fire separator
Machines traditionally employed for the large-scale recovery of microorganisms such as baker's yeasts and bacteria.

Micro-biological Engineering

by Teruhiko Beppu, Research Director of the
Department of Fermentation and Microbiology of the
Faculty of Agriculture at the University of Tokyo.

*If G. Surbled has been able to write that the
heirs of Pasteur were Japanese, this is
essentially because of their mastery of
strains, the decisive factor in biotechnology.
T. Beppu is a world-ranking expert in this
field. A University Professor, he works in
close collaboration with the largest (competing) companies in Japan. Selection and
screening are undoubtedly the two decisive
weapons in mastering the biotechnologies.*

The recent technological innovations in
basic biology that began with DNA recombination have aroused great expectations
for industrial applications of biological
capacities. Microorganisms have been
widely and successfully used for a very
long time in industrial fermentation procedures. The ease with which they may be
grown, the wide variety of different types
available, and the technological experience
acquired through using them are trump
cards that should ensure that they play a
crucial role as the "living tools" of biotechnology. Systematic identification or screening, which results in the discovery of
microorganisms with novel functions in the
natural milieu, and recombinant DNA techniques, which consist of rearranging genetic material, are essential research methods
opening up new applications for microorganisms.

Selecting and Screening Microorganisms

For thousands of years, man has been harnessing a limited number of microorganisms – yeasts, molds, and lactic ferments –
for industrial purposes. Modern man uses
a wide variety of microorganisms to produce antibiotics, amino acids, and primary
and secondary metabolites – in other
words, to produce a wide range of biological products used in the pharmaceutical,
food-processing, and chemical industries.
The range of microorganisms thus har-
nessed was increased by the discovery of
microorganisms present in the natural environment and characterized by optimal
response to particular applications. Screening, which consists of identifying and systematically selecting microorganisms from
among thousands of similar organisms, and
then studying their behavior so as to identify the particular microorganism most
adapted to the job in hand, may seem at
first sight to be a complicated and somewhat illogical procedure, but a number of
concrete examples prove that it generates
some very novel results. For example, in
1955, it would have seemed completely
illogical, given the state of biological knowledge at the time, to launch a program to
study microorganisms to produce glutamic
acid; today, it is an industry in full expansion. Nevertheless, research into new and
effective ways of isolating glutamic acid
produced by microorganisms, as well as
better screening techniques, has allowed
Japanese researchers to come up with a
fermentation technique with completely
new characteristics: by bringing into play
the permeability of cell membranes, glutamic acid is made to accumulate in the
cell by using cornybacteria. This one
example is a perfect illustration of the originality of this approach.

To carry out systematic identification of
organisms successfully, researchers must
be able to isolate and define their objectives, i.e., the search for new functions.
Here the quality of screening stretches the
boundaries of current knowledge in the
field.

Nevertheless, we should point out that the
wide range of microbial species is crucial to
successful screening. We know that the
quantity of air used by microorganisms living on a single hectare of fertile land corresponds to the quantity of air required by
many thousands of human beings. As this
implies, microorganisms are an enormous
mass with the ability to adapt to all sorts of
terrestrial microenvironments, including
earth, sea, thermal waters, and animal
intestines; this adaptability makes them the
most formidable varied species in the world
of living beings. They are able to grow and
develop in environments characterized by
extreme conditions (temperatures exceeding 70°-80°C, alkalinity greater than pH 10)
in which other organisms would not be able
to survive (figure 1). Many researchers are
now examining or screening bacteria to
obtain new, useful enzymes resistant to
very alkaline environments or very high
temperatures.

Systematic identification of microorganisms, a field with vast potential for new
discoveries, enables researchers to isolate
new genes and develop novel production
techniques.

Applications of DNA Recombination

Whereas the process of screening microorganisms depends to a large extent on
experience and accumulation of data by
trial and error, recombinant DNA techniques
were developed from the results of basic
research in molecular biology. Comparative
methodology is used to obtain a predetermined objective.

This method consists of inserting human or
animal genes in host microorganisms and
using their extremely fast rate of reproduction to produce proteins on a massive
scale. This method has the advantage of
giving practical results in a very short time.
In concrete terms, by taking a host-vector
system using bacteria from the large intestine, researchers have produced human
growth hormones or interferon in quantities
of up to 20-30% of the weight of the host
bacteria. To produce vaccines for the hepatitis B virus, since bacteria from the large
intestine were not suitable hosts, researchers now use yeasts. Massive formation of
heteroproteins in microbial cells can cause
formation of masses (figure 2); some
researchers have also encountered difficulties in extracting proteins without inhibiting
their activity, even after they have been
made soluble. It is therefore crucial that
researchers come up with a technique
applicable upstream of this process, which
would discover technical means compatible with basic theory concerning the structure of space in proteins, thereby enabling
us to deal with problems of extraction in
industrial production.

It is obvious that DNA recombination will
play an essential role in the development of
microbial cultures for industrial fermentation. We know that by making several

copies of plasmids to use as vectors, certain genes can be cloned, increasing the number of genetic copies in host cells and thereby multiplying the quantity of proteins (enzymes) for which these genes code. The process of biosynthesis of amino acids is thus enhanced and productivity increased. Researchers have also discovered specific regulator genes in actinomycetes, which are instrumental in producing many antibiotics. Their work has shown that if these genes can be inserted into other actinomycetes, they can be used to produce new and hitherto unknown antibiotics on a massive scale. By applying recombinant DNA techniques to bacteria selected by screening, researchers have not only succeeeded in obtaining large increases in production, but have even been able to discover new substances (figure 3).

All this leads us to believe that we shall very soon see significant development in the technological applications of microorganisms: reorganization of genetic information and association of screening techniques (which allow novel discoveries in terms of genetic information) with DNA recombination (which allows scientists extraordinary freedom of action).

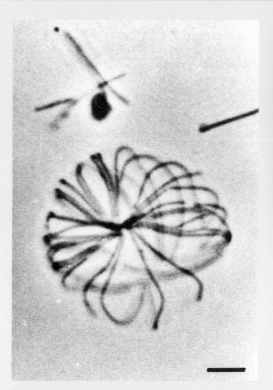

Fig. 1
Totally new anaerobic and thermophilic bacteria discovered in hot springs. These bacteria in the shape of a long fiber assemble to form a spherical structure.

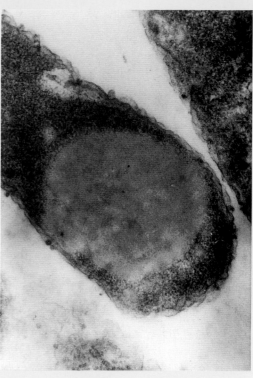

Fig 2
A large inclusive body of chymosine produced by the microbian cells of the large intestine (in cattle).

Fig. 3
By inoculating regulator genes taken from among the actinomycetes streptomyces coelicolor *into actinomyces* S. lividans, *we obtain the production of massive quantities of pigmented antibiotics that would not have been produced originally.*

Right: S. lividans.

Left: S. lividans *inoculated with regulator gene.*

An Obligatory Procedure

by Jacques Chardon, Director of Development and Industrial Fermentation, and Dominique Mison, Doctor-Engineer INSA, Production Department, Roussel-Uclaf.

Roussel-Uclaf, established in 1920 by Gaston Roussel, is one of the world leaders in corticosteroids and vitamin B_{12}. First specializing in fine chemistry, this company subsequently developed pharmaceutical products. Hoechst became its principal shareholder in 1974. Roussel collaborates with Transgène and INSERM in research on interferon gamma and with INSERM on human calcitonin, and in 1984 it signed an agreement with Sanofi for the production of interleukin 2.

For all of the biotechnologies, whether they be traditional or new, the obligatory procedure remains fermentation and bioconservation. J. Chardon and D. Mison describe this engineering of life.

The choice of a chemical compound to be produced by means of a biological process is primarily an economic gamble. In organic chemistry, one's first thought is to use chemistry whose potential has no longer to be proved. However, in some cases, when a synthesis is too complex, when a particular quality is looked for, or when the molecules are not stable, biological synthesis offers undeniable advantages.

About 192 fine chemical products, manufactured by 145 companies (located mostly in the United States, Europe, and Japan), are produced only by industrial fermentation,[1] while there are thousands of molecules that can be synthesized by microorganisms. This demonstrates the strength of economic selectivity.

Antibiotics, with over 100 original molecules, represent the largest group of pharmaceutical compounds produced by fermentation. But it is the food industry that provides the greatest tonnage and turnover for production by fermentation. This is not surprising, as fermentation has been used in food production since ancient times. This tendency can also be explained by the destination of produced foodstuffs, for which the natural qualities such as taste and aspect have to be better and better preserved.

In order to be competitive, the fermentation industry depends on three essential factors: the strain, the material (including the fermenter), and the process.

The Strain
The strain is the fundamental element. It is an absolute necessity for industry continually to improve the performance of their producing microorganisms; otherwise they would rapidly become uneconomic, as competition is very strong between companies. Many processes have to be abandoned because this principle has been forgotten, and, if it is not too late, it is very often expensive to overcome this handicap.

Another concern for production managers is the maintenance of sterility for their cultures, as the majority of fermentations are realized with very pure cell cultures. The effect of contamination is, of course, different according to the manufacture. As a consequence of lack of sterility, there can be, in some cases, a fall of 10-30% in production yield within one year, which is considerable. In serious instances of microphagial invasions, the whole production unit may be stopped for several months. Thus, it is essential that the material, and first of all the fermenters, be built so as to guarantee the purity of cell cultures.

The Fermenter
The fermenter is the second factor of great importance. Its capacity, its mechanical and energy performance, and its degree of

(1) D. Perlman, *ASM News* (1978).

strain preservation
and preparation laboratory

Fig. 1

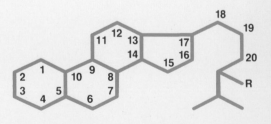

Fig. 1
A chemical reagent that has an affinity for the position 1 has a similar affinity for the positions 2, 6, 7, 11, 12, 15, 16. The enzyme can selectively react on only one position; this is the "regioselectivity effect."
The sites concerned by an enzymatic activity are
● *dehydrogenation on positions 1 and 2,*
● *hydroxylation on 11 and 16,*
● *cleavage of the molecule between the carbons 17 and 20,*
● *oxidation of alcohol in ketone on 3.*

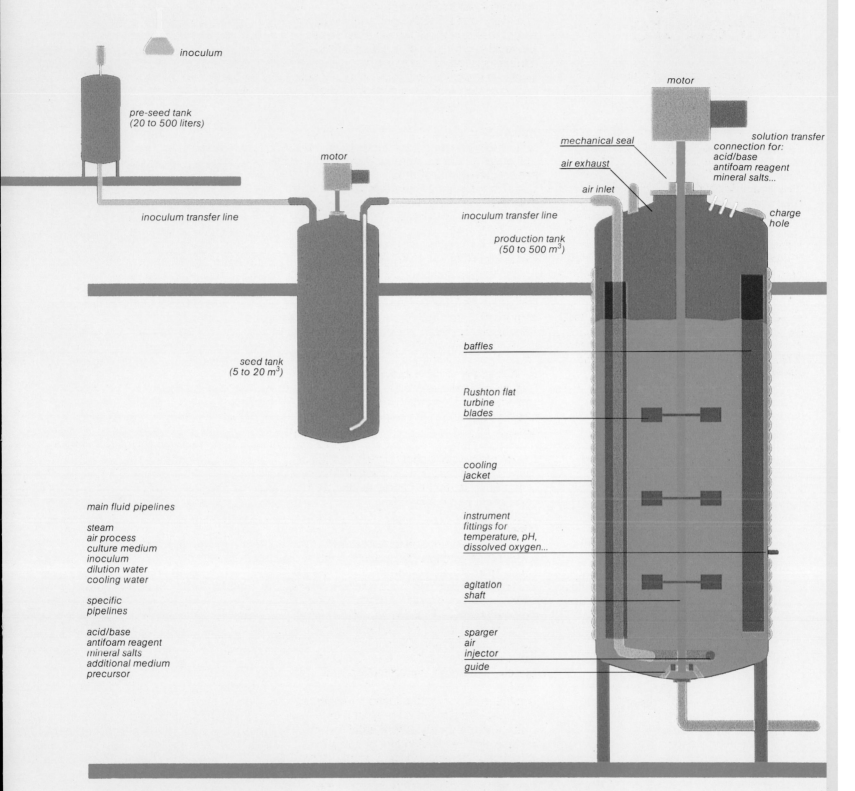

inoculum

pre-seed tank
(20 to 500 liters)

motor

inoculum transfer line

seed tank
(5 to 20 m³)

motor

mechanical seal

air exhaust

air inlet

inoculum transfer line

production tank
(50 to 500 m³)

solution transfer
connection for:
acid/base
antifoam reagent
mineral salts...

charge
hole

baffles

Rushton flat
turbine
blades

cooling
jacket

instrument
fittings for
temperature, pH,
dissolved oxygen...

agitation
shaft

sparger
air
injector
guide

main fluid pipelines

steam
air process
culture medium
inoculum
dilution water
cooling water

specific
pipelines

acid/base
antifoam reagent
mineral salts
additional medium
precursor

Starting from the laboratory inoculum, one or two stages of culture (two on the figure) are generally necessary to obtain the quantity and the quality of microorganisms with which the producer will seed. Insufficiently seeded, the lag phase will be too long. Overseeded, the fermenter risks not being able to follow the requirements of the culture in terms of oxygen transfer. The culture lengths are very variable from one fermentation to another: 24 to 72 hours for the development stage. For the metabolite production, for which one is trying to prolong to the maximum the biosynthesis, this prolongation is often obtained by adding nutritive elements to the medium (mineral salts, carbon and nitrogen sources). A precursor addition from which the metabolite biosynthesis occurs is often necessary. It is, for instance, the case of phenyl acetic acid for penicillin biosynthesis.

Most fermentations are made in aerated and stirred reactors. The design of the fermenter must take into consideration the need of the cultures (oxygen transfer and various nutriment feed) and the cell specificity: heat production, sensibility to shearing, and sensitivity to contamination. Agitation constitutes one of the major elements of this kind of fermenter. The number and the size of the turbine varies according to the volume of the fermenter and the power that one wishes to generate, generally 1-3 kilowatts per cubic meter. They are often flate blade disk turbine types. The drive shaft is maintained by a mechanical seal. The shell and sometimes the bottom of the fermenter is equipped with a cooling system by spiral heat exchanger. In the case of a particularly exothermic culture, it may be necessary to have additional exchangers in the fermenter. The fermenter must carry, finally, several holes permitting the connection to the piping as well as the connection of the probe, which permits automatic control of the fermentation.

automatization determine to a large extent the degree of expression of the power of the strain. The quality of a fermenter can be estimated according to mass transfer, homogeneity, heat exchange, and sterility. To function, the fermenter needs an important industrial complex. Steam, electricity, sterile compressed air, and water production alone can produce installations of a greater size than the fermentation workshop itself.

Yearly Production Costs in a Fermenter Process

Quantities	Percentage according to total costs of the product
Water (4.6 million m^3)	3-5%
Steam (42,000 tons)	11-13%
Air (110 million m^3)	5-7%
Electricity (15,700 megawatts)	5-7%
Total	24-32%

Size: 1,000 m^3.

Energy represents a large proportion of production costs. It is evident that productivity improvement is a major factor in remaining competitive on an international scale. The product obtained must then be extracted and purified, using chemical engineering techniques. Laboratories are also necessary for strain manipulation and analysis of the finished product. Finally, storage facilities must be available for raw materials and finished products. The effluent must be treated correctly. Social and administrative installations are also needed. One appreciates that all these facilities and the space they occupy represent very heavy investments, which must be made profitable.

Such complex installations cannot follow immediately upon a technical evolution. Before starting any new production, one must evaluate very carefully the risks involved.

The Fermentation Process
The fermentation process determines, on a scientific and experimental basis, the physical and chemical values conditioning the environment of the microorganism. Any fermentation process must include at least the following details:
• operational volume,
• composition of the culture medium,
• details on the quality of raw material and water supply,
• temperature and duration of sterilization,
• quality and quantity of the inoculum,
• pH value (acidity-alkalinity), temperature, aeration, rotation speed, during the different phases of growth and production,
• the nature and quantity of nutritive elements,
• minerals and other substances to be eventually incorporated during the process,
• duration of the production period.

It may be necessary to indicate difficulties that might occur (e.g., foaming) and the means to overcome them.

The evolution of processes has followed scientific and technical progress. During the 1970s, a great step was taken with the availability of sterilizable probes that became more and more reliable for measuring pH and dissolved oxygen.

This made it possible to use more sophisticated methods, permitting the precise regulation of these parameters and the control of respiratory phenomena of the cultures. It was during this period that the theory of mass transfer was developed – especially that of oxygen.

The availability of fixed enzyme probes has also helped to control the concentration of essential substrates of the product, by reducing the time necessary for analysis. This particularly concerns the observation of glucose consumption, by using glucose-oxidase electrodes in many methods where this substrate serves as a catabolic repressor.

The simultaneous adaptation of data processing and the introduction of microcomputers resulted in industrial automation, in turn helping to reduce considerably the costs of manpower.

The automation of operations as delicate as reactor sterilization or culture pilotage, on the basis of mathematical models, is now possible. The investigation of the procedure is generally carried out in the laboratory in small units. Specialized companies sell pilot material, which has now become very accurate in integrating technological progress long before the industrial stage.

Culture Media and Materials
The simplified chemical equation for a culture of microorganisms is written

carbon and energy source + oxygen + nitrogen source + growth factors → cells + product + carbon dioxide + heat + water.

The first part of this equation represents the culture medium, the second the products obtained by fermentation.

If at the stage of laboratory research there is no problem of supply or quality for the making of a culture medium, this may not be the same on an industrial scale. The standardized materials of classical microbiology are replaced by less expensive raw materials, coming very often from the food industry. This is the case, for example, of corn steep liquor (a by-product of corn in the manufacture of starch). This product advantageously replaces yeast extract, which is ten to twenty times more expensive and much employed in microbiology, being a good growth factor source.

Corn and soy meals are also a good source

of nitrogen. Carbon is often provided by by-products of sugar or milk (molasses, whey). More complex products can also be sources of carbon: native starch, various hydrolysates, sucrose, glucose. Other carbon sources widely used are fats, such as corn or soy oil or animal fats.

Inorganic nitrogen sources, such as ammonium sulfate, ammonium acetate, and urea, must also be added along with mineral supplies necessary for the cell metabolism (phosphorus, magnesium, oligo elements). Finally, there must be a system to maintain the pH (basis, different acids) and adjuvants (antifoam agents) within the limits of the process. Frequently, antifoam agents are necessary to maintain culture material at the most economic volume.

Materials represent a very important proportion of fermentation costs, being as great as energy costs. These two items alone represent about 50% of production costs.

Control of Fermentation
There are two kinds of fermentation products:
Primary metabolites:
These are molecules necessary for cell construction (vitamins, amino acids, enzymes). Primary metabolites are synthesized during the growth phase. During this phase, culture conditions and media must favor development of microorganisms and prevent any risk of catabolic repression by the substrate, or any feedback repression by the reaction product.

Secondary metabolites:
The synthesis of these products is independent of the growth period. This is the case for antibiotic production. The physiological conditions of the biomass must be such that the desired substance is produced, and it is then imperative to maintain these conditions as long as possible.

Two techniques are employed:
• the "batch culture," where the entire sterile culture medium is introduced in the fermenter at the beginning of the culture;
• the "fed batch," where only a part of the medium is introduced at the beginning and additions are then made at regular intervals or continuously.

This last technique permits rapid initial growth of the cultures. The control of the growth and production phase is thus easier, often permitting the maintenance of high productivity for at least 10 hours more.

Bioconversions
A bioconversion is a chemical reaction where microorganisms, or possibly animal or vegetable cells, play the role of the catalyst. Depending on the process, the microorganism used can be alive; a spore dried, free, or immobilized on a support; or simply an enzyme extract.

Bioconversions involve enzymatic reactions and fermentation:
• enzymatic reactions, in which the active enzyme requires the cell environment and machinery to function;
• limited fermentations, in which reactions are reduced to the transformation of the substrate to a closely related product, although normally permitting synthesis of complex molecules (antibodies, vitamins) from simple molecules (glucose, O_2, NH_4^+).

Thousands of bioconversions have been recorded. One of the oldest, without doubt, is that of the transformation of wine to vinegar by the bacteria *Acetobacter* and *Gluconobacter*. However, in industry the best known is that of steroid transformation. It is in this field that bioconversions have permitted the most spectacular progress, where classical chemical reactions are impossible, difficult, or unsuitable.

The reasons for which bioconversions prove to be useful are[2]
• their narrow position specificity or "regiospecificity" (figure 1),
• their stereospecificity, which may go as far as the recognition of asymmetrical molecules,
• the fact that they can be carried out on a molecule position that would not react, or only slightly, with a classical chemical catalyst.

One single bioconversion can replace several chemical steps. The work of D. H. Peterson and H. G. Muray (1952) concerning the 11 alpha hydroxylation of progesterone allowed the synthesis of cortisone in 11 steps instead of the 32 previously necessary, using diosgenin, which is a relatively abundant steroid plant extract. These considerable advantages, as we shall see, are moderated by the difficulties of the production procedure.

Bioconversions are generally realized in two steps, the first being the biomass production phase. During this period, the greatest possible cell density must be obtained, and above all the maximum enzyme activity. The composition of the culture medium and the growth conditions are therefore very important. Especially catabolic repression or "feedback" phenoma must be avoided. On the other hand, it is often important to induce enzyme formation by the addition of the natural substrate or an analogue in the culture medium.

For example, the acylase of *E. coli*, employed during the bioconversion of benzylpenicillin into 6 APA, is repressed by the presence of glucose, fructose, or glycerol. These substrates should be avoided and may be replaced by, for example, a medium containing glutamate, phenylacetic acid (as inducers), and mineral salts. It takes about 10 hours, at pH 7.0 and at 30°C, to synthesize enough enzyme to catalyze the desired reaction.

During the second step, which, in general,
takes place at the end of the growth period, the substrate to be transformed is brought in contact with the microorganism or the enzyme extract.

This manipulation can be realized either directly in the fermenter or in another reactor. The latter option is taken if the cells undergo a preliminary treatment, such as concentration, washing, modification of permeability, lysis (to recover the enzyme extract), or immobilization on a support. Substrate concentrations are very variable from one process to another. They must be adapted to the cell activity and take into account possible toxic or inhibitory effects.

Steroids present other peculiarities:
• They dissolve very badly in water. This necessitates their solution in a solvent as untoxic as possible to enzyme reaction. Acetone, ethanol, methanol, or dimethyl formamide is frequently employed.
• They can also form mixed crystals, mixing the substrate and product, impeding the normal process of transformation.

In spite of these difficulties, concentrations can reach up to several tens of grams per liter.

The reaction time varies, usually, between 1 and 3 days, depending on the initial quantity and the transformation capacity of the cells. Finally, the reaction conditions (pH, temperature, aeration) must be optimized in order to obtain the best enzyme activity. The presence of a releasing agent may be necessary.

Extraction
This is an indispensable, very often rather long, and delicate period, totally incomparable to the extraction of a product realized by chemical reaction, where the product concentration is often high and the objective is to remove traces of impurities. Here, on the contrary, the substance is lost in such a large volume that it must be concentrated, extracted, and finally purified. This is an ungratifying task, for the results are never totally assured. Any contamination or slight variation of the metabolism during fermentation might provoke the breakdown of results patiently sought after. If microorganisms are often excused for their frequent fluctuations in productivity, this is rarely the case with operations of chemical extraction, for which a higher yield is a prime objective.

A good extraction procedure must be capable of a great selectivity toward the required metabolites and insensitive to variations in the composition of fermentation broth. It must be not only profitable but also easy to work and to extrapolate to a higher volume.

Several steps are necessary to obtain this objective:
• The separation of microorganisms. Two techniques are generally employed: centrifugation and filtration, which both involve
costly installations, materials, and manpower.
• The lysis of microorganisms if the metabolite is inside the cells. The method selected depends on the volume to be treated, efficiency, cost, sensitivity of the product, degree of purity required (thermic or osmotic shock, pH extremes, shearing, ultrasound, crushing, enzyme attack).
• Several steps of concentration and purification. In many cases, for products that are hard to concentrate, ultrafiltration membranes or dialysis techniques are used. Purification may involve solvent extraction, chromatography, selective precipitation, or crystallization.
• The production, eventually, of a by-product that can be obtained by a chemical reaction.
• The final purification phase, such as sterilization just before the packaging process.

This description has outlined the great variety of bioconversions. It takes, generally, a big effort to set up, often more expensive and more hazardous than is the case for chemical procedure. It is for this reason that few bioconversions have been used in industry, at least until recently. It is quite possible that a deeper knowledge of microorganisms and their complex enzyme systems will allow better development in the future of these techniques.

(2) G. Nominé, and L. Pénasse, *Biofutur* (March 1983).

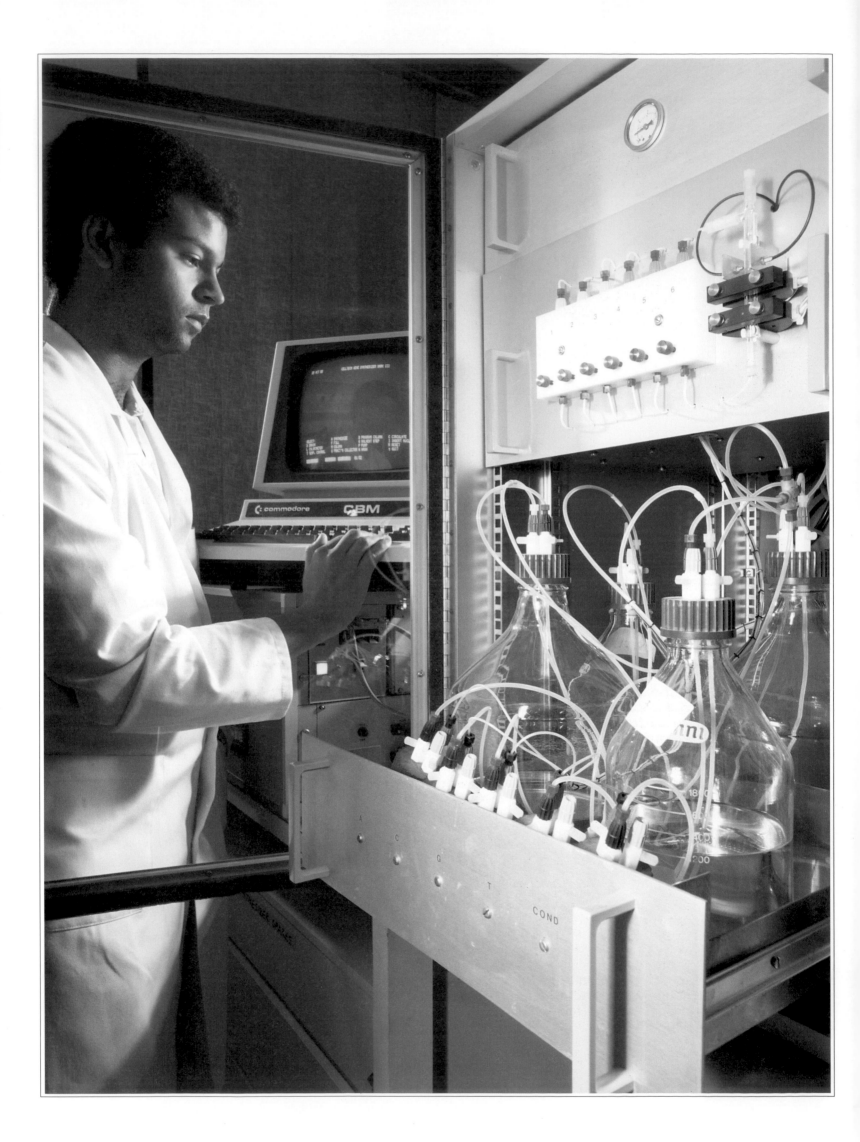

19

Black Boxes
and Chemical Robots

"Those who made money out of the first Californian gold rush, the investment analysts, were those who produced the picks and shovels," states Sam Eletr, the bearded Ph.D. with a biblical profile who is president of Applied Biosystems, the company that makes tools for genetic engineers. Applied Biosystems of Foster City, south of San Francisco, was born in 1981, but – uncharacteristically for the new crop of biotechnology ventures – has been making money since 1983.

The reason is that it took an instrument invented by the California Institute of Technology (Caltech) and developed it into a commercial product.

The instrument is an automated protein sequencer. Its unique feature is the sensitivity with which it can perform its analysis on microscopic amounts of protein. Its user just writes his requirements into a visual display unit and the machine does the rest. The basic technology of automatically working out the sequence of amino acids in a protein molecule has been available for nearly two decades. Per Edman had developed the first automated sequencer in 1967, but this instrument, using liquid-phase chemistry and a spinning cup, had several disadvantages: incomplete or secondary reactions, and sometimes loss of the sample, which became stuck to the glass walls during the course of the reaction.

It also required too great an amount of protein – up to several milligrams. To cope with these deficiencies, in 1971, R. A. Larsen proposed a solid-phase sequencer, whose essential difference lay in its method of immobilizing the protein (glass or polystyrene balls on which the peptide was fixed by covalence). But the quantity of peptide again remained too high. Caltech researchers led by Professor Leroy Hood developed a new microchemistry based on a gas-phase analysis that could handle as little as 1/500th of the amounts of protein previously needed: the reagents are introduced in the gaseous phase, and the support is a disk of very thin paper inpregnated with a polymer. The micro-

sequencer is electrical and can accept a wide variety of samples (glycoproteins, membranes, etc), and is 10,000 times more sensitive than Edman's device. Analysis is carried out by HPLC (High Performance Liquid Chromatography) and radioactive marking. The principle remains that of a series of chemical reactions that recognize the final amino acid (the "tail"), removes it from the chain, continues to the next, and so on. The limit is about 90 amino acids. Explains Sam Eletr, "A chemical reaction repeated 90 times over cannot remain chemically perfect; it loses 3 to 4% in accuracy each time. At a certain stage, the accumulation of these margins of error becomes intolerable and the process must be stopped."

These "molecular scissors," to use Professor Hood's own language, are now being sold by Applied Biosystems (under an exclusive license from Caltech), which can now claim 80-90% of the world market for these "chemical robots." Among other things, the microsequencer makes it possible to establish the sequence of the proteins coded by "rare-message genes," that is to say, genes that produce only an infinitesimaly small quantity of messenger RNA. To clone these genes, it is often difficult to obtain a probe of complementary DNA. However, if the amino acid sequence of a specific protein is known, one can, by consulting the genetic code, work back to the coding gene and, by a reverse process, fabricate fragments of this gene.

According to Sam Eletr, Beckman Instruments (now SmithKline Beckman), which had previously dominated the market for protein sequencers, when offered a license by Caltech, had said it would take three or four years to develop such a sophisticated instrument to the prototype stage. His company took a year. "Large companies cannot move as fast as small ones such as ours." But he has close associations with Becton Dickinson, which helps fund his research program. He also collaborates with an NBF (new biotechnology firm) called Genetic Systems Corporation in developing novel diagnostic tools.

The synthesis of genes
A Celltech researcher
prepares oligonucleotides
for the DNA synthesizer.

183

Leroy Hood

Eletr, who left a senior executive position with Hewlett-Packard to help found Applied Biosystems, used his former company's technology extensively in engineering the protein sequencer. His company has concentrated on the intricate system of automatic valves that must handle the flows of reagents and solvents without picking up traces of contamination. It also focuses on the reagents and solvents, specified by Caltech researchers, but made and purified by Applied Biosystems. Orders have begun to come into the industrial laboratories from the chemical and pharmaceutical industries (Du Pont, Eli Lilly, Hofmann-La Roche, etc.), as well as from the USSR.

Other chemical processes have made it possible successfully to synthesize proteins up to 40 amino acids in length (Applied Biosystems, SmithKline Beckman, Vega Biosearch). Once the decision has been made to modify an amino acid in a sequence for one or another application, the theory can be tested by synthesizing the peptide thus conceived. This procedure has already been put to work to produce synthetic vaccines against diphtheria and hoof and mouth disease.

Hard on the heels of the sequencer, however, is a less costly tool for genetic engineers with a much bigger appetite for reagents. This is the automated DNA synthesizer, which assembles and purifies pieces of DNA from 10 to 40 nucleotides in length (soon to be 100) using innovative chemistry developed at the University of Colorado, where Professor Marvin Carruthers, another of the company's principal consultants, is based. Eletr believes that revenues from sales of the reagents needed by his instruments will catch up with instrument earnings within a few years. The synthesizer, for instance, makes it possible to cut out a slice of gene, synthesize a new slice to modify three or four bases (which would change the structure of the protein coded by the gene), and then reinsert it in place of the old one – somewhat like splicing pieces of magnetic tape.

Making pieces of gene is not exactly a new idea in molecular biology. At the 1976 Congress of the American Chemical Society held in San Francisco, H. Khorana's team presented the total synthesis of a gene (the transfer DNA of alanine). And from the 1960s onward, it was thanks to patched-up fragments that it became possible to decipher the genetic code. In itself, the principle of synthesis is quite simple: it is a matter of joining the letters (nucleotides) end to end in the proper order so as to reconstitute a coherent sequence (the gene). In practice, the job is considerably more complicated. The first obstacle that must be overcome is the degeneration of the genetic code, which occurs when more than one codon can correspond to one amino acid. Furthermore, the components of DNA allow for chemical functions that can interact between themselves, and the nucleotides may inadvertently create "spoonerisms": the chemist must thus leave free the function he wants to see react while "tying up" the others, something that is far easier said than done. He must then liberate the nucleotides thus "protected." And finally, he must purify the end result – eliminating truncated chains and parasitical derivatives – and verify the sequence he has obtained.

It is thus not hard to imagine to what extent automation of the process can simplify the life of the researcher. The first companies to become involved in the automatic synthesis of genes lost their shirts, as did Biologicals Inc. of Toronto, which spent $3 million before finally throwing in the towel in 1983. With the Applied Biosystems machine, two, three, or four bases can be placed in a single position, which is important in view of the ambiguity of the reverse translation of a protein into DNA sequences. It was with the aid of such a machine that, in only two months, Marvin Carruthers succeeded in synthesizing the entire gene coding for human gamma interferon (500 bases). In a similar case, he synthesized sections of 100 bases each in a way so that he could "glue" them together, that is to say, with terminal sequences that would make it possible for a chain to link up with the one preceding or following it.

Development is now under way of a sequencer for DNA, whose chemistry still remains relatively complex. One system to analyze DNA has been developed in Japan by Akiyoshi Wada's team at the University of Tokyo. It consists of a machine to extract radioactively labeled DNA segments automatically, an automated procedure for the massive production of electrophoresis films (designed by Fuji Film), and computer programs for analysis (designed by Mitsubishi Knowledge Industry). The microchemical manipulator, controlled by the computer, was developed by a division of Seiko.

The Japanese will most certainly have to be taken into account, for here again there is a good chance that they will follow the same path that proved so profitable in the field of electronics: drawing their inspiration from Western inventions to develop smaller, simpler, handier and more attractive products, like the Walkman. In the field of "black boxes," a timid attempt has just been made by Nippon Zeon, established in 1982 as part of the Furukawa group. But research in Japanese laboratories is being accelerated, as they are well aware of the importance of these "chemical robots," which can become to biology what particle accelerators have become to physics. As Leroy Hood puts it, "Biology must evolve to the same stage physics was at twenty years ago, in that we can now conceive of the instrumentation required to enable it to take giant strides forward."

1. Protein sequencer
2, 6. DNA synthesizer
3. Inside a gene synthesizer
4. In Japan
Nippon Zeon's "compact,"
the first automatic DNA
synthesizer machine of this
type sold by the Japanese.
Worth remembering:
Applied Biosystem holds
70% of the market.
5. Peptide synthesizer

1

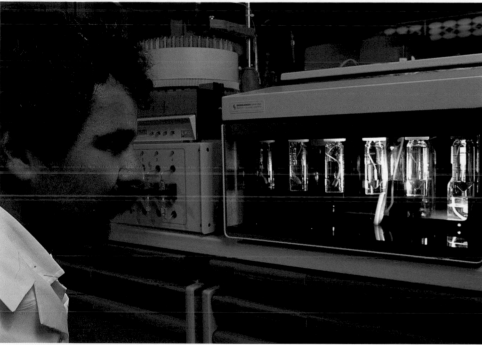

2

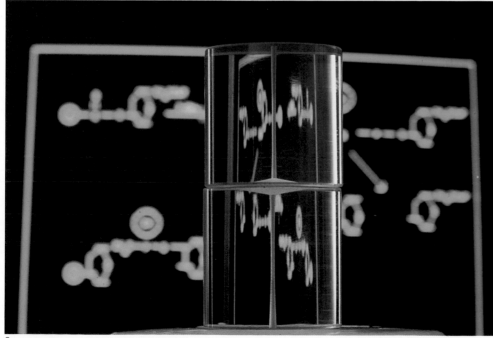

3

4

5

6

20

Patenting Life

A valuable bug was once stolen from a British laboratory in full view of the scientists – and there was nothing the embarrassed scientists could do about it. The incident occurred at the Lord Rank Research Centre of Ranks Hovis McDougall (RHM), where they were developing the biotechnology for turning plant starch into mycoprotein and edible protein.

A Russian visitor to whom the process was being described as a potential future customer for the technology asked whether he could see the bottle in which the organism was kept. His hosts saw no harm in him holding a code-labeled bottle containing the dried organism – until he removed the stopper, inhaled deeply, then blew his nose vigorously on a clean white handkerchief and tucked it away with a satisfied smile.

In 1972, Ananda Chakrabarty, then working for the US company General Electric (GE), invented a new organism that ate oil. It was not strictly the product of genetic engineering, being made neither by recombinant DNA nor by hybridoma technologies. Chakrabarty had taken four *Pseudomonas,* microorganisms with appetites for different fractions of crude oil, and had crossed them to create a stable hybrid bacterium with a taste for "whole crude." Laboratory trials showed that in oxygenated water this brand-new drug would digest about two-thirds of the components of crude oil, converting them mainly into microbial protein – food for aquatic organisms – and carbon dioxide. Since the appetite of the hybrid microorganism included a taste for the more toxic parts of the oil, GE thought that it might be a way of cleaning up oil spills.

At first the US Patent and Trademark Office refused to grant a patent for this new form of life. But the decision was overturned by the US Supreme Court in midsummer 1980, and in March 1981 GE was assigned US Patent No. 4,259,444 for "Microorganisms having multiple compatible degradative energy-generating plasmids and preparation thereof."

The basic issue on which the US Supreme Court eventually ruled in favor of GE was that the Patent and Trademark Office could not withhold patent protection for a genetically tailored microorganism merely because it was alive. The decision was crucial to the interest of the genetic engineers, who by then were inventing many new living organisms by splicing and resplicing genetic material.[1]

In fact, on the basis of the US Supreme Court ruling, the genetic engineers obtained their first patent three months before Chakrabarty's patent was granted. US Patent No. 4,237,224, granted Stanley Cohen and Herbert Boyer, described a "Process for producing biologically functional molecular chimeras." It appears to establish a basic claim to the gene-splicing invention of recombinant DNA technology.

Cohen and Boyer, at, respectively, Stanford University and the University of California at San Francisco, assigned their US patent to Stanford's Office of Technology Licensing. In August 1981 this office announced the terms under which it was willing to grant the use of Cohen and Boyer's invention – one they believed was as basic to genetic engineering as Bell Telephone Laboratories' patent on the transistor had proved to be to semiconductor development. The terms of the Stanford license are that, for an initial nonexclusive license fee of $10,000 and an annual fee of $10,000, the techniques described by Cohen and Boyer can be used in research. Academic scientists are not required to make these payments. The companies for which the license was intended include most if not all of the new biotechnology firms (NBFs), as well as many multinational groups in pharmaceuticals, agriculture, and food.

The license terms had been carefully calculated to net a large proportion of commercial genetic engineers. Royalties on product sales required by Stanford University would vary from 1% on the first $5 million of earnings to 0.5% on sales of over $10 million. They are relatively low for patent agree-

Sealed ampules
Patents means royalties, particularly to the pharmaceutical industry. To what extent can plants and microbes be patented? Beyond what threshold does the genetic engineer "create"?

(1) In 1973, a convention, signed at Munich and entering into force on 1 June 1978 in eleven European countries, established definitively the principle of patent rights to certain life forms.

187

1. Stanley Cohen
2. Herbert Boyer
3. César Milstein

1

2

3

ments – although against this must be set the fact that they are nonexclusive. Nor does the patent claim to protect for organisms or plasmids with properties, as the Chakrabarty patent claims.

As a further inducement to companies not to contest the patent, Stanford University said that companies that signed a nonexclusive license agreement before 15 December 1981 would be allowed to offset the annual subscription against future royalties. This helped to forestall accusations that the California universities were out to make a quick killing. Cohen and Boyer, meanwhile, had waived their personal rights to patent income other than that for more research.

Some seventy-three companies applied for a license before the deadline. Many were small research companies – new biotechnology firms – which took the view that the payments were modest compared with the cost of research they were funding based on the techniques covered by the patent, and more modest still in relation to the profits they expected successful research to yield. They believed it was up to big business to contest the patent – as GE had done in establishing the patentability of living organisms. But the applicants also included chemical and pharmaceutical giants – American Cyanamid, Allied Corporation, Bristol-Myers, Burroughs Wellcome, Corning Glass, Du Pont, Eli Lilly, Hoechst, Hoffmann-La Roche, Monsanto, Schering Plough, Texaco, and Upjohn. By 1984, Stanford University was reported to have collected nearly $3 million in licensing fees and royalties from the first Cohen and Boyer patent.

In fact, the first challenge to the patent came from the US Patent and Trademark Office itself. In July 1982 it withdrew a second patent application by the same pair of scientists only two weeks before it was due to be granted. This one covered the product of gene splicing, whereas the first covered basic techniques. Stanford University was closely relating the two patents in negotiating its licenses; together, they promised to give Stanford University monopoly rights over a wide range of products arising from recombinant DNA research.

It was withdrawn on suspicion of flaws in the original gene splicing patent – suspicions publicized in a book published by Cold Spring Harbor Laboratory, a US research center, that was based on a private conference held in October 1981. According to a lawyer with Exxon Research and Engineering, Albert Halluin, the inventors had failed to deposit plasmid pSC101 – the vector crucial to their gene splicing experiment – with the American Type Culture Collection until six months after the patent was issued. Halluin also cast doubt on whether the vector can be made in the way described in the patent, citing one of the inventors as admitting such doubts in print in 1977.

In other words, the inventors may have failed one of the main tests for the granting of the monopoly that accompanies any patent – namely, that the recipe shall permit anyone skilled in the art to repeat the experiment. One further argument was that there may have been prior disclosure as a result of articles written after a private meeting in 1973, before the patent application was made.

After two years of legal arguments – including a claim from a third scientist to be a coinventor, which the patent examiners rejected – they finally assigned Cohen and Boyer's process patent to Stanford University in August 1984. Between them, the two patents on process and product give Stanford University a proprietary position in the new biotechnology of recombinant DNA, or gene splicing, until 1997. But the legal tussle teased out enough issues and inconsistencies to ensure that there will be many more battles once profits are being made from the gene splicing technique.

Hybridomas

Cohen and Boyer recognized from the start the profound commercial importance of their experiments. In contrast, the two scientists who discovered the second big new technique of biotechnology – the cell fusion or hybridoma technique – recognized only its significance for medical science and made no effort to protect their discovery.

Celltech owes its existence to the failure of the Medical Research Council in Britain to obtain patent protection for the discovery of the hybridoma cell fusion technique for producing monoclonal antibodies. The technique was invented by César Milstein and Georges Kohler at the MRC's Laboratory of Molecular Biology, Cambridge, in 1973. As Nicholas Wade described it in Science in 1982, "In place of the riot of different antibodies raised naturally to a given antigen, the hybridoma technique makes available a constant pure source of a single antibody. Of such techniques are revolutions made."

British university and national laboratory inventions are normally patented by the National Research Development Corporation (NRDC), a government agency that in 1982 formally became part of the state-owned British Technology Group. The NRDC has frequently been blamed for failing to patent this discovery for the United Kingdom. It is even said that its merger into the British Technology Group came about because of the Prime Minister's annoyance at the failure to protect this Nobel Prize-winning invention.

"With hindsight, one can certainly say that if the right steps had been taken it could certainly have been possible to patent the process," Gerard Fairtlough, chief executive of Celltech, said in August 1982. In the event, no one can get a patent because no one made an application before details of the technique were published. As Fairtlough put it, "Somewhere, something went wrong."

What apparently went wrong was that Milstein sent a copy of his paper to a Medical Research Council (not an NRDC) official, but went ahead nonetheless with publication in the scientific press in August 1975, thereby establishing the claim to a scientific discovery. The official eventually passed a message on to the NRDC – thirteen months later, in September 1976, with a note apologizing for the delay.

The Wistar Institute in Philadelphia subsequently obtained a US patent for a "Method of producing viral antibodies" and a "Method of producing tumor antibodies," describing the procedures leading to the production of monoclonal antibodies against viral and tumor antigens. According to the Wistar Institute, the Milstein publications were reviewed by the US Patent Office before any patents were granted.

Celltech, founded with a unique exclusive

agreement with the Medical Research Council in 1980, promptly set about combing the MRC Laboratories for any genetic-engineering inventions with commercial potential and transferring them to its own laboratories in Slough, England.

The Patent System

"When an inventor is granted a patent he makes a Faustian bargain with the State. The inventor discloses his manner of new manufacture in his patent specification, which is published by the Patent Office and available to all the world. In return, the State grants him a 20-years[2] monopoly in the manner of new manufacture disclosed in his patent, after which it is available to anyone."

That is how Alistair Kelman, a London lawyer specializing in the production of "intellectual property," began to describe the system to a conference on investing in biotechnology in London. The subsequent discussion indicated a high level of interest in the protection of biotechnology's inventions.

The alternative to patent protection is the "trade secret." The classic example of a well-kept trade secret is the formula for the liqueur Benedictine, made by French monks. But, should the secret escape, the courts do not line granting an injunction preventing its use, but prefer to order damages, leaving the defendant free to use the information. This legal fact, coupled with modern surveillance and bugging equipment, greatly reduces legal protection for trade secrets. The US Freedom of Information Act is another route to disclosure of trade secrets when, for example, the secrets have to be revealed to licensing organizations such as the Food and Drug Administration (FDA).

Patents are a particularly useful form of protection for the drug industry, which is one of the most important outlets for the new biotechnologies and their newly created forms of life. For the same basic reason, patents are also good protection for the (human) food industry. The reason is the cost and time it takes to steer a new product through the regulating system for new drugs or foods. It is not possible to design a way around a patent and thus avoid meeting the full cost and delay of the safety regulations. Dr. Jack Edelman, director of research at RHM in England, estimates that his company has a lead of five to seven years "because we have had to go through that period of experimentation, safety evaluation, for the regulatory authorities." Any competitors would be "really quite mad if they didn't come to us to license technology, which we have patented," Edelman believes. The RHM mycoprotein organism was patented before the Chakrabarty hybrid bacterium was invented. It was patented as a new plant.

(2) Seventeen years in the United States and Canada.

Codex Borgia
The Aztec Goddess Mayahuel seated on an agave cactus below a jar of pulque decorated with paper streamers and arrows. This was a time when patents on life belonged only to the gods.

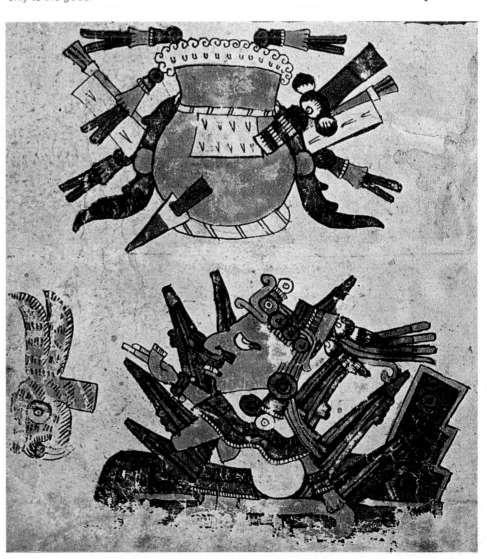

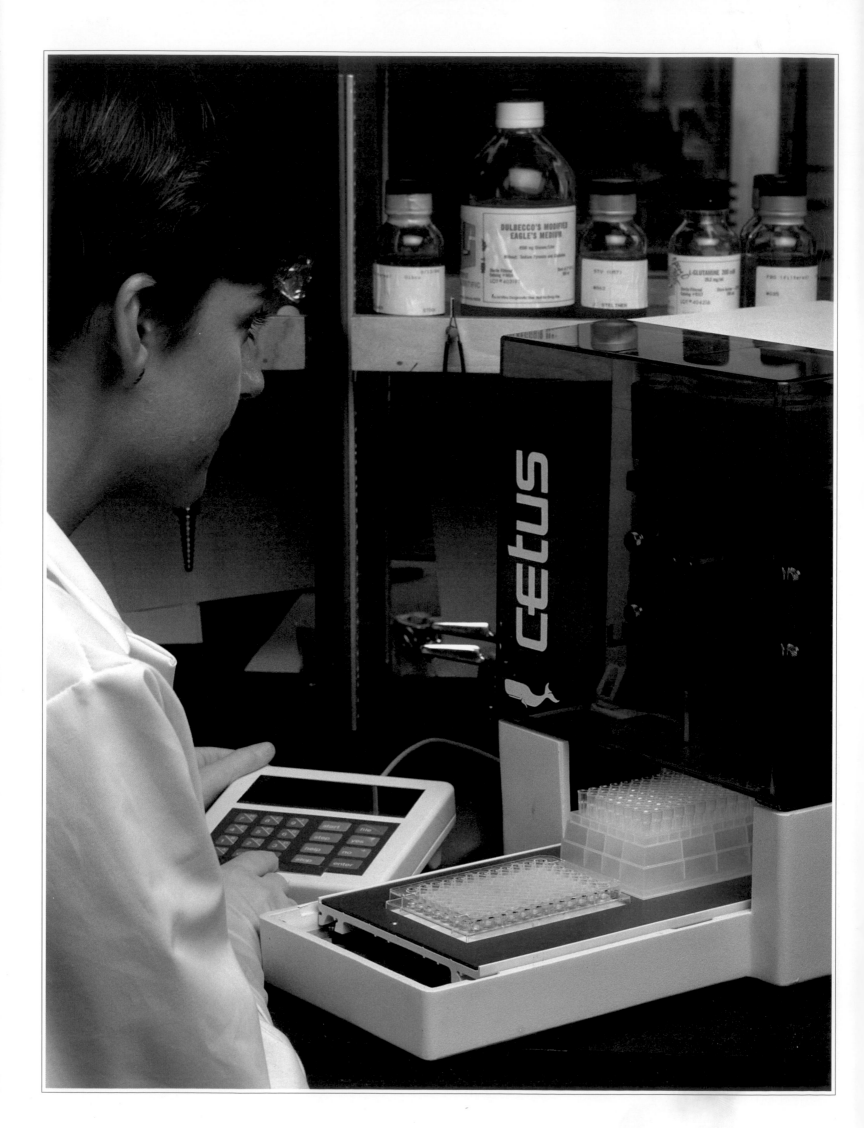

21

Risks to NBFs

The NBF (new biotechnology firm) is at risk from the moment it is born. It is trying to sell brainpower, an ability to do something no one else can do, or no one has recognized the value of doing. As in any research area, the risks of failing to do this successfully are high.

When Lord Rothschild, the Cambridge scientist, persuaded his family bank to set up a venture capital trust specifically to invest in biotechnology, his eye was on the investment opportunities of companies like Genentech, with their dramatically fast rise in market value. But his scientific training warned him that the claims of such companies would need very careful examination. The Genentech example was encouraging many others to overprice themselves.

One of the first NBFs to fail was a "toolmaker" called Bethesda Research Laboratories, founded to make the novel reagents needed by the genetic engineers. It built up a successful business in Bethesda, near Washington, DC, supplying the vast research campus of the National Institutes of Health (NIH). Its corporate plan in 1980 called for a steady expansion of its own research program in genetic engineering, funded out of the sale of its biological reagents. Then abruptly, and to the consternation of its bankers, the company began to recruit scientists in droves. Its founder talked publicly of making his company "the IBM of biotechnology." Soon it was deep in financial trouble, with no income to pay its large team of scientists. Its bankers refinanced a much smaller company with a new chief executive, but by the mid-1980s Bethesda Research Laboratories had been bought by a bigger biotechnology group.

One may confidently forecast that this will be the fate of many, probably the vast majority, of the NBFs. They attracted investments with claims that seemed incredible, and encouraged their patrons by fulfilling some claims even sooner than they were forecasting. But those early fulfilments – such as the cloning of genes for human interferon and insulin – were only the first step on a long and hazardous development path needing many, quite different skills from those with which the company was founded.

For some entrepreneurs the thrill is in the starting, the growth of a new-born company into one with an income of, say, $50-$100 million a year. They sell out at this point, perhaps to a big firm that needs a ready-made pocket of technical expertise to support a new venture or for diversification. They may reinvest in a new start-up.

Other entrepreneurs dream of growing into a big company, rivaling established firms in the field. Genentech threw down this challenge to other NBFs when Robert Swanson announced that his company would move from contract research and development for other companies into the manufacture and marketing of its own pharmaceuticals, starting with t-PA, human growth hormone, and gamma interferon.

Biogen was one company that responded to Genentech's challenge. Its chief executive, Walter Gilbert, propounded a strategy for focusing its debut as a drug company on cancer drugs such as gamma interferon and interleukin 2. His idea was that the US specialists in cancer treatment, who were his primary market, were a sufficiently well-defined community to be reached by quite a small sales force, commensurate with a small firm such as Biogen. But by late 1984 the company was failing to find enough financial support for all the ambitious opportunities Gilbert had opened. It was obliged to trim back, and Gilbert himself left the driving seat he had occupied so conspicuously since the company's founding.

The surprising thing is not that NBFs should have setbacks and sometimes founder, but that far more have not failed. An important factor has been the ingenuity of the companies in finding fresh sources of venture capital to refinance their programs. Many American NBFs found temporary relief in limited research partnerships, for example.

Quality control
Beta seron (beta interferon obtained by genetic engineering) being tested on a Cetus machine.

For the investor the problem is to recognize in good time when an NBF is overreaching, when its ambitions are racing ahead of its achievements and its ability to bring in more cash.

Cetus – the Pioneer

Cetus claims to be the first of the NBFs – founded in 1972 – before the term "new biotechnology" was invented, or indeed before Boyer and Cohen invented the first genetic-engineering technique. Its founders were Dr. Ronald Cape, a molecular biologist, and Peter Farley, an entrepreneur businessman who has since left the company.

Cetus pioneered the idea of a company adept at recognizing opportunities arising at the frontiers of the biological sciences, targeting those opportunities, and raising risk capital to turn them into new technology for industry. It set itself up in Berkeley, near San Francisco, and close to the California universities, which pioneered genetic engineering. A decade later it became one of the first to run into serious financial problems, caused partly, at least, through overdependence on two large research sponsors who withdrew their support. Ironically, the research in each case had been a success.

The successful research for which Cetus was most widely known was a biotechnology process that used a natural (not genetically engineered) enzyme to convert glucose to the sugar fructose and yield alkene oxides as a valuable by-product. It promised a cheaper route to a sugar that is sweeter than fructose. The sponsor was Socal (Standard Oil of California). But in 1982, Dr. Cape pointed out that the cost of bringing this new sweetener to the market was on the order of $150 million. Socal decided to drop the project, having spent about $8 million with Cetus on research and development. But Dr. Robert Fildes, chief operating officer, points out that the fructose technology is ready and waiting for a change in the market and was not abandoned for technical reasons.

Dr. Fildes, a scientist with top executive experience with such companies as Glaxo, Bristol-Myers, and Biogen, joined Cetus to reorganize a company badly shaken by Socal's decision and also by Shell Oil's plans to pursue interferon – the second success story – alone. Instead of a bioscience research company, he and Dr. Cape refocused on health care, with the aim of bringing novel pharmaceutical products to the market itself. By 1984, more than 70% of its activities were devoted to human health care, 20% to novel agricultural products, and only 10% to enzyme biotechnology processes for the food and chemical industries. Many research projects funded by Cetus itself had been abandoned. "Cetus intends to be a totally integrated human healthcare company in North America," Dr. Cape says. "We will create new products in the laboratory, develop large-scale production methods, complete necessary testing to gain regulatory agency approval, and then manufacture and market these products."

Unlike the next two NBFs we describe, Cetus is not advised by a panel of eminent scientists. Dr. Cape himself remains the principal link with academia.

As a contract research company, Cetus was accustomed to developing specialized instrumentation. In 1983, it decided to add the manufacture of advanced tools for bioscience to its portfolio of targets, creating Cetus Instrument Systems. Dr. Fildes forecasts that this NBF's first $10 million-a-year market from bioscience will come from the production of this laboratory tool.

Late in 1984, Cetus found itself a new partner in W.G. Grace & Company, the chemicals group, to develop, make, and market novel products for agriculture. Bob Fildes says that the new joint venture, Agracetus, "represents the realization of our strategy to join with an established leader in the agricultural field to complement Cetus's technological skills with financial resources and strong marketing and distribution capabilities."

Genentech – the Leader

Genentech, by any standards, is a phenomenally successful new venture in high technology. Its track record of scientific achievement has no rival anywhere in the world. It has served as the model for well over 200 other NBFs created since 1975 to exploit the new techniques of molecular biologists.

It was the idea of a young entrepreneur, Robert Swanson, who had been helping Citibank to invest venture capital in starting companies. He decided to set up his own firm to exploit the new technology.

Swanson recruited Herbert Boyer, who with Stanley Cohen had filed the first patent on gene splicing. In the next year or so, by early 1978, the partnership of a pushing entrepreneur and a scientist (who, the academic world believed, was heading for a Nobel Prize) raised $1 million in venture capital from such firms as International Michel and Monsanto, both seeking to diversify from traditional business sectors. By its fifth birthday in 1981, it had published 36 scientific papers and was earning wide praise for the quality of its science. Far behind were the days when a potential Swedish client had to send air tickets to California before Swanson and colleagues were persuaded to journey to Stockholm to discuss the idea of genetically engineering human growth hormone.

Genentech pioneered the idea of a high-powered panel of scientific advisors drawn from the disciplines it wished to exploit. This novel approach to "technology transfer," which offered academics a stimulating new challenge and unprecedented financial rewards for success, was widely imitated by other NBFs.

Genentech has been associated with most of the more highly publicized "targets" of genetic engineering: human insulin, growth hormone, and alpha interferon.

In 1983, it earned $47 million, almost all in contracts for research or development. Late in 1984, the high-technology investment specialists Hambrecht & Quist Inc. were forecasting revenues of about $114 million for 1986, of which half would be revenues from genetically engineeered products.

These analysts point to three products that they expect to bring in the first substantial product revenues. Significantly, they do not include human insulin, Genentech's first big scientific success. One is tissue plasminogen activator (t-PA), a substance that rapidly biodegrades blood clots and promises to be a powerful new drug for the treatment of thrombosis. Although other NBFs have also claimed t-PA

1

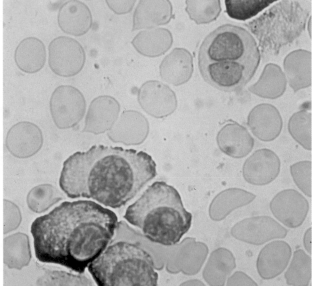

2

3

4

1. *Weissman, Davies, Flavell*
The Board of Directors of Biogen: left to right, Charles Weissman, Julian Davies, Richard Flavell.
2. *Monoclonal antibodies*
Starting in 1979, Hybritech was the first company to sell monoclonal antibodies.
3. *Ronald Cape*
Founder and President of Cetus.
4. *Clinical tests*
The production line for beta interferon at Cetus. This was the first of the company's therapeutic products to be subjected to clinical tests.
5. *Cell culture*

5

a target, Genentech is believed to have a three-year lead on what could be one of the biggest cardiovascular drug successes. It acquired an exclusive license to the research of Dr. Desiré Colicu at the University of Louvain in Belgium. The big advantage foreseen for t-PA over other clot-digesting agents available or under development is that it is highly specific in attacking only the clot.

Another early revenue earner is expected to be human growth hormone (HGH), the drug with which Genentech plans to make its own debut as a pharmaceutical company. Although developed originally under contract to Kabi Vitrum, HGH will be sold in the United States as Protropin, made by Genentech, while Kabi Vitrum will market the drug in Europe and Sumitomo in Japan.

The third of the initial revenue earners is expected to be gamma interferon, believed the most potent of the three genetically engineered human interferons. Genentech has raised $55.6 million through a financial arrangement called a limited partnership to fund the development stages of both HGH and gamma interferon. Interferon is expected to absorb more of the funds because it has a broader spectrum of therapeutic use, although it is not expected to go on sale until late in 1986.

One of the highest public tributes of Genentech's phenomenal record of technical achievement appeared in the scientific journal *Nature,* where many of the "breakthroughs" of molecular biology – starting with the double helix of Watson and Crick – have been recorded for posterity. Late in 1984 this normally sober journal heralded Genentech's latest scientific paper as "a technical triumph without parallel."

Celltech – a UK Copy

The idea of a British attempt to emulate the activities of Cetus, Genentech, and many other American NBFs was born in 1980 in the Belgravia offices of a government agency called The National Enterprise Board. Its principal architect was an industrialist, Gerrard Fairtlough, whose experience had ranged from a Cambridge University degree in biochemistry to being managing director of Shell Chemicals UK, part of the Royal Dutch-Shell group, in the mid-1970s before he joined the agency.

One of the roles of the agency – now renamed The British Technology Group – was to seek situations where the government might help to launch a new industrial initiative. A report from the government's Advisory Council on Applied Research and Development had urged strongly that such initiatives should be taken in the case of biotechnology. But the report coincided with a growing skepticism in government of the value of government intervention in industry.

Fairtlough and a couple of aides achieved two considerable coups. They put together a "blue book" outlining how the United Kingdom might compete with American NBFs, proposing launch capital of £12 million, and persuaded four private investment houses in the City of London to put up more than half of the cash. Confronted with what was manifestly a display of great entrepreneurial initiative in the public sector, the government agreed that it should become the fifth and biggest shareholder in the new company, called Celltech.

The second coup was to strike a deal with the Medical Research Council, the government-funded agency supporting academic medical research in Britain. This agency's own facilities included the Laboratory of Molecular Biology in Cambridge, whose activities since it was set up have earned eight Nobel Prizes. The MRC assigned Celltech rights of first refusal for exploiting ideas and inventions in the new biotechnologies, such as monoclonal antibodies and genetic engineering. Its chief executive, Sir James Gowans, joined Celltech's board of directors.

Celltech has built its business very substantially on the back of this close relationship with a world-famous medical research program costing about £120 million a year. In particular, it has set out to exploit the technique of making monoclonal antibodies discovered at the MRC's Cambridge laboratory in 1975, for which César Milstein and Georges Kohler received Nobel Prizes in 1984. By the end of 1983, Celltech has investigated the possibilities of at least ten of the MRC's research centers, and had begun to develop six of them in its own rapidly growing laboratories. It was also establishing world leadership in the "mass production" of monoclonal antibodies, using large-scale cell cultivation techniques of its own invention, which involve fermenters of up to 1,000 liters capacity.

In 1983, the Boots Company entered into a joint venture with Celltech to develop the first fruits of monoclonal antibody discoveries. These will be new medical tests for diseases such as chlamydia and respiratory syncytial virus, and for conditions such as infertility in both humans and cattle.

In 1984, Celltech embarked on a second joint venture with a big company, this time with Air Products, UK offshoot of a US industrial group with $1.7 billion sales that year. Celltech was seeking ways of developing the process-technology side of a business. It can see some of its research leading to a need for bioreactors and other novel process technology of substantial scale, perhaps running continuously for many weeks.

Transgène – a French Venture

Robert Lattès, a banker with a background in mathematics and nuclear physics, founded Transgène in 1980. Lattès, a director of the French state-owned Paribas Bank, brought together two eminent French scientists, Professor Pierre Chambon of the Louis Pasteur University in Strasbourg and Dr. Philippe Kourilsky, director of the Pasteur Institute's molecular biology laboratory, as Transgène's leading scientific advisors. Its laboratories are in Strasbourg, whose connections with bioscience can be traced back to Pasteur's discovery of the role played by yeast in making beer – in 1867. It has assembled an international team of scientists, mostly recruited outside of France, and established close links with universities and institutes in France, elsewhere in Europe, and in the United States.

Transgène is owned by five big companies, two being finance houses (Paribas and the insurance group Assurances Générales de France), and three being industrial groups (BSN-Gervais Danone, food and beverages; Elf-Aquitaine, oil and pharmaceuticals; and Moët-Hennessy, alcohol and perfumes). Together, these five raised about $10 million to launch the NBF and support it for up to five years.

Transgène is operated as a contract research company, independently of its industrial shareholders. Indeed, it stresses the security with which it shrouds research undertaken specifically for clients, refusing to divulge details even to board members where there may be a commercial conflict of interest. Its income from such research contracts has grown steadily from 18% of its operating costs in 1981 to 64% by 1983. It has publicly claimed three genetic-engineering successes: in cloning gamma interferon, under a contract for Roussel-Uclaf; in cloning a vaccine that may protect against rabies virus; and in cloning a blood factor, factor IX, used to treat some hemophiliacs, under a contract for Mérieux, a subsidiary of Rhône-Poulenc.

A substantial part of the company's workload comes from independent programs. These programs have been discussed internally – they are conducted with the company's own funds and will be offered to industries after a breakthrough has been achieved and patents have been applied for.

The company claims that the fact that all its shares are owned by five major groups avoids the need to overpublicize its research in order to keep raising more money, as has happened with many American NBFs. The five original shareholders will be expected to meet the next round of finance, if and when its development plan requires it.

There are, obviously, many other NBFs, both in the United States (Genex, Hybritech, Calgene, Sungene, etc.) and (more timidly) in Europe – France and Great Britain, but also Switzerland and the Netherlands. Now more cautious, most specialize in what businessmen term "a narrow niche": research on oncogenes, nucleotides, collagen, etc. To them the future is uncertain. But without them biotechnology would undoubtedly not have experienced the explosion we have seen, nor, from the very inception of genetic-engineering techniques and monoclonal antibodies, would they have been able to collect so much capital.

1. *Transgène*
The Board of Directors: left to right, Philippe Kourilsky, Jean-Pierre Lecocq, Pierre Chambon, Étienne Fisenmann.
2. *Celltech*
Celltech, established in Great Britain to exploit Milstein and Kohler's invention. In this picture, Malcom Rhodes controls microbe fermentation in a pilot reactor.
3. *Sungene*
Many new firms, particularly in the United States, are devoted to research on plants and seeds. Here, a researcher studies the DNA of a plant. After centrifugation, the DNA is concentrated in strips in a solution and the researcher shines ultraviolet light on the plastic tube, making the DNA visible. It is then extracted with a sterile syringe.

195

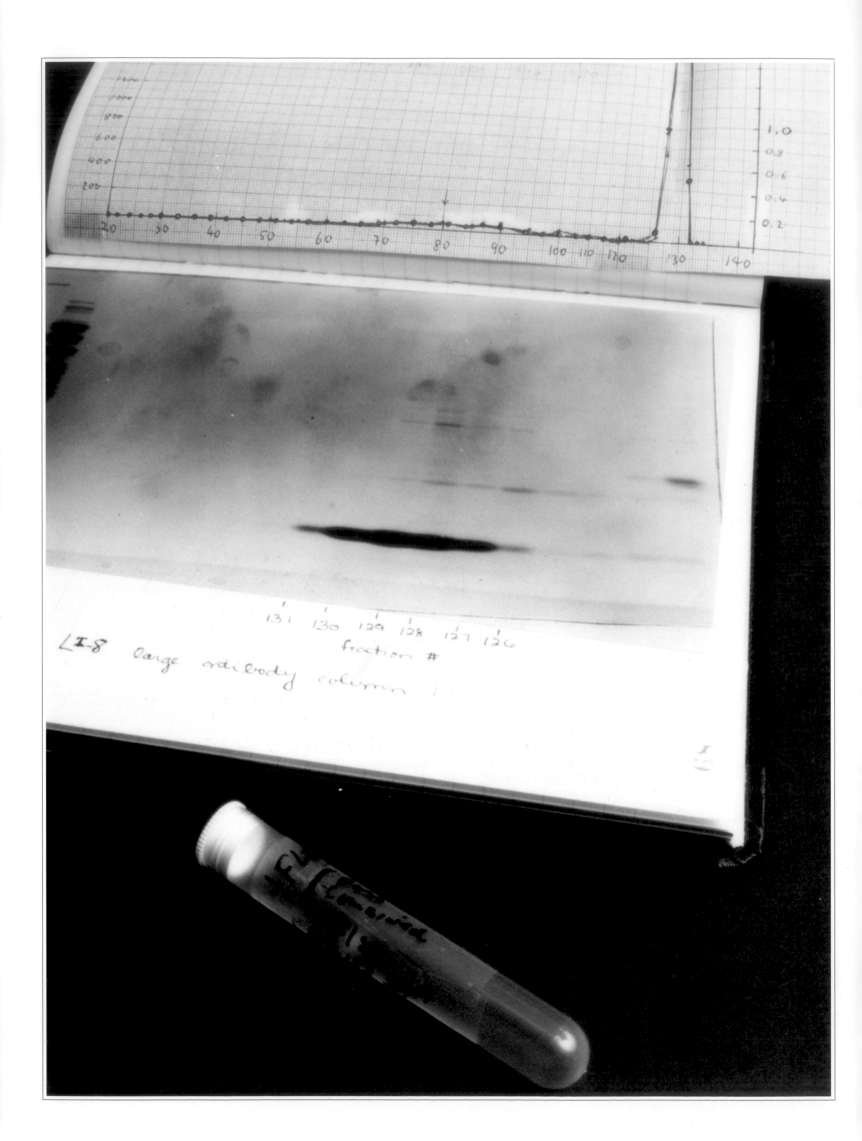

131 130 129 128 127 126

fraction #

LI-8 large antibody column

22

Big-Company Strategy

Big companies, it is sometimes said, were slow to recognize the importance of scientific events of the early 1970s in molecular biology and were, therefore, upstaged by entrepreneurs and their new biotechnology firms (NBFs).

It is true that big companies tend to react more slowly to events than small ones, if only because they have commitments to targets – technical as well as financial – and because many professional reputations may be committed to established patterns. But it is not true to say that big companies were unaware of the excitement of the new science. Many were cautious because they had already had uncomfortable experiences with trying to exploit biotechnology.

In the early 1960s, the Western world awakened to the threat of an uncontrolled explosion of population in the developing countries. New systems of birth control, such as the Pill, had only just been invented and were far fom being widely accepted. The world seemed to face an acute shortage of food. Science invented ways of "factory farming" nutritious organisms – bacteria, yeast, microfungi, algae, etc. – as potential foodstuffs. Private industry invested hundreds of millions of dollars in what was then known as biotechnology. These companies mounted a prodigious technical effort to solve problems that the manufacturing industry had never encountered before.

The Swiss pharmaceutical company Hoffmann-La Roche took a different approach to the long-term promise of biotechnology as it unfolded at the laboratories of molecular biology in the 1960s. In about 1970 the company, flush with profits from its great commercial success with the first tranquilizers, invested in two academic institutes of molecular biology, independent of its corporate research and development activities. Although geographically close to corporate R&D centers in Basel, Switzerland, and Nutley, New Jersey, these institutes were closer intellectually to academia, from which they attracted research workers for newly unfolding

areas of bioscience such as the mechanism of the immune system. Scientists of the caliber of Georges Kohler and Niels Jerne, who shared the 1984 Nobel Prize in Medicine with César Milstein, came to work at the Basel Institute of Immunology, for instance, so high was its scientific standing. By the late 1970s, these two institutes were valuable bridges between the company and the universities in the vanguard of genetic-engineering and cell fusion techniques. Roche was one of the first industrial patrons of Genentech, the California research company started in 1976.

Pharmacia, the Swedish pharmaceutical group formerly called Fortia, is a good example of a well-established company that sees clearly the central importance of biotechnology for its future and also the value of small start-up companies. It simply creates its own NBFs.

Pharmacia is a research-based company of about 4,500 with headquarters at Uppsala, a university town near Stockholm, where it has long-standing links with academic bioscience. One-sixth of its own staff is engaged in research and development, which absorbs about 14% of the company's sales. Yet Gunnar Wessman, its chairman, cheerfully confesses that "we have a not-invented-here syndrome – most of our ideas come from outside." Pharmacia tries to stay close to the science of about fifty universities worldwide. Internally, too, it works hard at efficient technology transfer, and a new product will often draw upon the technical resources of the entire group. Its corporate motif is the drop of blood dangling from the tip of a needle. Everyone in biotechnology talks to this company.

Other executives credit Wessman as the man who alerted the 74-year-old company to the true value of its "sleeping assets." Wessman, formerly chief executive of Perstorp, the Swedish chemical group, joined Pharmacia's board in 1973 and became chief executive in 1980. He piloted its first public issue in 1973 and became chief executive in

197

1

2

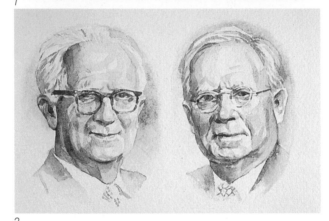

3

4

1. Fritz Hoffmann-La Roche
The founder of the company
bearing his name (seated)
with Emil C. Barell, who
continued his work for over
a quarter of a century.
2. The Pedersen brothers
Thorvald (1887-1961) and
Harald (1878-1966), who
founded Novo, in Denmark.
3. Marcel Mérieux
In 1897, in Lyon, Marcel
Mérieux, in the course of his
work on tuberculin and an
antitetanus serum, founded
the institute that bears his
name.
4. A. Beckman
The founder of Beckman
Instruments, whose motto
was "There is no substitute
for excellence".

1980. He piloted its first public issue in 1981 – until then the biggest for a Swedish company. Wessman puts a lot of faith in two factors: Pharmacia's well-honed skills at turning science into cash, backed by a corporate small-company approach to exploiting science quickly.

Small business units "lead to faster growth because every unit wants to be bigger." He sees no limit to the size a company can attain in this way. He has subdivided Pharmacia into fourteen small companies. "But you have to ensure that the growth is evenly spread." If it is not, it is time to rearrange your units or hive something off, he adds.

Wessman likens growth in such a company to the way cells grow and divide in a living organism. By watching it carefully, management knows when to feed nutrients to a slow-growing cell, he says. Pharmacia is investing heavily in management training – an activity it dignifies with the term "personal and business culture."

Behind these small business units stands a well-tested faith in university connections and efficient technology transfer from academia, particularly the Swedish universities. In the 1960s it picked up the work of Professor Arne Tiselius, a biochemist at Uppsala University who earned the Nobel Prize in Chemistry in 1948. It turned the professor's invention of electrophoresis, as a novel way of separating large molecules like proteins, into laboratory tools – for which the university was among its first customers. Then, as the requirements of biotechnology became more exacting, it developed the tools into pilot and then production equipment.

Pharmacia counts itself particularly fortunate in having in Uppsala, 2 kilometers from its headquarters, the vast university Biomedical Center, a union of three departments beginning in the 1960s, with a staff today of 2,000. But Stockholm's famous Karolinska Institute is also very important to Pharmacia, Wessman stresses. Back in Uppsala, the company has a joint venture called Bionova with the Wallenburg Laboratories in fundamental research in molecular biology, the basic science of genetic engineerring. This gives the company first option in exploiting any discoveries.

Dr. Johansson, who describes himself as "a frustrated university dean," heads a corporate R&D effort, employing about 600 and costing over £30 million in 1984 – nearly five times as much as it spent in 1980. About 20% is spent on what they call "explorative research" to open up new opportunities involving immunology (vaccines), peptides, or free radicals, for example.

The biotechnology of Pharmacia began in the 1940s with a substance called dextran, cultivated by fermentation, which the company purified to pharmaceutical standards for intravenous use as a substitute for blood plasma. Although, to this day, this is still its main use, Pharmacia has employed dextran as the raw material for nearly 200 different ion exchangers used to separate and purify biological materials. The purity of these polysaccharides is the basis of widespread confidence in what executives claim is the widest range of separation products available from any company.

In the late 1950s, the company began to engineer experimental separation systems and their controls for the big molecules of biotechnology. A decade later it moved into production systems.

In 1982, Pharmacia introduced its most sophisticated separation system, FPLC (fast protein liquid chromatography). It can cut separation times from hours to minutes, which is especially important for fragile or labile biomolecules. It was a wholly in-house development for both the chemistry and the electronics. FPLC has already found acceptance both in biotechnology research and in purifying such early products of genetic engineering as interferon.

In speaking of biotechnology, one tends to forget the "peripheral" industries, those that supply equipment of every kind and whose contribution is of capital importance. Pharmacia and Alfa-Laval are good examples, as is Chemap, the Swiss company that leads in sales of fermenters (50% of the market), or, in the same sector, Electrolux, Biolafitte, and John Brown, which helped in the construction of ICI's Billingham facility. Equally in the picture are companies making yeasts and ferments of various kinds: fast-starting strains, lactic ferments (for which the American market is shared by Miles and Hansen Ball). The Eurozyme Company, established in 1981 by Air Liquide, has developed a freeze-dried lactic ferment that permits continuous fermentation. The Danmark Protein Company (a subsidiary of Kabi Chemie AG and Danish Cheeses) is the leader in treating whey by ultrafiltration; of course, many other examples could be mentioned.

Five markets seem exceptionally promising for the immediate or very near future: red gold (blood derivates), estimated at $1 billion; green gold (seeds), worth about $30 billion; vaccines ($750 million); diagnostics and reagents ($300-$500 million by 1990); and enzymes, with a market of $500 million.

It is difficult to make a distinction, as is often done, between the strategies of the pharmaceutical, chemical, food-processing, petroleum, and other affiliated industries. How, in fact, can one compare the strategy of SmithKline, for instance, with that of Merck, Sanofi, or Takeda? The former originated in a Philadelphia chemist's shop opened in 1830 by the son of an immigrant, John K. Smith, and its industrial adventure only really began in 1932, when it launched benzedrine. Or, where the chemical industry is concerned, how can Monsanto's ominidirectional strategy be compared with that of ICI, which bets heavily on long shots for the future, or with Sumitomo, which concentrates on development, or, in food-processing, with Ajinomoto and Suntory's decision to get involved with immunology, while General Foods "industrializes" traditional techniques? As for the major oil companies – Shell, Standard Oil of California, and Exxon – little can be said, as in most cases they have opted for secrecy.

Where big companies are concerned, pharmaceutical firms benefit from long-term research and tried-and-true experience in fermentation techniques and chemical companies are used for "creating" molecules. In the energy field, companies find it prudent to prepare less costly alternatives in case of a crisis. And as Campbell, General Foods, Ajinomoto, and Kyowa Hakko have become aware, the food-processing industry may indeed be the first to become concerned with new techniques. But biotechnological research is also of interest to companies in other industrial sectors: metallurgy (Inco); electricity and electronics (Emerson, General Electric); glass (PPG); engineering and construction (Lafarge-Coppée, Fluor, Coppers); cosmetics (Revlon, with its research company, Meloy); etc.

Most of the major companies concerned have invested heavily in research centers and institutes: Monsanto some $150 million in its Saint-Louis facility; Du Pont $85 million for its Wilmington Center; Exxon $22 million for Exxon Biomedical; Searle $15 million for its High Wycombe pilot plant in Great Britain (it has other research centers in Skokie, Illinois, and Sophia Antipolis, France); Elf-Aquitaine $10 million for the Labège Center; Ciba-Geigy $19.5 million for its research center in Basel.

Biotechnology in Human Health Care

Biotechnology, as we understand the term today, really began with the discovery and production of penicillin. The significance of this was marked by the award in 1945 of the Nobel Prize in Medicine not only to Alexander Fleming, the bacteriologist who discovered the antibiotic powers of the mould fungus *Penicillium,* but also to the biochemist Boris Chain, who recognized the significance of the original observation, and to Howard Florey, who planned the wartime research and manufacture of penicillin.

In 1940s, when Chain first demonstrated the remarkable healing properties of penicillin, Britain was in no condition to mount the industrial effort needed to exploit the discovery. But the government let Florey tour US industry appealing for help. The response was immediate and immense. While back in Britain penicillin was being grown by surface culture in bottles and milk churns, US industry developed the vastly more productive technology of deep culture. A single 1,200-liter fermentation vessel could make more penicillin than all the "bottle plants" Britain had built. Distillers and Glaxo swiftly adopted the new technology. Yields improved at a phenomenal rate, from a maximum of 10 units in the first deep culture fermenter to 150,000 units or more a few years later.

These events marked the turning point of traditional industrial microbiology into an advanced manufacturing technology. Its achievement, however, was not confined to the fermenter, but extends to the separation and purification of antibiotics, different in each case.

Nowhere else than in Japan did penicillin mark such a decisive industrial turning point, for immediately after the war, in a completely devastated country, several companies turned to the production of penicillin. Only four of these finally survived, including the "candy maker" Meiji Seika, the country's leading producer, which has also become the world's second largest producer of antibiotics, and Kyowa Hakko. Kyowa's is a story worth telling: In 1936, three Japanese established a cooperative to produce alcohol, Kyowa-kai. During the war, at the government's request, Kyowa made isooctane (aviation fuel) from butanol, drawing on its experience in the field of fermentation. The war over, the main factory at Hofu turned to the production of spirits (gin and shochu – grain alcohol), and soy-based foodstuffs, but also butanol and acetone. Kyowa started to develop penicillin in 1947, and in

Hamsters
Flies, mice, monkeys, hamsters – for various reasons, researchers have their own favorite animals for experiments. The Japanese firm Hayashibara, for its part, has chosen hamsters for the production of lymphokines.

1949 the Kyowa Sangyo company assumed its present name. Subsequently, in 1951, an agreement was signed with Merck for the production of streptomycin. Kyowa Hakko's great turning point came in 1956, when S. Kunoshita discovered the famous fermentation process for the production of glutamic acid by controlling the metabolism of a certain microorganism. And it was also that year that Shigetoshi Wakaki, collaborating with Tohju Hata of the Kitasato Institute, discovered an important anti-cancer agent, mitomycin C. Today, Kyowa Hakko is, along with Ajinomoto, one of the leaders in biotechnology in Japan, with a highly diversified business extending into food products, pharmaceuticals, petrochemicals, fertilizers, ethanol by fermentation, etc.

Biotechnology already exists as an industrial activity. One authoritative estimate suggests that there are about 100 products of biosynthesis established in the market today, with sales that totaled $9-$10 billion in 1980-1981. Many of these are competing with the products of chemical synthesis: for example, lactic acid, riboflavin, and various steroids and amino acids. These 100 different products are being manufactured by about 125 companies worldwide. Almost half of these companies are based in Europe, the rest mainly in the United States and Japan.

The first products of the new biotechnologies, such as genetic engineering, began to reach the market in the early 1980s. In the 1980s, many major

*1, 2. **Yesterday... and today**
A shop producing aspirin at
Rhône-Poulenc about 1918
(1), and contemporary
installations for the multiplication of cells and viruses
at Rhône-Mérieux (2).*

1

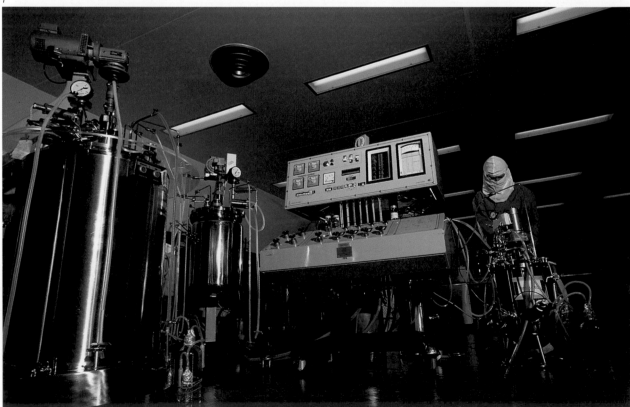

2

pharmaceutical companies acquired part or all of smaller companies with specialist skills in the biosciences. Du Pont bought New England Nuclear, for example, one of the biggest companies specializing in medical diagnostics. Schering Plough Corporation paid $29 million for the DNAX Research Institute, a California NBF. Most were content with a shareholding, however, that would give them a "window" on the new developments.

Agriculture

The pattern of big-company strategy toward biotechnology's impact on health care is being repeated for the agricultural industries, although the time scale is expected to be longer. Unilever, the Anglo-Dutch agrofood group, was among the first to put a big scientific effort into the possibilities of radically modifying plants through the new biotechnologies. As we already saw in our chapter on plants and seeds, the number of companies engaged in research is enormous. The Meccas for such research are unquestionably the University of California at Davis in the United States, the University of Ghent in Belgium, INRA (Institut National de la Recherche Agronomique) in France, and the Plant Breeding Institute in Great Britain.

The potential market is far larger than that open to pharmaceuticals. But research is long and costly, and sometimes the competition with chemicals is stiff indeed. Certain sectors, however, such as pesticides, may in the long run prove extremely profitable, first of all because there are a large number of insects that are unaffected by chemical pesticides, and second because certain pests are beginning to develop a resistance to insecticides and fungicides. Tate and Lyle have initiated a development program for several organisms isolated by the Glasshouse Corps Research Institute, including micromushrooms, to fight against certain insects. Dr. Stephen Lisansky sketches a comparison of the respective costs of the new chemical products and biological agents: "While a new chemical pesticide may cost $12 million to develop and require annual sales of $30 million to break even, a biological agent may be discovered for the sum of $0.4 million and require sales of only $0.6 million to show a profit."

In addition to the purchase of the major seed companies, which we considered in chapter 12, we must also mention Röhm and Haas, which is involved in research on hybrid wheat and is financing research on insect-resistant plants at the University of Ghent; Shell, which has bought out Nickerson; ICI, which has bought Garst Seed; Union Carbide, which has purchased Keystone Seeds; Elf-Aquitaine, which has bought Rustica;... and this is only a beginning.

An Original Example of Diversification

Considering the biotechnologies and big-company diversification, imagination has moved into the driver's seat: traditional, prudent or resolute, audacious or fearless, narrowly or broadly aimed. We have watched these strategies unfold in the United States, Japan, and Great Britain. But in France we must mention the original initiative taken by a cement-manufacturing concern, in which bio-

technological activities may play an increasingly important part.

It was in 1833 that two brothers, Edouard and Léon Pavin de Lafarge, opened a limestone quarry in the Rhône valley. One hundred and fifty years later, the Lafarge group had become a leader in the building-materials industry, essentially manufacturing cement. As a result of the energy and cement crises in the early 1970s, the company's president, Olivier Lecerf, decided to diversify into another business (outside of the building industry) that, in the long run, could absorb the shock of any new crisis and spread the risk over several sectors.

The working group met in 1978 to establish a list based on priorities drawn up in Germany, Japan, and the United States as well as in France. Several sectors could be distinguished: *computerization* (office, communications, robots) – crowded, this sector required too many specialized personnel and a sustained marketing effort, something that was not Lafarge's strong point; *ocean-bed mining* (polymetallic nodules) – only a change in the international law of the sea would make this worthwhile; *space exploration* – "flying cement" would be a strange combination; *new energy sources* – too distant from Lafarge's current preoccupations; in the field of the *biotechnologies,* on the other and, Lafarge had many assets – a proven capacity for managing long periods of time and heavy investments, energy, and continuous processes.

But the biotechnologies constitute a wide field. The narrowing down process was resumed: "We thought as businessmen," says P. Juston, the firm's information manager. "We wanted to settle on a sector which was still open to competition (which, for instance, excluded the pharmaceutical industry), but which was already well-established (which excluded sea-bed mining). Why not then get closer to those already on the biotechnological scene?"

It was then that Lafarge turned to Baron Coppée's Belgian company, which, like most family-owned firms, was suffering a bit from obsolescence, but which had many years' experience in the field and was the first during the 1950s to derive value from sugar residue (molasses), thus engaging in biotechnology before the word was even invented. In its 130-year history, Coppée had passed from coal and steel to engineering and then to fermentation. The company has two factories in northern France: Orsan in Nesles, which makes glutamate, and Eurolysine in Amiens, founded jointly by Orsan and Ajinomoto in 1973. The Lafarge-Coppée marriage was celebrated in 1981. The bet was successful inasmuch as Orsan's profits rose by 25% in 1983 and by 50% in 1984, and it began to produce another amino acid, aspartic acid. Where lysine was concerned, Orsan and Ajinomoto extended their lease by starting up another plant in the United States (Heartland Lysine in Iowa). With the two largest plants in the world for the production of glutamate and lysine, Lafarge-Coppée has become world leader for lysine and aspartic acid and among the frontrunners with respect to glutamate.

A new turning point came in 1982 with the purchase of a 10% interest in the Claeys-Luck Company. After three years of observation and apprenticeship, Lafarge-Coppée decided to get into the market for seeds. In the spring of 1984, the company bought a small hybrid corn firm (Wilson Hybrids), again in Iowa, the heart of the American "corn belt." In December of the same year, it bought out Celanese subsidiaries: Harris and Moran Seed, one of the world's top ten companies for vegetable and flower seeds (with installations in the northeastern United States and in California), and Celpril, a specialist in the coating of seeds, which can pave the way to genetic engineering. At the beginning of 1985, Orsan and Claeys-Luck together founded a subsidiary, Hybrinova, which concentrates on the extremely promising market for hybrid wheats. Bernard Collomb, president of Orsan (as well as general manager of Lafarge), puts it this way: "These development should be viewed in a planetary perspective where the demand for food is growing rapidly, though in the case of certain major crops a surplus is grown in our industrialized countries. World population forecasts lead us to expect that, particularly in Southeast Asia, a considerable improvement in agriculture will be necessary to meet the demand for food. It is reasonable to think that, despite the sociological inertia prevalent in agriculture, an important role will be played by improved plant species. And it should be remembered that 80% of the present world market for industrial seeds, estimated at $10 billion, is in the developed countries (roughly 40% in North America, 30% in Europe, and 10% in Japan)."

1, 2, 3, 4, 5. **Sodium glutamate**
The production of glutamate at Orsan: fermenters (1), the ultrafiltration shop (2), continuous crystallization (3), and packaging line (4). Also, the fermentation process for sodium glutamate as shown in a drawing from Ajinomoto (5).

1

2

3

4

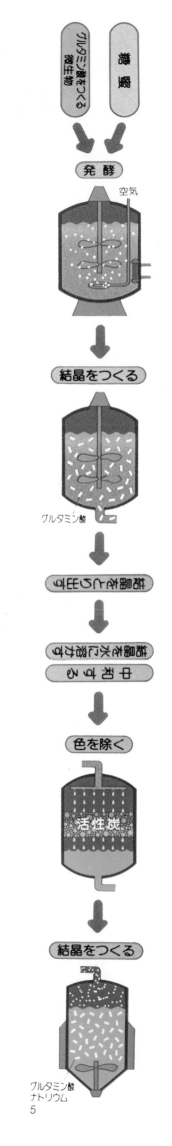

糖蜜　　グルタミン酸をつくる微生物

発酵　空気

結晶をつくる

グルタミン酸

塩酸や水に溶かす

塩酸や水に溶かす中和する

色を除く　活性炭

結晶をつくる

グルタミン酸ナトリウム

5

Corporate Strategies for Diversification

by Bernard Langley, responsible for the ICI Group's interactions with universities.

After degrees in physiology and chemistry and a PhD in Organic Chemistry at Cambridge University, Dr. Langley did postdoctoral work at medical schools in Cleveland and at Yale before returning to a research fellowship followed by a lectureship at the University of Leeds. He joined ICI in 1956 and became Biochemistry Manager in 1965. He joined the Policy Group of the Corporate Laboratory in 1975, where he plays a part in the company's general strategy. What are the attractions and dangers inherent in diversification in the "life industries"? Have not major companies – Corning Glass and Volkswagen – shrunk from the quicksands opening before them? Strategies vary, but they all have points in common, which B. Langley attempts to describe here.

Biotechnology, however one defines it, will affect almost every human activity – our health, food, fuel, waste, and virtually every biological process or interaction with our environment. It also proffers such prospects for controlling biosystems that we may choose to harness them for doing things now done only by physics and chemistry.

If all this is true, if the science is burgeoning and the applications are wide – and if even only some of our present hopes are realized – one might well ask: Why do companies need a strategy? If several of the geese will lay golden eggs, why worry which bird to feed? But not all the geese will lay. Some of the eggs may be small, leaden, or both. Some may require platinum eggcups. Many of the best eggs will come not from today's geese but from their distant offspring in future generations.

Biotechnology has been oversold as a way of making money. Commercial time scales have been unrealistic, despite the fact that many of the scientific time scales have shrunk beyond expectation due to the imaginative vigor of researchers. Overall the technology sparkles, but it is far from clear who most stands to gain what, and when. There will be no problems with scientific interest; there may well be with bank interest.

None of this should be surprising. Something similar must have happened with every potential step change in technology. Today we may be greedier and expect a faster pace, but the bigger the change – and this one ultimately will be very big – the bigger have always been the problems, delights, and opportunities. Above all, a step change in science and technology inevitably brings opportunities for business change, for diversification.

An Enabling Technology

It is argued that a new technology like biotechnology will have profound effects on society. When was this last claimed? What was the topic? What happened?

The most obvious comparison is with electronics. Few technologies can have grown so fast or have had substantial effects on such a wide range of human activities. At first sight they have much in common. Both stemmed from basic science, from what would now be called "curiosity-driven" rather than "strategic" research. Both are still dependent on high science, which would in any event have been self-sustaining. But each has received early and continued stimulation from farsighted scientific entrepreneurs who saw what use could be made of the research.

Although each is dependent on frontier science, in both fields this science has been of a type that readily stimulated public excitement and expectation. Each required long years of dedicated research by a small number of enthusiasts, often in disciplines only indirectly related to the main line. There have been mammoth jumps in both fields, many of them recent. Some of the basic science, say that involved in constructing lasers or producing insulin from bacteria, has not only allowed the unthinkable but has moved from laboratory to production scale in a very few years. Both technologies belong in commerce to massive industrial enterprises, but equally both have given rise to hundreds of small specialist companies. Both technologies are fast moving and nowhere near their maturity. Both are potentially so broad in their applications, and likely to make such changes when applied, that they have caused all near them to sit up and take notice. They are broadly alike in so many ways that perhaps more can be learned, especially in the context of diversification, by looking for points of difference.

Both are essentially "enabling" technologies; portfolios of techniques that relate to very-wide-ranging activities and problems. But what exactly do they enable and who wants to be enabled? Although at the arcane research end one scientist's fast computer is equivalent to another scientist's gene-splicing system, the similarities attenuate as one approaches the market. Can one ever envisage the biotechnological equivalent of a domestic calculator, to say nothing of a color television, hi-fi, or cordless telephone? Is there any biological equivalent of automation? Can one see any biotechnological products being mass-produced in small units and used on a scale that will cause the dramatic drops in production costs analogous to those for microchips?

The impact of biotechnology will be different. Different types of money will change hands and over different time scales. For example, one can see biotechnology bringing vast social benefits to the hungry and sick in poor countries, but at the same time one cannot, alas, see the wealthier governments underpinning this work, as they now do electronics for their own defense.

Those who need a speedier return must go for today's best, in say drug research, agrochemicals, or plant breeding. Even here the real returns will require a long haul. And in most areas biotechnology lags twenty or more years behind electronics. So the longer-term returns will not be made in today's world, but in the very different one resulting from several decades of the silicon chip and its congeners.

Major Issues In Diversification

Let us examine the key factors involved in diversification. Some are peculiarly important to biotechnology, but most govern all business diversifications.

Who will pay? Some of the best biobusinesses (for example, pharmaceuticals) are so well understood that there is no problem at least in deciding who *should* pay. But even here there are problems – who will pay for Third World ills, for example? In a similar vein, the veterinary and agrochemical equivalents are so cost-effect dominated that a new product had better either be satisfying a desperate need or be of very high potency. At the large-tonnage end, the problems of rewarding capital are much greater and become immense when, say, dealing with waste disposal. Few biotechnologists have found much truth in the English saying, "Where there's muck, there's brass."

Where is the added value? The real costs are so large that those initiating diversifications of this sort must analyze carefully at what stage the real cash markup occurs. If it is your own process, well and good. If you have to enable somebody, enable yourself. Otherwise, go for the product. The history of the chemical industry, for example, is one of continued and widespread diversification. Most of its more profitable ventures, from drugs, agrochemicals, and fertilizers to paints, explosives, films, and even recording tape, have come from downstream moves toward the market. Some chemical companies do indeed make a reasonable living from supplying intermediates, but it is hard to see this being a prime target for biotechnology at present.

Diversification or diversion? Although those under severe threat have little choice, all those others in, say, the agrochemical or pharmaceutical businesses, no matter how prosperous, need every last man in their present research programs. You cannot enter this new field with a man and a boy. Biotechnology will not be the answer to every biological mediator problem. Unless you can expand, what will you stop to start this?

What is the entry fee? It may seem quite cheap to start a small biotechnology laboratory, but what if it is successful? What will it cost to make and market the product? If the diversification is into a new business area, how well do you understand that business?

What is closest to the present business? There is great merit in first testing the biotechnological water by applying it to peripheral extensions of your existing business. Infinitely better judgments can be made both of opportunities and threats. Masterly examples are Lilly's collaboration with Genentech on insulin, and Unilever's work on cloning oil palms. Small diversifications have led to massive business changes. Almost all the chemical industry has grown in this way.

How patient are the backers? The real returns from production biotechnology, as distinct merely from research, will be slow to accrue. It will be largely a twenty-first-century affair. Anything other than experiments to prove some scientific point are

rather hard to do in practice. Those involved in translating such observations into, say, the reproducible production of homogeneous protein from an "engineered" organism on a pilot scale need patience themselves, as well as from their financiers.

Timeliness or promise? In the United Kingdom the best-known science-funding agency, the Science and Engineering Research Council, judges all research grant applications first by these two criteria. They can scarcely be bettered in that context or in the context of diversification. Nobody doubts the promise of biotechnology – what of its timeliness?

Will the problems it aims to solve still be around by the time the biotechnological solution has been worked out? Other technologies will not stop. For example, organic chemists are currently making huge strides in achieving regio- and stereo control in stoichiometric and even in catalyzed reactions. They are using solid-phase synthetic systems to make ever-more complex molecules. Fermentation biosyntheses and transformations may not always be the best way in future of making relatively small amounts of complex molecules, just as they can rarely compete with chemistry at present in producing large-tonnage commodities.

Nor is biotechnology itself the only thing that is fast moving. The ICI single-cell protein plant is the largest continuous monoculture fermentation unit in the world and generally regarded as a technical triumph. It was a massive enterprise and took over ten years from conception to full-scale production. Unfortunately for ICI, during that decade the price of its starting material rose severalfold, and the price of soya protein, with which its product competed, fell to an all-time low. Eventually the technology should pay handsome dividends for a variety of purposes – but not on the time scale first envisaged. Since biotechnology is really for the future, diversifications involving it now should be aimed at solving problems for the future. Fortunately for the practitioners, such problems abound.

Is there national interest? There is always a big attraction in getting the largest banker to prime the pump – providing always that his demands are not excessive or restricting.

Although, fortunately, there is too little military interest in biotechnology to warrant any direct equivalent of what governments have done for electronics or aerospace, most have indeed recognized that the subject is sufficiently important to warrant special help. Probably the paramount example is that of Brazil, which is pioneering the massive production of fermentation ethanol as a motor fuel. This is effectively the biotechnological equivalent of the coal-based chemistry done in Germany in the 1930s

and being done in South Africa today. Other predominantly agricultural countries are realizing that this new technology may alter the relative value of their crops. The Russians have taken a special interest in single-cell protein production from waste to relieve their protein supply problems.

Sometimes national concern for the present hinders progress for the future, as, for example, when EEC glucose prices are maintained at 2.5 times the level elsewhere. However, in general, technically advanced countries are naturally seeking to underpin both academic and industrial research in biotechnology to maintain national prosperity.

Which Industry?
No industry will be more affected by the changes biotechnology may cause or by the opportunities and challenges of diversification that it proffers than the chemical industry. It is large and, at its best, highly productive. Its success is traditionally closely coupled to advances in basic science. It already makes much of its profit from pharmaceuticals, agrochemicals, and materials made for biological uses, and is anxious to develop such interests. In its primary role of converting one material into another, it already supplements chemical with biochemical processes. It is anxious to improve its understanding of biological systems, especially in their interaction with chemicals, both to improve the efficiency of those it makes with this intent and to diminish the effects on the bioenvironment of those it hopes will be biologically inert.

But other industries will be loath to accord it the lion's share. Healthcare, agriculture, and the production of food and drink are, in effect, solely concerned with biological systems from start to finish. As such they will be profoundly affected by the greater understanding of biology and the newer opportunities for modulating biosystems that will stem from biotechnology.

Let us take an imaginary example, the future production history of an alcoholic drink. The cereal crop might be from cloned stock, conceivably with a designed nitrogen-fixing symbiont. The plants would be protected until harvest with the newer panoply of agrobiochemicals. The grain would be collected and any malting or similar process biochemically controlled. The more subtle aging of some of the grains before, and indeed of the beverage afterward, would be monitored by biosensors. Whatever fermentable oligosaccharides were needed would be provided by enzymatic hydrolysis. The fermentation would be continuous and automated. The product might be freed from clouding agents by supported enzymes. The waste would be disposed of by a specific fermentation system. An enzymatic detergent would be used to wash the glasses.

Clearly there will be room, besides those

concerned in primary production, for countless makers of specialized fermentation and separation plants and monitoring and analytical equipment.

Pharmaceuticals and Agrochemicals

The pharmaceutical business, closely followed by its agrochemical equivalent, will be the earliest and largest beneficiaries of the new biotechnology. This is even more obviously true if one broadens the definition, to rename them health care and agricultural aids. Why is this so? Consider some of the more obvious characteristics, particularly of the pharmaceuticals sector.

1. They are fast growing and highly profitable. Only electronics and computing are in the same league.

2. They have not always been in this position, nor was it always obvious, even within the chemical industry, which mostly spawned both, that this would be the case.

3. Together with electronics, they stand as shining examples of the conversion of high science into hard cash.

4. They can flourish without oil, coal, or mineral abundance, or an agriculturally luxuriant climate. They are especially attractive to those countries with neither.

5. It now costs very much more to prove a drug or pesticide is safe and useful than it does to discover it in the first place.

6. Most of the big companies have grown slowly and been in this business for a long time. Products of this sort do not develop, register, and sell themselves in 80 countries without the continued and updated attention of many different types of experts.

7. All of its progress starts from laboratory discovery. This, although difficult, is sufficiently challenging to attract scientists of very high caliber to the industries. Everything depends initially on the interaction of groups of talented and unusual people.

8. Although mature, neither business is senile. There is continued need for renewal and improvement of the products in both fields. Many of the major human disorders are still largely unaffected by chemotherapy.

9. The prosperity of both industries has fed back to enliven and in part endow those biological and chemical fundamental sciences on which they depend. However, in contrast to many other parts of science, on which other industries depend, these areas are themselves in a state of high and self-sustaining ferment.

10. Because of the supreme interest in health, esoteric and expensive techniques can be used to maintain or restore it. Only "defense" is comparable as a stimulus to the pioneering of scientific techniques,

which are then invaluable for other purposes.

11. Commercial competition is fierce. The costs of health care are scrutinized by all governments. Veterinary and agrochemical products are, of course, sold purely on a cost-effectiveness basis without emotional overtones.

12. The industries are luminous examplars of the advantages of selling not processes, methods, intermediates, or techniques, but the complete formulated products for the eventual customers. Future products will be complete "treatment packages" of, for example, diagnostics or seeds.

13. The businesses peripheral to both areas – diagnostics, medical aids, dental products, therapeutic diets, seeds, plants, soil and crop treatment techniques, and equipment – are mostly very different from the core businesses of pharmaceuticals and agrochemicals. Almost every aspect from innovation to cost structure and marketing is different. Some of the primary practitioners will broaden their interests. Some will reject them as initially less profitable. There is room for biotechnologists to start in these areas on the ground floor.

Clearly there are many ways in which biotechnology, in both the enabling and progenetive senses, will have a huge impact on both pharmaceuticals and agrochemicals in the medium term. In the long term, it will change some parts of their business beyond recognition.

Diagnosis or Cure?

The balance between preventive maintenance and breakdown repair in any field depends on what is possible and what each costs. In extreme situations, where the cost of failure is prohibitive, as with nuclear power and in aerospace, diagnostic techniques are stretched to the ultimate. In other cases, such as human medicine and road transport, things are more evenly balanced. In general, the developed world is moving steadily toward preventive maintenance, especially in medicine. In any field, preventive maintenance first requires selective and sensitive techniques for monitoring what is happening. Then one must define normality and finally correlate abnormality with later events. In recent years, biotechnology has furnished a huge portfolio of techniques for probing complex biosystems. Monoclonal antibodies, gene probe analysis, enzyme analysis, enzyme-bound electrodes, and the like have been used to devise an impressive array of diagnostic methods. As their biochemical origin would imply, many are inherently both selective and sensitive. When coupled with electronic methods for shifting the signal-to-noise ratio and computational methods of pattern recognition, they herald a quantum leap in diagnosis.

The leading edge of this work, true to tradi-

tion, is, of course, in human medicine. But the applications are enormous in veterinary medicine, animal breeding, forensic and security work, and throughout the whole domains of agriculture and environmental control. Within biotechnology itself, of course, these techniques, by helping the measurement of nutrient and metabolite levels, will permit much more precise control of fermentations, leading to greater efficiencies and, possibly, the production of hitherto unobtainable products.

What Is Special about Microbiology?

The main realities of applied microbiology are well known, but some deserve emphasis, since they determine so many biotechnological opportunities.

1. The techniques for carrying out contained microbial fermentations are well established.

2. Biochemists argue that, in the wake of recombinant DNA discoveries, microbial fermentation should be the preferred way of carrying out any biosynthesis or biotransformation on any scale.

3. Although such contained fermentations are commonplace, for high-cost materials the plants are as yet too costly and the processes too inefficient to be used for many commodities or low-cost manufactures.

4. Even though traditionally contained, or at least controlled, large-scale fermentations (such as are used in the preparation of food and drink) could benefit from modern biotechnology, there is every excuse not to change low technology, customer expectation of tradition, little patentable property, or vast capital in existing plants.

5. Even where microbiological processes occur in the field, they can be influenced by modern biotechnology. Apart from the well-known suggestions for adapting nitrogen-fixing bacteria, there is talk of releasing other special types to improve soils, remove waste, or aid fish farming. Although the organisms cannot be contained, there are growing prospects of modifying their metabolism for our benefit by exogenous chemicals, as is now done with soil nitrification inhibitors and ruminant growth promoters. There is another side to this coin, what one might term negative biotechnology. Thus it is important when developing new agrochemicals to ensure their residues have no harmful effect on useful soil organisms.

The reason why the solution to the genetic code was welcomed as such a momentous discovery was, of course, that the system is ubiquitous in all living things. The mainline process whereby information, as DNA, ends up as protein happens throughout nature. It follows that genetic engineering and other new biotechniques can be applied in principle throughout the plant and animal kingdom.

Removing Impediments

Virtually all the biotechnological opportunities I have mentioned have been concerned with "go" – with the many new things that can be made or done, or old things made or done better. But what can biotechnology do to remove or reduce the many "stop" signals that hold us back – for example, in toxicology and ecology?

Accidents, side effects, and occasional malpractice have together given rise to much public concern and suspicion. In turn, this has provoked legislators to devise ever longer and more elaborate safety tests for new and old products. Not infrequently these are exaggerated to the point of scientific futility to stay popular alarm. The pharmaceutical and agrochemical businesses have borne the brunt of this legislative efflorescence. Safety testing in both fields accounts for a huge and growing proportion of the total invention and development costs for any new bioregulatory substance. Nor are these industries the only ones affected. Materials are not necessarily devoid of biological effect just because they are intended for nonbiological purposes. Although the tests required for most such materials are, not unreasonably, not so stringent as those demanded for drugs and agrochemicals, their associated costs often restrict innovation. Where novel substances are intended for use in food or drink, there the testing is so expensive as to have crippled work on new sweeteners, flavors, and preservatives.

The help in devising simpler and more meaningful tests that is coming from molecular biology should in effect release the brake from many business areas, including several apparently remote from the mainstream of biotechnological advance.

Most of the enabling techniques for studying the detailed interaction of chemicals with living systems are highly relevant in toxicology and ecology. The ability to seek fine metabolic or biological differentials between species will be invaluable. The possibility of being able to probe for chemical damage to DNA itself should make screening for genetic damage and potential carcinogenesis – two of the largest and most emotionally loaded toxicological problem areas – infinitely simpler and, above all, more scientifically satisfying.

What Can We Own?
Those investing in new or old technologies want, above all, to be able to ensure a return on their money. Biotechnology will inevitably contribute immeasurably to human welfare, but how can those investing huge sums in research and development be certain of getting their money back? Nobody will spend on this scale if anyone is free to imitate or even use his invention with negligible outlay. Patents and intellectual property are key issues.

Each fresh spurt of potentially profitable scientific advance has caused lawyers to grapple with new problems. The development of industrial fine organic chemistry for dyes, pharmaceuticals, and agrochemicals was retarded in many countries until substances could be protected by patents as such and in relation to new uses. Comparable problems are arising today in relation to the protection of computer software.

Many key aspects of the patenting of biotechnological inventions are still being worked out by legislation and court practice. Their resolution will be crucial to those seeking to extend research in their own areas of biotechnology, let alone diversify into others. For example, the decision to allow the patenting of new plant varieties has provided an enormous stimulus to commercial plant breeders.

Around the Corner
Only those at the frontier can make much valid comment, but those much farther back at headquarters, wondering where to put their money, where to diversify, can ask how to find out, how to ensure they can capitalize on the most relevant future science.

Strategy is one of the words of this chapter's title. In one common usage, "strategic research" implies that it is applications driven, in contrast to "basic research," which is curiosity driven. If that is the case, then those contemplating future business strategies in biotechnology should realize that all the initiating discoveries in this field came from the basic rather than the strategic end, from penicillin to restriction enzymes. In both cases, the discoveries were soon used for strategic purposes. In some cases, practitioners of basic science themselves soon moved to a strategic mode.

The point is made here in conclusion only to remind those unfamiliar with the problems and profits involved in diversifying into a fast-moving scientific field that they must be ready for the unexpected. It is more like buying a greyhound than a truck. The scientists work for interest alone. So do the bankers, but the word, as implied in the introduction, has different connotations.

To be successful in diversifying into a field such as biotechnology, especially in its present early, formative stage, one must be ready for almost anything. If unfamiliar with past in-house needs, with truly speculative research, one must either acquire expertise of one's own, or know where to get it. The main thrust of the work will be in academia for many years. Even large industrial enterprises, which traditionally fund the bulk of the research they need, themselves find that close collaboration with universities is essential in a field such as biotechnology. While recognizing that academics provide the mainspring for most that is truly novel in science, industrial collaborators should not complain if many academic ideas seem naive in the commercial sense. At a time when most national paymasters are daily urging their academic beneficiaries to work in more commercially "relevant" areas, research teachers can be excused for trying to get near the marketplace. This phase will pass. But only when those with a real understanding of the history of past successful commercial diversification based on pure science have their say. No subject more obviously requires clear debate between Academia and Industry. That many future profitable diversifications will flow from biotechnology there can be little doubt. In the meantime, by all means select the goose to lay the golden eggs. But don't force-feed it too hard.

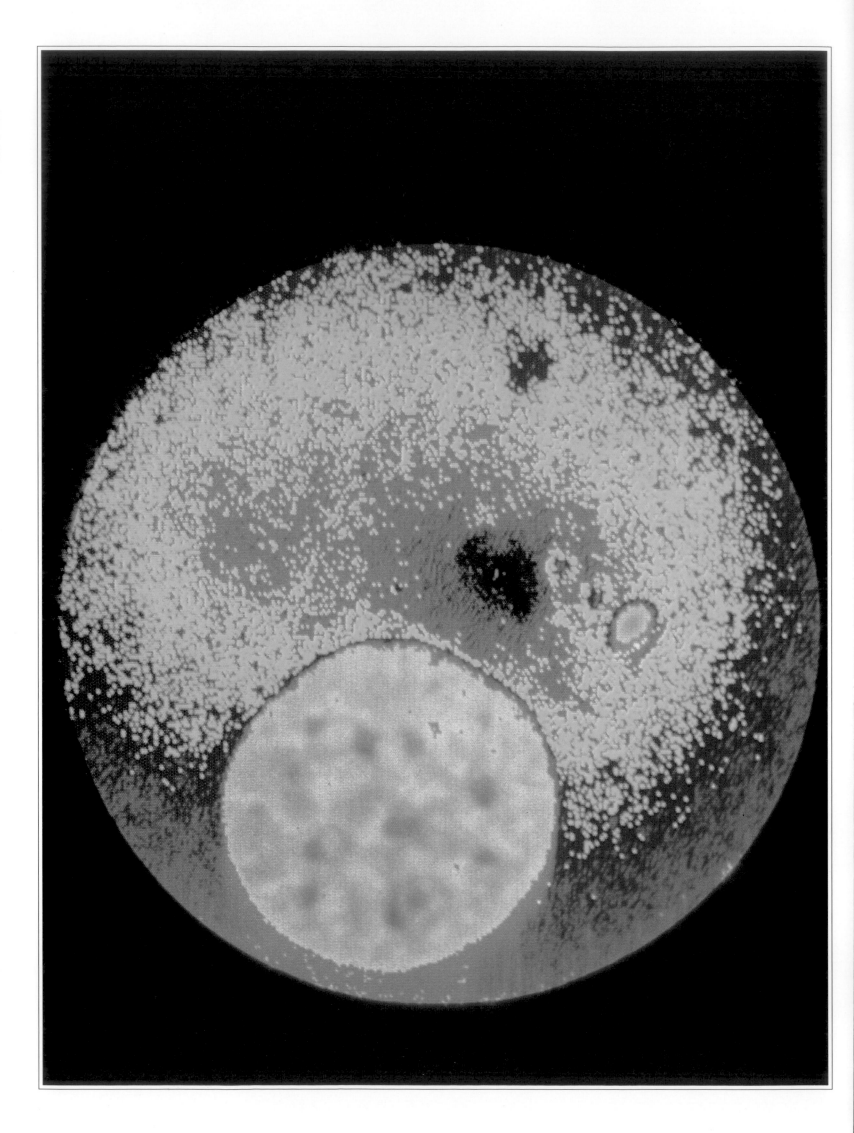

23

Planning: The State

"To European ears, the song of the cricket heard so clearly on autumn nights is but a pleasant sound. The Japanese for their part hear a spoken message in this song. Autumn days are getting shorter, it will soon be winter: the Japanese thus feel they are being told something about the impermanence of this world. It is through their sensitivity that the Japanese confront the essence of reality. They are endowed with a special insight that intuitively perceives the very heart of a phenomenon. On the other hand – and this dates as far back as anyone can remember – they have never been very gifted with respect to abstraction and analysis, and even less so for induction and deduction." It was in such terms that in May 1982, Yasuichi Nagano, honorary professor at the University of Tokyo and president of the Franco-Japanese Biological Society, welcomed the French delegation to Tokyo.

The global approach is, in fact, Japan's strategy. Like France, the land of Pasteur, Japan is a country of ancestral traditions where biotechnology is concerned. But even more than France, it is a land of pragmatism, accustomed to adapting itself to new situations and of profiting from its own techniques. A company such as Mitsui Toatsu, which is today becoming involved in the biotechnologies – as is Mitsui Petrochemicals – is part of a sogo shosha (which we might translate as a consortium), closely linked, as are Sumitomo, Mitsubishi, and so many others, to Japanese history. Mitsui started off as a family business: it originated, over 300 years ago, when Hachirobei Mitsui opened a fabric shop. He was also a moneylender, and initiated a policy of fixed prices and cash payment. Today, Mitsui is a giant conglomerate of nearly 1,000 companies in a wide variety of sectors, some of which are open to the biotechnologies. And if, since the beginning of the Meiji Era (1868), the government has been encouraging innovation in fermentation techniques, this is essentially because of the huge income it derives from sales of saki.

It is against such a backdrop of tradition and adaptation that the biotechnologies developed in Japan, with the role of innovator held by businessmen in the food-processing, chemical, and pharmaceutical industries (Kyowa Hakko for fermentation, Tanabe Seijaku for immobilized enzymes, and Mitsubishi Chemical in founding the first fundamental research institute devoted to the life sciences in 1971) rather than by the universities.

Yet it would be dangerous to ignore Japan's university researchers (particularly the microbiologists), whom Nobel Prize winner Severo Ochoa considered among the best in the world. In 1953, to support this research effort, the University of Tokyo even established an Institute of Applied Microbiology. "Nevertheless," states W. Yamaya of Mitsubishi Chemical, "research and development in Japan is supported mainly by private industry, not only in the case of biotechnology but in anything concerned with advanced technology. The government willingly opens its mouth, only reluctantly its coffers."

While it is true that the Japanese in most instances prefer an industrial patent over an article in a scientific journal, the government has nonetheless supported and coordinated the nation's effort. As the result of two reports by the National Council on Science and Technology (in 1971 and in 1980), in 1981 the Ministry of International Trade and Development (MITI) launched a ten-year program to develop three essential sectors: gene splicing, large-scale cell culture, and bioreactors. MITI is assured of the close collaboration of fourteen companies in pursuing this program, to which it has devoted over $100 million. Five other government agencies are financing research in the biotechnologies: in 1983, they received the sum of $67 million from the government. Alongside this public effort we might mention the activities of the Life Science Committee of the Keidanren (the largest industrial association in Japan) and those of Bidec (Bioindustry Development Center), a private organization founded under the auspices of MITI and with a

Simulation
"What does the earth resemble most?... A simple cell" (Lewis Thomas).

209

membership of about 140 companies. Furthermore, everything is being done to ease legislative regulations (the volume of fermenters, safety conditions, the state monopoly on rice, etc.).

The Japanese start off from a pragmatic view of their strengths and weaknesses. Saburo Okita, a former minister, stresses the fact that the Japanese are not infected with "staromania" and that they more often talk of an entire company rather than a single individual as responsible for developing some new process. Others tend to evoke the rigidity of a system that forbids a university researcher from working awhile in industry and then returning to the university. At Sumitomo Chemicals, Rintaro Ishiwata says, "We have but a single Nobel Prize winner, Kenichi Fukui; the Americans have a slew of them." At Hayashibara, Sumio Kamiya says, even more mysteriously, "We are small in the good slots, good in the small slots." And he tells the following story: Three businessmen are in an airplane that goes into a dive. The Frenchman commits his soul to God and starts to pray. The Japanese opens his mouth, but the American quickly gags him with his hand. "No you don't," he says, "you're not going to tell us your business philosophy!"

The Japanese business philosophy is, in fact, a strategy with several stages. The first of these consists of "buying time" (T. Noguchi at Suntory), that is to say, buying foreign patents, and counting on the second stage, production, while doing everything possible to cut prices. In addition, Japan's protectionist policies force foreign businessmen to adopt a policy of signing commercial agreements with Japanese companies if they wish to enter the market. The second stage is the so-called *feedback effect*, that is to say, sending Japanese researchers to Europe and the United States to assimilate the latest discoveries in basic research (this process is sometimes but rarely employed in the opposite direction, and we might mention such French scientists as G. Gelff and P. Monsan, who worked with S. Fukui at the University of Tokyo). The third stage consists of organizing a potential for *rapid reconversion* made necessary by the policy of life employment, which is commonplace in Japan. In 1981, fewer than 10 companies employed more than 10 researchers in genetic engineering. A year later, 52 companies employed 60 or more. The fourth stage is to recruit *more and more women* determined to prove their competence in a field not necessarily favorable to them: Suntory has 40 women among its 100 researchers, and 2 women manage Mitsubishi Kasei's 2 research departments.

United States

If we consider only the number of patents filed, the United States would rank second, after Japan. But it is the most advanced country from the standpoint of basic research in genetic engineering. An imposing report published in the United States by the OTA (Office of Technology Assessment) in January 1984 on the commercialization of the biotechnologies and international competition in this field concludes by stating that the United States is the world's leading investor in basic research, but that its financial effort is quite limited with respect to applied research, considered the province of

industry. In no other country in the world can there be found a research budget comparable to that of the National Institutes of Health (NIH) in Bethesda, near Washington, DC, for its program in support of biomedical research. It was on the basis of this financial support for basic research that the commercial potential of molecular biology became apparent and that an alliance was formed between university researchers and businessmen, leading to the establishment of the NBFs (new biotechnology firms). The United States (leader in antibiotics, hormones, and blood derivatives) has not developed a governmental program per se. But university research is supported by federal allocations, frequently using the NIH as an intermediary, and many programs receive state support, while encouragement for innovation assumes the form of tax advantages. Pressure from ecologists also assumes greater violence and impact than in any other country and frequently hampers the speedy development of a technological breaktrough. Someone like Jeremy Rifkin, who waves the flag of the impending Apocalypse at every new experiment, is a figure who could exist in no other country, and with whom the government as well as universities and business must continuously cope.

Europe

In Europe, the hour of mobilization against the two superpowers has rung, but the situation differs widely from country to country. In some countries, we cannot speak of the government's global strategy, since the focal points are the big, forward-looking companies: in Switzerland, the three great pharmaceutical firms (Ciba-Geigy, Sandoz, and Hoffmann-La Roche); in the Netherlands – despite the establishment in 1981 of a government coordinating committee for biotechnological research – Gist-Brocades and Akzo; in Italy, Montedison; in Denmark, Novo; in Sweden, Pharmacia and AB Fortia (despite the existence of the government pharmaceutical company KabiVitrum AB, which owns 50% of Kabigen AB); in Belgium, the Union Chimique Belge and the Solvay Institute.

In West Germany, the government's policy was seriously put to the test by the agreement signed between Hoechst and the Massachusetts General Hospital of Boston ($50 million over ten years to support the research of H. Goodman and his team). Back in 1980, Bayer had allocated $300,000 to MIT, but this time the enormity of the sum was such that there was an outcry against the flight of German capital. Yet since 1974 the government has granted huge credits, mainly to the research centers of Gesellschaft für Biotechnologische Forschung (GBF) and to Jülich. But even if the credits allocated have been larger than in France or Great Britain (Albert Sasson says ten times as much), there is still a lack of qualified researchers in biology, a situation that has not changed since the scientific brain drain that occurred under Hitler.

France for its part enjoys an appreciable reputation with respect to genetic engineering, molecular biology, biochemical engineering, and even the culture of plant cells. But, paradoxically enough, it seems underprivileged with respect to microbiology. Drawing his inspiration from reports written in 1979 by F. Gros, F. Jacob, and P. Royer, and in 1981 by J.-C. Pelissolo, Minister Jean-Pierre Chevène-ment in July 1982 announced a plan of mobilization for the biotechnologies that was largely the work of Pierre Douzou, an enthusiastic researcher at the Rothschild Institute and advisor to the government. Stress was laid on three main sectors: basic research on microorganisms, cells, and enzymes; biotechnological reactions, that is to say, everything having to do with selection and screening, control systems, extracting, and purifying procedures; and development and application in pharmaceuticals, food processing, the environment, and energy. The aim of the plan was to raise France's share of the world market for biotechnology from 7% in 1980 to 10% in 1990. Even in 1980, three public organiza-tions (CNRS, INSERM, and INRA) had decided to affiliate and to establish, along with the Institut Pas-teur, the 3G group (Groupement de Génie Généti-que). Among the leading French companies involved in the biotechnologies, many are state owned, and thus by right benefit from government aid: Elf-Aquitaine, Roussel-Uclaf, BSN-Gervais Danone, Rhône-Poulenc. And some of them hold majority shares in still other companies (Rhône-Poulenc in Institut Mérieux; Sanofi, an Elf subsidi-ary, in the Institut Pasteur-Production).

In Great Britain, the English equivalent of Doctor Douzou is Ronald Coleman, "the nation's chemist," who became the government's scientific advisor on biotechnology in 1982. In April 1980, a lengthy report had been presented to the government by a group of researchers presided over by ICI's former director of research, Alfred Spinks, that proposed government financial aid, the creation of university-industry liaison committees, and an easing of exist-ing regulations. Two years later, exactly, R. Coleman was the source of a national plan aimed at making the best use of the nation's traditional skills in bio-chemistry, molecular biology, and fermentation. According to the Spinks report, in fact, British industry at that time had the largest fermentation capacity in Western Europe. Furthermore, did not Britain's laws authorize such American companies as Eli Lilly and GD Searle to manufacture their first product resulting from genetic engineering in Great Britain? Priorities were drawn up: enzymatic engi-neering, diagnostics, applications in agriculture, equipment, and instrumentation. Coleman's team also called for the gradual evolution of a biotechno-logy aimed at producing high-quality products for advanced sectors to a biotechnology capable of mass-producing goods of less exalted value.

But even before publication of this government plan, it was due to state intervention that the first new biotechnological firm, Celltech, was estab-lished in 1981, with the National Enterprise Board providing 44% of its initial capital. Celltech enjoys a priviledged relationship with the Plant Breeding Research Institute and the laboratories of the Impe-rial Cancer Research Fund, and above all with the Medical Research Council.

As we have seen, strategies vary enormously from one country to another, from a national effort for some to the encouragement of private initiative for others. Sometimes, there is a combination of both: a complex fabric woven between the public and private sectors, with a certain feeling of unease when they intermesh too closely, as we shall see in the "dangerous links" between town and gown (that is, business interests and universities). About twenty different reports on the biotechnologies blossomed between 1976 and 1982, in Japan, the United States, Canada, Australia, West Germany, Great Britain, France, Belgium, and the Nether-lands. But any "crystal ball" predictions in a sector as volatile as the industry of life must perforce be constantly put in doubt, if not patently disavowed. All countries are aware of this, particularly the busi-nessmen, who at times impose a watch-and-wait attitude on their respective governments. Yet major trends are now coming to light, with the United States in the lead for genetic engineering and Japan for certain fields of application. Does not Europe risk paying dearly for its hesitation and los-ing control of entire facets of the world economy? Satellite countries or a third Great Power? Only the future will tell.

1. *Family portrait*
Johann Rudolf Geigy Gemuseus (1733-1793) surrounded by his family. It was in 1758 that he opened the hardware store on the Freie Strasse in Basel, which became the ancestor of the House of Geigy.
2. **Origins**
Like Asahi Chemical, Sumitomo Chemical forms part of a group whose origins date back more than 400 years, to a period in the sixteenth century when the brother-in-law of the founder, Riemon Soga, invented a revolutionary method of refining copper, immortalized in this woodcut.
3, 4. **The art of wielding chopsticks**
From an age-old culinary tradition (3) to sorting of spermatozoa, as seen by the cartoonist Loup (4).

1

2

3

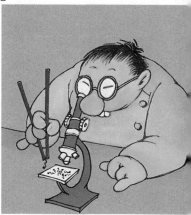

4

The Big Two: Japan and the United States

by Olivier Fond, PhD, Nancy University, graduate of the Company Management Institute of Nancy.

Despite his youth (he was born on 20 January 1960), Olivier Fond has already made his mark. In 1981 he received an award from Elf for his study of the biotechnologies in Great Britain and an award from the France-Japon Association for his report on bioindustry in Japan. He did his tour of duty at the Renewable Resources Laboratory of Purdue University in the United States.

Based on his personal experiences and conversations he has had with researchers in both Japanese and American universities and industries he sketches the means available to these two great powers in bioindustry, and the strategies they have chosen to adopt in placing their pawns on the world chessboard. On the sidelines stands Europe: has it a third way, or is it merely a spectator in this new division of the world?

Biotechnology is, after the computer revolution, the last great technological revolution of the twentieth century. Political and economic pundits all over the world have recognized its importance, and their awareness can be seen in the increasing concentration of capital and companies seeking innovation and diversification in this industrial sector. Some governments have even drawn up specific blueprints to encourage and structure development of this new sector.

Japan and the United States are the unquestioned leaders in the field, because of the enormous financial resources available in both countries for research, production, and marketing of products, in a context where new patents are registered every day.

However, the development of the bioindustrial sector in these two countries took place against two very different backdrops. Their very different economic and social characteristics have conditioned the strategies adopted.

Factors Affecting the Development of the Bioindustrial Sector in the United States and Japan

The three main ingredients of America's success are a spirit of enterprise, a tax environment favoring investment and company creation, and an excellent foundation in the life sciences.

The American businessman's spirit of enterprise is grounded in his character, honed by a particular type of university training, which encourages individuality and competition, essential characteristics for launching new companies. Since 1976, more than 100 venture capital companies have been created in the United States to exploit breakthroughs in genetic engineering, all specializing in research for new products and the associated processes.

The American tax structure favors the creation of new companies by means of special deductions for venture capital and research contracts. In venture capital companies, investors looking for profitable investments provide capital and become shareholders. In research contract companies, investors finance a research program in return for an option on new technologies or the profits from patents taken out as a result of the research they have financed. In either case, typical investors may be either companies such as banks, insurance companies, or pension funds, anxious to invest their capital in rapidly expanding sectors allowing substantial profits (from 1 to 5 times the initial investment), or large companies with plans to diversify into the biotechnological sector by participating in biotechnological research. The financial resources committed to these new companies are considerable: Biogen's market capitalization in January 1985 represented $140 million, while Genentech and Cetus were worth $460 million and $215 million, respectively.

On the flip side, many shares have actually lost ground since coming onto the stock market. Agrigenetics was the first company to benefit from research contract financing, to the tune of $55 million. A great many companies have followed suit, and many have taken up both options: in 1981, Genentech won research contracts worth $90 million, while Cetus had contracts amounting to $75 million. Today these forms of investment appear less attractive, since in 1984 Biogen won only $28 million worth of contracts instead of the $60 million it had originally projected.

These new companies, most of which concentrate on genetic engineering and monoclonal antibodies, have made a handsome contribution to technological innovation in the field through basic or applied research, technology transfer with large companies, development of new markets, and the beneficial influence of competition.

The large companies involved in biotechnology may be divided into two categories: pharmaceutical companies interested in its expansion potential for their current activities, such as Eli Lilly, Abbott Laboratories, and Schering Plough; and chemical or petrochemical companies, such as Du Pont, Monsanto, Standard Oil, Exxon, Shell, and Dow, which are looking for areas of diversification.

Today, more than 200 companies are working in this sector, which is expected to account for a total turnover of $25 billion in the United States by 1989.

The larger American companies have registered a slower response time to the biotechnologies than have the new companies. They began by signing contracts (venture capital or research contracts) with the newer companies: Eli Lilly with Genentech on insulin, Monsanto with Biogen, Standard Oil with Cetus, Schering Plough with Biogen, and Abbott Laboratories with Amgen. This type of investment represented more than $120 million in 1982.

But it was not until much later that the big companies got around to setting up their own research centers; the most typical scenario is still a mixed strategy, combining internal development through company research laboratories with financial participation in venture capital companies.

The new companies have chosen a strategy of short-term development designed to produce immediate income, which will then be used to finance long-term research. In contrast, the large companies, except for the contracts signed with the new companies, are more oriented toward long-term research. The development of biotechnology is still at a very early stage, and competition is fiercest at the research stage and for preliminary industrial development; here the new companies play a major role. However, in the future, the situation may well change in favor of the large companies when competition shifts to mass production and marketing, where these companies will be able to use their experience of industrial processes and existing distribution networks.

In the meantime, complementarity between the two types of companies through vertical integration from basic research to product distribution promotes fast development and commercial applications of biotechnology in the United States.

The third important factor in this development is substantial financial support of basic biological research provided by the federal government since World War II. The United States has the highest research budget in the world: in 1983, $511 million was earmarked for basic research in biology, as against only $6.4 million for applied research. This means that the United States has many high-level molecular biologists and immunologists, who have excelled in the creation of new companies and early applications of this research in the pharmaceutical industry.

Another important factor is that innovations spread quickly to the industrial sector through the many contacts between industry and academia, either by means of research contracts or consultancy agreements, while many academics at MIT or the University of California at Berkeley, for example, have even created their own companies, some right on campus.

In Japan, the bioindustrial revolution developed along different lines and in quite a different spirit than in the United States. The main factors in this development were Japanese team spirit, which favors large companies, the coordinating role played by the Japanese government between the different branches of industry, and an emphasis on applied rather than basic research.

In very many industrial sectors, the Japanese economic miracle may be explained by the cultural underpinnings developing a spirit of cooperation between citizens. One of the greatest contributions of Japanese culture to economic life is the famous social consensus, which unites government, political parties, industrialists, and employees around the primary objective, i.e., economic development by whatever means possible. This situation encouraged growth of large companies, which are often the main or even sole center of interest of their employees. In contrast, the creation of venture capital companies has been very limited. The first such company was created in 1982. As an illustration, during that same year, investments in this type of company represented $84 million in Japan as against $5.8 billion in the United States. The Japanese MITI (Minister of Industry and External Trade) now encourages venture capital. The Japanese financial market is also structured differently. Companies acquire funds by means of bank loans

financed by household savings, which in Japan represent a significant 18% of net disposable income, instead of selling shares on the stock exchange.

The Japanese biotechnological industry is built on traditional dietary habits based on the use of fermented products: shoyu, miso, natto, and saki. It has spread to the pharmaceutical and food-processing sectors with the objective of producing refined substances with high added value. This strategy stems from Japan's lack of natural resources: the country is obliged to import much of its primary materials, mostly from neighboring Asian countries (starches and molasses from Thailand, the Philippines, and Indonesia). Reliance on imported raw materials is often risky and may even prove to be a source of conflict, depending on political developments in supplier countries. This dependence is illustrated by the case of the Kyowa Hakko Company, a leading world producer of amino acids. The company imports 150,000 tons of molasses each year as a fermentation substrate. It is now experimenting with cultivation of sorghum, with the assistance of the Japanese government, the aim being to reduce its dependence on imports. In this major industrial project, sorghum would be used as a raw material in the production of ethanol. Given its special situation, Japan has chosen to make the most of the resources drawn from raw material imports, in other words, to maximize added value. This strategy increases the Japanese added value content for equal turnover of finished products, and reduces its dependence vis-à-vis imports. Moreover, the high level of Japanese technology shelters Japan from competition if supplier countries decide to process their own raw materials.

The companies currently working in biotechnologies may be categorized in the following way. In the first category are food-processing companies that have specialized for many years in fermentation technology. Ajinomoto and Kyowa Hakko are the most well-known; both are now moving into the pharmaceutical sector. The second category consists of pharmaceutical companies that already possess good technological foundations and powerful research means: Tanabe and Fujisawa. The last category consists of groups such as Mitsubishi and Mitsui that are primarily interested in finding areas of diversification.

Since the 1950s, the Japanese government has worked through MITI to foster and, to an extent, control an industrial policy based on tax incentives and subsidies, while also encouraging contacts between business, academics, and public laboratories. MITI recognized the importance of the biotechnological revolution and made it a national priority, financing several 10-year programs: $43 million for bioreactors, the same for DNA technology, $20 million for cell cultures, and an impressive $150 million for synthesis of secondary metabolites from

plants. Of this work, 10% will be conducted in laboratories attached to the various ministries, and in particular in the new scientific city of Tsukuba. The other 90% will be done by 14 selected, competitive companies, with the government handling contacts between them and encouraging competition. To sum up, we might say that Japan relies on subsidies and tax incentives to promote investment in research and development and production, whereas tax policy in the United States tends to encourage the creation of capital.

In Japan for many years both government and industry tended to favor applied research, and excelled in the art of quickly applying and improving the results of basic research carried out in other countries. Thus, Japanese companies can marshall a great many competitive industrial processes, long experience, and already trained personnel.

Bioindustrial Profiles of the United States and Japan

In the United States, biotechnological development has had its greatest impact in the pharmaceutical sector. There are a number of reasons for this:
● DNA research, which revolutionized modern biology, began in medical research centers;
● the experience of pharmaceutical companies enabled them to make quick use of this technique in commercial applications;
● pharmaceutical products are characterized by high added value, yielding high profits to finance costly research.

Companies involved in this sector are either new companies or large companies, some pharmaceutical, others not, which are often associated with the new companies through venture capital or research contracts. The main applications involved are insulin, interferons, antibiotics, diagnostic kits based on monoclonal antibodies, and new vaccines. For example, Eli Lilly has formed an association with Genentech to produce a new bacterial insulin, and another with Bioresponse Inc. for monoclonal antibodies. Merck is associated with Biogen and Chiron for hepatitis B vaccines, and Bristol-Myers with Genex for interferon.

Among companies that have recently diversified into pharmaceuticals, we find Du Pont associated with Whitehead Institute in molecular genetics and with Biotech for marketing monoclonal antibodies and diagnostic tests, while Shell is associated with Cetus in work on interferon. Under the terms of these contracts, the new companies generally grant exclusive rights to the resulting processes to the larger companies.

In 1984, 62% of the 219 companies whose commercial applications are known were working in the pharmaceutical sector, 28% in the veterinary industry, and 24% in agriculture (total exceeds 100% since some companies are working in more than one sector). By 1995, sales are expected to reach $4-$5 billion a year in pharmaceuticals, with a similar figure for agriculture and the veterinary sector, as against $35 million at the present time in the pharmaceutical industry and a mere $2 million in agriculture.

Most investment in the veterinary sector was provided by the new companies, whereas in agriculture investment came mainly from the large agrochemical companies. The predominance of new companies in the veterinary sector is explained by the fact that the clinical research required is less time consuming and less costly than in the field of human health. The market is also smaller and therefore more accessible to small structures, while the distribution system is rather different. Applications include the development by Genentech of a bovine and porcine growth hormone. Genentech and Molecular Genetics Inc. are involved in the production of vaccines.

The application of genetic engineering in the agricultural sector is proceeding at a rate that would have been impossible to imagine some years ago. The main stakes are plants resistant to drought, pesticides and herbicides, and plant/bacteria associations to produce plants that manufacture their own pesticides or prevent formation of frost. The large companies in this sector include Dow, Monsanto, Pfizer, Rohm and Haas, and Standard Oil, while Calgene and Plant Genetics Inc. are typical of the new companies working in the field.

Production of special chemicals (ethanol, organic acids, food additives – amino acids, vitamins – and enzymes), wholly or partly by fermentation, is also undergoing considerable commercial development in the United States, although it has been well established for some time in Japan and Europe. In 1984, the breakdown for fermentation products was as follows: antibiotics 63%, ethanol 21%, organic acids 8%, amino acids 5%, enzymes 3% – a good illustration of the importance of this sector. Since 1979, ethanol, used as a gasoline substitute, has taken over 5% of the American oil market (1.9 billion liters) thanks to major fiscal support. The war against lead emission should benefit this sector.

In the field of amino acids, the development of aspartame, a low-calorie sweetener consisting of aspartic acid and phenylalanine, produced by G. D. Searle, has benefited from the popularity of low-calorie soft drinks in the United States. G. D. Searle products are sold exclusively to manufacturers of sparkling beverages: Coca-Cola, Pepsi-Cola, and Seven-Up. The production of fructose syrups (4.4 million tons in 1984) was also encouraged by beverage manufacturers, but further growth may be restrained by the difficulties of adapting this syrup for the food-processing industry, because of its liquid state and also for reasons of purity.

In Japan the bioindustrial sector is quite different. Japan now dominates the world market for amino acids and new antibiotics (cephalosporins). It also has wide experience in biological processes and cell culture. Among the amino acids, lysine, which is used as a protein additive in animal feed, and glutamic acid, used to enhance flavor in the food-processing industry, take the lion's share of production. Kyowa Hakko and Ajinomoto are the main producers. These companies, which started off in the food-processing sector, have been making amino acids for some time. Tryptophane, an animal feed additive, is produced by these two companies and also by Showa-Denko, a company originally involved in the chemical sector. Likewise, aspartic acid, an ingredient of aspartame, is now produced by the chemical company Mitsubishi, although the main producer is still Tanabe Seijaku, a pharmaceutical company. Development in this sector takes the form of an increase in the number of amino acids produced by fermentation and an improvement in existing processes in order to reduce cost prices, thereby rendering the market for animal feed or food additives and flavorings more accessible. Products include phenylalanine, threonine, serine, and cysteine.

The production of antibiotics accounts for the second-largest industrial sector in Japanese biotechnology, with pharmaceutical firms such as Meiji Seika, Fujusawa, Takeda, Shionogi, and Kyowa Hakko. Companies like Ajinomoto, Sankyo, and Toyama Chemical have also diversified into this sector.

Despite past weakness in basic research and molecular engineering, Japan has no intention of missing out in the race to find industrial applications for genetic engineering and monoclonal antibodies. This is why Japanese companies, both pharmaceutical and recently diversified companies, have made considerable investments in the field. Many of them have signed joint research and development contracts, sometimes in association with public research institutes: Yamanouchi Pharmaceutical and Ajinomoto in the production of thrombolytic agents, Asahi Chemical and Dainippon Pharmaceutical with the University of Tokyo, Green Cross and Toray Industries in interferon research, Ajinomoto and Takeda in the development of interleukin. The associations encourage development by sharing financial means, skills, and existing industrial structures. They resemble the hookups between American research companies and large companies, although in this case integration tends to be horizontal rather than vertical.

Since 1981, Japan has led the world in the introduction of new pharmaceutical products (30% of the world total). Japanese companies have tended to concentrate their efforts on monoclonal antibodies and the development of biotechnical processes, an essential step in industrial application of research results and anticancer therapies.

Commercial Strategy and Apportioning the World Market

The development of biotechniques is still very much in the research stage. However, it will not be long before it reaches the marketing stage, and competition for the world market is expected to become much fiercer.

We have seen that in the United States, the large companies will continue to hold their own through sales muscle and existing distribution networks. Contracts with the new research companies enable them to find applications quickly for innovations and to win sales on the world market. At the present time, this mutually beneficial association gives American industry a considerable advantage. However, there is some uncertainty as to the way in which the two types of companies will evolve, and whether they will be able to maintain their strategic advance.

The main element in future development of the new companies lies in their ability to attract capital to finance their expansion and to place their products on the market while maintaining their technological lead, so as to avoid being swallowed up by the large companies. They must learn to produce and distribute their own products to provide higher cash flow, but also because they need to recognize market trends and study consumer response to new products. Nevertheless, it is not certain that they will be able to come up with the financial resources needed to carve out a place for themselves on the world market beside the large companies with their highly developed marketing structures. As an illustration, some large companies employ more than 500 people in the marketing department alone, whereas total salaried staff at Genentech, the largest research company, does not exceed 500 people.

In 1984, many new companies did attempt, at enormous cost, to become involved in the production and marketing end of the industry. This experiment resulted in considerable deterioration of their financial situation. Their ratio of available assets to short- and medium-term debts fell quite significantly. During 1984, this ratio fell from 6.6 to 3.7 for Genex, 14.8 to 5.2 for Hybritech, and 5.6 to 2.3 for Monoclonal Antibodies. Nevertheless, given that the average ratio for chemical companies is around 2, this is hardly catastrophic. In 1984, Biogen, Monoclonal Antibodies, and Genex lost money, the latter because of the high costs involved in opening its own plant. Only Genentech and Hybritech made money, although their profit margins were low (3-4%).

Breakthroughs are more difficult in some markets than in others. The market for human therapeutic products is the most difficult to break into despite its size, because of the high costs of clinical studies and sales licenses, as well as the competition created by large, already established companies. The market for in vitro diagnostic reagents is far more easily accessible, but is bound to become quickly overcrowded.

Geographically too, some foreign markets are much more difficult to break into than others. Although the Japanese government refuses to recognize that it discriminates against foreign companies, the Japanese market does seem to be a closed one, owing to customs procedures and the complicated approval system that products must negotiate before they can be sold in Japan. Foreign companies wishing to do business in Japan must create their own subsidiaries or set up joint ventures with already established Japanese companies. Many American research companies have signed contracts with Japanese companies, which usually provide for joint research development associated with an exclusive license for the Japanese company for production and marketing in Japan. These licenses usually include marketing rights for other parts of Asia – Korea, Taiwan, and Southeast Asia – thereby taking advantage of the experience and distribution circuits of Japanese companies in this part of the world. Thus, Green Cross has an exclusive license from Biogen to produce the hepatitis B vaccine on the Japanese market, while Mitsubishi Chemical has the same type of contract with Genex for albumin serum, and Yamanouchi with Schering Plough for interferon. Daiichi will develop a blood diagnostic test made by Genetic Systems, and will produce and market the product in Japan, Korea, and Taiwan, with Genetic Systems receiving royalties.

Up until the 1970s, Japan was notorious for its economic isolationism. Today, faced with the enormous obstacles to be cleared in order to tap the pharmaceutical market in Western countries, Japanese companies do not have adequate distribution networks or sales forces to conduct an aggressive sales policy. To compensate for this weakness, these companies are entering into joint ventures with American or European companies so as to use their systems of national and international distribution. Ajinomoto is associated in France with Orsan (Lafarge-Coppée group) for the production of lysine (Eurolysine), and this association will soon lead to construction of a new lysine plant in the United States. Takeda has signed an agreement with Abbott Laboratories, Fujisawa with SmithKline, Shioniji with Merck for hepatitis B vaccine. Likewise Tanabe Seijaku and Marion Laboratories are about to form a joint venture with a view to developing, patenting, marketing, and even manufacturing on American territory some of the pharmaceutical products manufactured by Tanabe.

In the meantime, Japan and the United States lead the world in the field of biotechnology, in both research and industrial applications. However, over the next decade, they will only be able to maintain their competitive edge if they reinforce their mastery of biotechnological processes and continue to innovate in genetics, immunology, and the culture of organisms.

Japan is ideally placed to develop processes, thanks to its wide experience, but must try to encourage basic biological research in its public and private laboratories. The United States has a strategic advance in genetics and immunology, but is relatively weak in biotechnological engineering. The American government envisages developing applied research in microbiological engineering, separation, and purification, as well as in process optimization and control. Its aim is to increase the number of process engineers, who may eventually be in short supply, and to encourage contacts between process engineers and biologists. Two centers for research in microbiological engineering are to be developed, one at MIT, the other at Purdue University, with a budget of $6 million a year. At this stage of industrial development, it is unlikely that we shall see the creation of new research companies, since the costs of industrial-scale equipment for improving processes can only be borne by large companies anxious to develop their own systems.

24

The "Dangerous Links" between Town and Gown

It's a scenario that could have been entitled *For a Few Billion Dollars More*. A cartoon that appeared in an American newspaper summarized the situation best – it showed a university researcher, test tube in hand, being lassoed by a dollar in the form of a double helix. For it is from a new angle that the problem of the relationship between the universities (or, more precisely, the centers for basic research) and industry resulting from the development of the biotechnologies must be viewed today, in Europe, in the United States, and even in Japan.

Major inventions in the field of electricity (dry cells, light bulbs, motors) were and are immediately exploited by businessmen: the Siemens, Edisons, and Westinghouses of the world. Those in electronics (vacuum tubes, radio, transistors, the computer) came into being in the industrial laboratories of the Marconi Company, Bell, or IBM. Important firsts in biotechnology were for their part mainly achieved in public research centers on the West Coast of the United States or at the Medical Research Council in Great Britain. The first-born of the NFBs (new biotechnology firms), Cetus, quickly gained the cooperation of Nobel Prize winners and researchers at the University of Wisconsin, while the most famous (Genentech) was established on the initiative of university scientists. To stay at Biogen, Walter Gilbert gave up his position as professor at Harvard. Stormy debates have followed one another. At MIT, for instance, David Baltimore has assumed direction of the Whitehead Foundation and the heat is on. Edwin Whitehead, president of an important company (Technicon), decided to grant MIT researchers recruited for a foundation he had decided to establish an initial sum of $20 million, an annual income of $5 million, and an endowment of $100 million on his death. But the question that has arisen is, Can MIT tolerate, within its body, an organization over whose recruiting and active research it has absolutely no control?

Another "scandal" arose on the West Coast of the United States when the powerful Allied Chemical Company decided to support the research of a professor at the Plant Biology Laboratory of the University of California at Davis, Raymond C. Valentine, and Valentine then founded Calgene... with Allied owning 20% of its stock. As Valentine holds the position of "counselor" responsible for the experimental Department of Agronomic Sciences, the dean of the College of Agriculture and the Environment has voiced his fears that, as Valentine's advice must be "neutral and objective," there may be some conflict of interest if he must keep his company's research a secret.

The problem has thus been raised: Is there not a serious risk of diverting public funds and using them for private purposes? Does this not authorize the creation of autonomous enclaves of power within universities, one of whose most precious privileges is the free circulation of ideas between scientists? Does this not mean concentrating research on commercial goals and abandoning work on products that are an obvious necessity (such as vaccines), but are far less profitable? Or "witholding" an article in a scientific journal until a patent has been obtained? The zealots of closer collaboration state that the taxes taken from profits made as the result of the development of new technologies enrich the public treasury, that the time of application – and therefore the availability to the public – of necessary products is abbreviated, that, in view of international competition, anything is fair, and that this form of "cooperation" may ease somewhat the financial embarrassment of the universities (due to dwindling government grants and the effects of inflation). But above all, opponents are faced with set circumstances: the size of the markets, which are opening up, and the siren song caressing the ears of researchers on the lookout for a few extra dollars.

In West Germany, Hoechst was the target of fierce criticism for not having generated its inventions at a West German university. Hoechst defended itself by arguing that biological research dried up

"Around the DNA molecule"
A sculpture by Bror Marklund erected at the biomedical center of the University of Uppsala, with which the Swedish company Pharmacia maintains close ties.

217

during the 1930s, mainly as the result of the brain drain to the United States, and that it never recovered. Furthermore, the German university tradition, which insists on working in highly compartmentalized areas without any interconnection, hardly encourages the transfer of technology. The outcry was equally loud on the far side of the Atlantic: in a period of stiff international competition, why institutionalize "leaks" to Europe? Despite all the noise, the agreement was finally signed. The shock was undoubtedly beneficial, as it encouraged the association of other companies with West German universities (such as BASF with the University of Heidelberg), an influx of foreign capital (like Schering, which financed a new institute in Berlin), and an effort to attract scientists of international reknown to West Germany (like G. Kohler, who left the Immunological Institute of Basel to take up the direction of a department at the Max Planck Institute).

The situation is quite complex in Europe. In France, for instance, several university centers collaborate generally more closely than one might expect with industry: the University of Compiègne with the Daniel Thomas team; Strasbourg with Pierre Chambon, one of the founders of Transgène; Toulouse with Pierre Monsan, who founded BioEurope with Jean-Bernard Borfiga thanks to the financial support of, among others, Roussel-Uclaf and Sucre Union. But the mentality of public research organizations such as INRA (Institut National de la Recherche Agronomique) often still remains conservative, showing a reluctance to engage in "dangerous links."

Great Britain, for its part, has definitely decided to encourage closer collaboration: in 1981, to this end SERC (Science and Engineering Research Council) named a directory, which collaborates closely with the Department of Industry, the Agricultural Research Council, and the Medical Research Council. Twelve companies (including Beecham, Glaxo, Wellcome, Shell, and Unilever) participate in this effort. A Biotechology Center has been organized in Glasgow in close association with that city's university as well as the University of Strathclyde. Imperial Biotech has been founded by the Imperial College and Technical Development Capital Ltd. to manufacture enzymes and other proteins and sell them to industry. In April 1982, University College received $7 million from Sandoz for research on the biochemistry of the brain. A company like ICI considers its relationship with the university so important that it employs a full-time man for this purpose, Bernard Langley.

In the United States, all configurations are in evidence. Agreements are signed between two institutions, as in the case of Monsanto and the University of Washington in St. Louis (which also signed agreements with Mallinckrodt); between an institution and an individual, as in the case of Du Pont de Nemours and Philip Leder at Harvard; between individuals and institutions, as in the case of the founder of Petrogen, Ananda Chakrabarty, and the University of Illinois. University researchers sit on the boards of directors of NBFs or act as advisors to companies, as do Leroy Hood (Caltech) and Marvin Carruthers (University of Colorado) for Applied Biosystems. And universities or research institutes themselves found NBFs – among them Engenics, the creation of Stanford University and six compa-

nies (Bendix, Elf-Aquitaine, General Foods, Coppers, Mead MacLaren Power, and a Noranda subsidiary, Paper), and Salk Institute/Industry Associates, established by Salk and Phillips Petroleum. Sometimes the situation becomes contradictory: Harvard, for instance, refused Mark Ptashne's proposition to found a company to exploit the discoveries made by researchers at the university. This "hesitation waltz," with one step back for every step forward, reflects a certain uneasiness.

Dr. Sydney Brenner, director of MRC's Molecular Biology Laboratory in Cambridge, was himself the center of a controversy that came to the ears of the Prime Minister: it resulted from the "off-hand manner" in which British universities had judged the commercial importance of the hybridoma technique and monoclonal antibodies, which the Americans were the first to patent. In 1982, in the course of a lecture at the University of Wisconsin, he set himself apart from those who considered the movement of researchers from the university to industry as a "sellout": "I can foresee," he said, "that when all this uproar about the 'gold rush' is over, this change will be considered as a major contribution of university science to American and European industry.... Naturally enough, the university departments do not want to be used as a 'clone service.' The effort to understand the fundamental phenomena of life must be continued, while at the same time liaison groups should be established to serve as 'bridges' for the development of products and their exploitation throughout the world."

In most instances, the universities most committed to cooperating with industry are those geographically closest to the NBFs and the industrial empires: Basel and Zurich, in proximity to Biogen, Hoffmann-La Roche, Sandoz and Ciba-Geigy; Uppsala, in proximity to Pharmacia; Strasbourg, in proximity to Transgène (not to mention the United States, where the major focal points are Silicon Valley on the West Coast and Route 128 on the East Coast).

At the crossroads are the NBFs. A modern researcher can no longer be a half-mad scientist lost in a dream world. A businessman in the biotechnologies must today invariably read *Nature*. The frontiers between town and gown (and between basic and applied research for that matter) are no longer as sharply defined as they once were. And all this is in large part due to the creation and proliferation of these small companies with their host of university researchers and Nobel Prize winners.

The NBF is one way of assuring the "transfer of technology" between two worlds frequently foreign to one another. In 1983, an editorial in *Science* defined the NBFs as "important driving forces for progress, crucial to establishing and maintaining a fast pace in the evolution of biology and its applications." Biogen remains the prime example, imprinted with the personality of the man who for long years presided over its destiny and created its image, Nobel Prize winner Walter Gilbert. During the first five years of Biogen's existence, Gilbert obtained credits of $183 million. With its international team of scientists recruited in the United States, Great Britain, West Germany, Switzerland, and Belgium, Biogen created a miniacademy of biotechnology, holding frequent meetings to com-

1. Walter Gilbert
2. James Watson and Sydney Brenner
Two of the most famous researchers in molecular biology, the American J. Watson, who presides over the destiny of the Cold Spring Harbor Laboratory, and the Briton Sydney Brenner, who presides over those of the Medical Research Council in Great Britain.

pare their latest results. Investors in Biogen considered the NBF as an "interface": the universities invented new tools for biotechnology; Biogen's job was to ferret out these tools and show its investors how they could be used commercially.

Julian Davies, associate professor of microbiology at the University of Geneva and Biogen's dynamic director of research, urges the company's scientists to use up to 20% of their time pursuing their own projects – quite apart from the goals set by Biogen. He encourages them to work when they want to and when it suits their experiments ("Microbes don't wait"), keeping the labs open to them twenty-four hours a day. In addition, he favors researchers maintaining personal contacts with the universities, with a degree of freedom any large corporation would undoubtedly consider intolerable.

The situation is entirely different in Japan. There is little or no uneasiness. The separation between the public and private sectors is less distinct to the extent that the large companies participate in national programs. They collaborate among themselves and with the universities but exclusively in matters concerning basic techniques. But this does not prevent them from being fiercely competitive. Some university researchers, like T. Beppu, act as counselors for a considerable number of them, while others, like H. Saito, act in the triple role of university researcher, "advisor" to companies, and "advisor" to the government. But here again tensions can arise as more and more industrial concerns establish their own research centers. One of the most flagrant examples is Mitsubishi Kasei, whose Mitsubishi Chemicals Company created a Basic Research Institute not always appreciated by the university with which it competes, enjoying as it does better financial backing and an outlook free of traditional mandarinism. For researchers in industry, it is a way of escaping the "principle of seniority" that prevails in Japanese business.

The development of the biotechnologies has obliged researchers to question the validity of established structures: the constant round robin between the university and industry acts as a mental stimulus, however trying it may be. Competition has now arisen in a field formerly free of conflicts of interest. But after all, did not Pasteur himself set an example when, while at the University of Lille, he worked for local businessmen? Can the industry of life be expected to avoid the problems of life?

1. **Uppsala**
Established in 1477, this university has seen within its walls the most famous researchers of their time, from the botanist Carl von Linnaeus to the astronomer Anders Celsius and Nobel Prize Winner in Chemistry Arne Tiselius, inventor of electrophoresis.
2. **University-Industry**
In buying up the DNAX research firm, the Schering Plough Company assured itself of the aid of such prestigious researchers as (from left to right) Charles Yanofsky and Nobel Prize winners Paul Berg and Arthur Kornberg, all three from Stanford Universy.

Conclusion

"Explore the musical mysteries of the cell," said the 1985 ad in the very serious scientific journal *Science* to promote the sales of a record, "DNA Music," based on the four notes, C,G,A, and T, of the genetic alphabet. The fact that this musical message from our genes has become a consumer product is significant: life is henceforth to be sold as a gadget, a sure sign that an industry has come into its own.

But what industry? An industry based on what the English call "enabling technology" – a technology that makes it possible to do everything from extracting chewing gum from plants to having bacteria manufacture artificial snow. Improving life through bacteria, shortening working hours thanks to microorganisms, a job with a future for dynamic microbes: such are the advertising slogans that are beginning to appear in the press. After the robot, the microbe.

By the year 2000 we may well smile at all this as fads, as ephemeral as the first New York nightclub of the "bio" era, named Interferon. Just as we smile today at the Abbé Nolet's "electric kiss," the "miracle" of radio, G.K. Chesterton's statement that the electron was "pure abracadabra," or H.G. Wells's prophecy that radio would only reach "a phantom army of inexistent listeners." Simple fads, or as yet enigmatic symptoms of a pivotal era whose real nature no one can guess?

And what about man? Has not man himself become a laboratory animal on a planetary scale? Over the centuries, from Epicurus, Plato, and Euclid to Galileo, Descartes, Leibniz, Robespierre, and Napoleon, men of science, war, or government have guessed that, behind the "tiny trickle of feeling," as the economist Edgeworth put it, there lay a program and a code. The program came to light in 1953, the code deciphered less than ten years later. The DNA that turns in the opposite direction, jumping genes, silent genes, the paramecia that cast doubt on the universality of the code, and other troublemakers today and tomorrow will but throw back the frontiers of exploration and explanation. Like even the most outstanding accomplishments – the DNA probe or monoclonal antibodies – every new discovery in molecular biology is a hook thrown into the universal soup to lure and catch the great enigmas. What is it that distinguishes a man from a chimpanzee? We finally know the answer: a 1% difference in DNA bases.

The *London Times* shows no hesitation in writing that DNA has become the "evolutionary clock." Why? In a quiet laboratory at the University of Uppsala in Sweden, a researcher, Svante Pääbo, is in the process of cloning the DNA of a human being, certainly nothing extraordinary in 1985. What is exceptional, however, is that the DNA in question comes from the flesh of a mummy. It is the genetic program of a dead man who lived in Egypt 2,400 years ago and is now being revived in a test tube in the middle of Scandinavia. The independent life of the genes, well after the death of the man (like the "immortality" of the cancer cells from a long-buried patient), is enough to make one dream. But scientists are not there to dream. A full year before this, at the University of California at Berkeley, A. Wilson was able to clone the DNA of a quagga – a species of animal assumed to be extinct – and Russell Higuchi that of a mammoth, dead some 40,000 years earlier!

Were these simply attempts to reconstitute a zoo of extinct species or to have a closer look at one of Tutankhamen's favorites? Not at all, but rather an effort to explore evolution, as one might explore the mechanisms of the cell, of birth, of aging, of cancer, of immunology, and of death.

There is nothing neutral about this investigation, as its ultimate aim is control and regulation, everything involved in "switching on" and "switching off." It is vertiginous, if, like W. Yamaya of Mitsubishi Chemical, we keep in mind that the information contained in DNA is 10 billion times 100 billion bits (the basic unit of data processing), compared with several tens of thousands of bits in a dictionary and a scant 1 billion in the most sophisticated computer imaginable.

The industry of life creates a change of perspective. Things change first of all for the engineer, used as he is in mechanics, electronics, or even in the nuclear field to scaling up from a model to full size: a living organism reacts differently if its environment is changed, and it is impossible to go from in vitro to in vivo. The engineer is consequently forced to do things in reverse and design large-scale machines and then scale them down to produce the best possible reactor for his experiments. The perspective has also changed for investors and businessmen, who for the first time are now forced to define narrowly the impact of products that do not yet exist and whose future costs are unknown. It also changes for academics and politicians, who no longer know how to train young people or in which direction to point them. And, above all, it changes for mankind as a whole, those "robots blindly programmed to transport and preserve those egotistical molecules known as genes," as Richard Dawkins defines his fellows in *The Selfish Gene*. "Curiously enough," says S. Guttmann of Sandoz, "discoveries in molecular biology and genetic engineering have always begun with the inhibitory factors of a system (repressor, inhibition of renin, etc.) rather than with activation. This may be a natural tendency of the human mind." Enzyme-feedstock, antibody-antigene: the key-lock or lock-bolt analogies are numerous in molecular biology. And men, ensnared by the most contradictory "bioideologies" (sociobiology or behaviorism, holism or reductionism) undoubtedly have trouble identifying themselves with those thousands of the Metropolis (the cells) at work in their own bodies. The youngsters of the "Me Generation" are faced with the ultimate quandary: When the "me" disappears, what remains? Listening to the music of one's own genes, perhaps, while escaping into the wonderful narcissism of gymnastics, aerobics, jogging, or other forms of bodybuilding. The researcher observing life in a test tube is like the spectator at a peep show, isolated behind his one-way window.

Oddly enough, these anxieties are reflected in debates on what is rather pompously known as The Ethical Question: Will we not witness the birth of monsters in our laboratories? Will we not show a tendency to stifle an incipient Beethoven in the womb because he shows a slight hereditary defect? Will we not change the planetary balance by creating invulnerable bacteria that will kill off all life? But since Asilomar in 1975, a conference organized by the scientists themselves to discuss the dangers of genetic engineering, the debate has taken a wrong turn. No institution, no law is capable of imposing an ethic on a society that has lost sight of the meaning of the word. The debate actually started when the first nomad decided to settle down, when the first farmer created the first tool, and maybe even before this. Perhaps as far back as Genesis, when God ordered men to "have dominion over the fish of the sea, and over the fowl in the air, and over every living thing that moveth upon the earth." And it is exactly a debate of this kind that both scientists and politicians avoid, for it raises questions that are not the province of science but of quite another sphere, and that science alone is powerless to solve.

GLOSSARY

Active area: Area on the enzyme where biochemical reaction takes place.

Adenine (A): One of the four letters of the genetic alphabet; nitrogenous base and the essential component of nucleic acid.

Adsorption: The superficial penetration of a gas or a liquid into the surface of a liquid or solid with which it is in contact.

Aerobic: Term applied to a microbe whose activity depends on the presence of oxygen.

Allele: Applied to the varied forms that may be assumed by the same gene situated at the same "locus" (level) and that confers different characteristics on the species concerned.

Amino acids: The "molecular building blocks" that make up the proteins. There are 20 different kinds. The sequence of amino acids is the linear order by which they are linked within the protein.

Antibodies (or immunoglobulins): Proteins produced by the blood of mammals that are mobilized in response to an attack on the body by a viral, bacterial, or tumoral aggressor.

Antigen: Substance foreign to the body that arouses the immunity defense reaction. It stimulates the production of antibodies capable of reacting specifically with one or another antigen.

Autotrophe: Applied to a microorganism capable or nourishing itself by synthesizing the elements necessary for its life (the contrary being a "heterotrophe").

Bacteria: Procaryote unicellular microbes averaging one micron (one-millionth of a meter) in size. They assume various shapes: spherical (cocci), elongated (bacilli and colibacilli) and spiral (vibrions and treponemes).

Bacteriophage: A virus that attacks bacteria. The lambda bacteriophage is frequently used as a vector in recombinant gene experiments.

Base: Substance capable of reacting with acids to form salts. The "letters" of the genetic code, A, G, T, U, are called "bases."

Biocatalysis: The acceleration of biochemical reactions by enzymes.

Biochemistry: The chemical reactions taking place in a living oganism.

Biochip: Electronic chip using biological molecules, and which can replace semiconductors in integrated circuits.

Bioconversion: Chemical conversion making use of a biocatalyst, that is to say, a catalytic agent produced by an enzyme.

Biomass: All organic material developing as the result of photosynthesis.

Biopolymers: Macromolecules, such as proteins, nucleic acids, and polysaccharides, whose molecular mass is many times that of monomers (simple molecules).

Bioreactor: Apparatus in which bioconversions take place.

Biosensors: Electronic systems using biological molecules to detect specific substances.

Biosynthesis: The production of a chemical substance by a living organism.

Cal: A grouping of undifferentiated plant cells, the first step in the regeneration of plants from a culture of tissues or cells.

cDNA (complementary DNA): The DNA complementary to a messenger RNA.

Cell fusion: The formation of a hybrid cell, produced by fusing two cells of different species.

Chimera: Name for any new organism produced by genetic engineering.

Chromosome: The physical structure containing the genes. Each species has a specific number of chromosomes – 23 pairs in the case of human beings.

Clone: Line stemming from a single ancestral cell.

Codon: Group of three bases in DNA or RNA that determines the position of an amino acid in "building" a protein.

Cofactors: Complementary molecules necessary within the framework of an enzymatic reaction.

Cytoplasm: The part of the cell outside of the nucleus.

Cytosine (C): A base, one of the four letters of the genetic alphabet.

Dicotyledons: Plants born with two leaves – soybeans, for example.

Diploid: Said of a cell or organism in which all types of chromosomes are represented twice. All cells in higher species, except gametes (sexual cells) are diploids.

DNA (deoxyribonucleic acid): The basic biochemical component of the chromosomes and the support of heredity, which contains the sugar deoxyribose.

Endorphins: The "natural opiates of the brain," peptides acting against pain through the endocrinal and nervous systems.

Enkephalins: Small peptides in the brain having almost the same effect as endorphins.

Enzymatic engineering: Techniques used to reinforce the action of enzymes.

Enzymes: Proteins that act as a catalyst in biochemical reactions. They are produced by living cells.

Eucaryote: Cell possessing a nucleus isolated from the rest of the cell by a membrane: yeasts, protozoans, fungi, and certain algae.

Exon: Gene expressed in the form of a protein.

Factor VIII: Protein that coagulates blood platelets and is used to treat hemophiliacs.

Fusion of protoplasts: Fusion of two cells whose walls have been eliminated, making it possible to redistribute the genetic heritage of microorganisms.

Gene: The basic unit of heredity, which plays a part in the expression of a specific characteristic. The expression of a gene is the mechanism by which the genetic information that it contains is transcribed and translated to obtain a protein.

Gene amplification: Increase, within a cell, of the number of the same gene. Amplification may be spontaneous or induced.

Genetic engineering: A technique used to modify the genetic information in a living cell, reprogramming it for a desired purpose (such as the production of a substance it would not naturally produce).

Genome: The genetic identity card of an individual: the combination of chromosomes constituting the hereditary material.

Genotype: The genetic makeup of an organism.

Glycoprotein: Protein to which groups of sugars become attached.

Glycosylation: The attachment of groups of sugars to a molecule, such as a protein.

Guanine (G): Base in the genetic code.

Haploid: Referring to cell or organism in which each chromosome is present but a single time. This is above all true of the gametes (sexual cells).

Heterotrophe: Applied to microorganisms that do not know how to synthesize the metabolite essential to their life and that require organic compounds, growth factors, vitamins, etc., to grow.

Heterozygote: A diploid cell (with a double set of chromosomes) whose paternal and maternal alleles are different.

Homozygote: Diploid cell whose paternal and maternal alleles are identical.

Hormones: The "chemical messengers" of the body. A hormone is secreted by an endocrine gland (internal secretion) and carried by the blood to the cells, tissues, and organs, on which it execises a specific action.

Hybridoma: Hybrid cell resulting from the fusion of a myeloma cell ("immortal") and a lymphocyte (cell producing an antibody).

Immunodepression (or immunosuppression): State in which the body's immunological defenses are weakened.

Immunotoxin: Molecule toxic to the cell, attached to an antibody.

In vitro: Biological reactions taking place in a laboratory, in an artificial system – outside a living being.

In vivo: In this case, the experiment takes place inside a living cell or organism.

Intron: A "silent" gene, which does not express itself to code a protein.

Isoglucose: Syrup with a high fructose content.

Karyotype: The characteristic arrangement of an individual's chromosomes, arranged in pairs.

Liposome: A little "bag" consisting of a membrane of lipids, used to contain substances to be transported within the body.

Lixiviation: Extraction of the soluble components of, for instance, an ore.

Locus: The location (or rather level) of a gene on a chromosome.

Lymphocyte: Cell that plays an essential role in immune mechanisms. Nearly one-fourth of the white cells are lymphocytes.

Lymphokines: Proteins that play an essential role in immune defenses.

Lysis: The destruction of organic elements.

Meiosis: Cellular division, with a passage from the diploid stage to the haploid stage during the formation of sexual cells.

Metabolism: The combination of physical and chemical processes that occur in cells and organisms, with the production of metabolites or energy.

Mitosis: Cellular division at the end of which each cell has the same number of chromosomes as the original cell. This takes place in several phases: anaphase, metaphase, prophase, and telephase.

Molecular weight: The sum of the atomic weights of the atoms constituting a molecule, expressed in "molecule-grams" (or "moles").

GLOSSARY

Molecule: A group of atoms united by a chemical bond.

Monoclonal antibodies: Highly specific antibodies derived from a single cellular clone.

Monocotyledons: Plants born with a single leaf, such as cereals (wheat, barley, rice, etc.).

Multigene: A single characteristic expressed by several genes.

Mutagenesis: The induction of genetic mutation by physical or chemical means to obtain a characteristic desired by researchers.

Mutation: Fortuitous or induced event leading to the hereditary alteration of the genotype.

Nodule: Part of the root of a plant where nitrogen-fixing bacteria live in symbiosis with the plant in question.

Nucleic acids: Organic acids of the cell nucleus consisting of sequences of bases. There are two kinds depending upon the kind of sugar of which they are made: ribose (RNA) or deoxyribose (DNA).

Nucleotide: One of the four bases (A, C, G, and T) to which a sugar and a phosphate are attached.

Oligonucleotides: Short segments of DNA or RNA.

Oncogene: The gene of cancer.

Organic (compound): Molecules containing carbon.

Organite (or organelle): Differentiated cellular component exercising a specific function, such as the mitrochondria (where ATP, the "fuel" of living beings, is manufactured) or the chloroplasts (where photosynthesis takes place).

Peptide: Short chain formed of a succession of amino acids.

pH: Abbreviation for "potential hydrogen." It is a measure of acidity or alkalinity of a solution, with a scale ranging from 0 (acid) to 14 (alkaline), and which reflects the activity of the hydrogen ion in the solution. Below 7, pH is acid; above that, it is alkaline.

Phage: Abbreviation for "bacteriophage."

Phenotype: The observable properties of an organism resulting from the interaction between its genotype and the environment.

Photosynthesis: The reaction by which plants fix the carbon dioxide in the atmosphere in the presence of light and by which sunlight is converted into organic substances.

Plasmid: The circular chromosome of bacteria that reproduces itself independently of the principal chromosome. A choice vector in genetic engineering.

Polysaccharide: A long chain formed of sugar molecules, like starch or cellulose.

Probe: DNA or RNA sequence marked by radioactivity or fluorescence and allowing genetic screening.

Procaryote: Cell or organism without a nucleus separated by a membrane. The chromosomes are bathed directly in the cytoplasm. The best example are the bacteria.

Promoter: Sequence of DNA giving the starting signal for transcription.

Protein: Macromolecule consisting of amino acids connected by peptidic links.

The "building blocks" of living beings.

Protoplast: Cell deprived of its cell wall.

Protozoans: Eucaryote microorganisms, such as amoeba and infusoria.

Regeneration: Process of obtaining (in the laboratory) an entire plant from a single cell or a tissue.

Replication: The formation of two new strands of DNA from existing DNA permitting the reproduction of an identically new cell as the result of the division.

RNA (ribonucleic acid): Basic biochemical component of the chromosomes and support of hereditary containing ribose sugar.

Semisynthesis: Frequently used synonym for bioconversion: a chemical reaction in which biological processes play a part.

Substratum: Substance on which an enzyme, for instance, exercises its action.

Thymine (T): Base in the genetic code.

Transcription: The first stage in the expression of a gene by means of messenger RNA.

Translation: Second stage in the expression of a gene: a synthesis of the amino acids in a specific order, corresponding to the sequence of the nucleotides of a messenger RNA.

Uracil (U): Base in the genetic code.

Vector: A DNA molecule used to introduce foreign genetic information into a host cell.

Virus: An infectious agent comprising but a single type of nucleic acid, DNA or RNA enclosed in a "cloak" of proteins or protein lipids (capsids). The virus lives as a parasite on the living cells and will only multiply within them. Viruses are classified in several groups depending on their structure or properties: adenovirus (virus with DNA), retrovirus (with RNA), etc.

WORKS CITED

CHAPTER 1

René Dubos: *Pasteur and Modern Science* (Anchor Books, Doubleday, 1960, pp. 36, 38)

ARTICLE 1

Works

Antibiotics and Other Secondary Metabolites – Biosynthesis and Production (1978) edited by R. Hütter, T. Leisinger, J. Nüesch, and W. Wehrli, Academic Press, London.

Economic Microbiology – Vol. 3: *Secondary Products of Metabolism* (1979) edited by A.H. Rose, Academic Press, London.

The Future of Antibiotherapy and Antibiotic Research (1981) edited by L. Ninet, P.E. Bost, D.H. Bouanchaud, and J. Florent, Academic Press, London.

Biotechnology – Vol. 1: *Microbial Fundamentals* (1981) edited by H.J. Rehm and G. Reed, Verlag Chemie GmbH, Weinheim.

Bioactive Microbial Products: Search and Discovery (1982) edited by J.D. Bu'Lock, L.J. Nisbet, and D.J. Winstanley, Academic Press, London.

Overproduction of Microbial Products (1982) edited by V. Krumphanzl, B. Sikyta, and Z. Vanek, Academic Press, London.

The Chemistry and Biology of Antibiotics (1983) edited by V. Betina, Elsevier Scientific Publ., Amsterdam.

Biochemistry and Genetic Regulation of Commercially Important Antibiotics (1983) edited by L.C. Vining, Addison-Wesley Publ., London.

Biotechnology of Industrial Antibiotics (1984) edited by E.J. Vandamme, Marcel Dekker Inc., New York.

Articles

T. Aoyagi and H. Umezawa (1981): Production of pharmacologically active agents of microbial origin, in *Advances in Biotechnology* (M. Moo-Young, C.W. Robinson, and C. Vezina, eds.) Vol. 1, pp. 29, 33, Pergamon Press, Oxford.

J. Berdy (1974): Recent developments of antibiotic research and classification of antibiotics according to chemical structure, in *Advances in Applied Microbiology* (D. Perlman, ed.) Vol. 18, pp. 309-406, Academic Press, New York.

J. Berdy (1980): Recent advances in and prospects of antibiotic research, *Process Biochemistry, 15,* 28-35.

M.E. Bushell (1983): Microbiological aspects of the discovery of novel secondary metabolites, in *Topics in Enzyme and Fermentation Biotechnology* (A. Wiseman, ed.) Vol. 6, chap. 3, pp. 32-67, Ellis Horwood (a division of Wiley), Chichester, NY.

A.L. Demain (1981): Industrial microbiology, *Science, 214,* 987-995.

A.L. Demain (1982): New applications of microbial products, *Science, 219,* 709-714.

A.L. Demain (1984): Capabilities of microorganisms (and microbiologists), *Basic Life Sciences, 28,* 277-299.

R.L. Hamill (1977): General approaches to fermentation screening, *The Japanese Journal of Antibiotics, 30* – Supplement, S 164 – S 173.

L.J. Nisbet (1982): Current strategies in the search of bioactive microbial metabolites,

Journal of Chemical Technology and Biotechnology, 32, 251-270.

L. Pasteur and J.F. Joubert (1877) : Charbon et septicémie, *Comptes rendus des Séances de l'Académie des Sciences, 85* (3), 101-115.

P. Schindler and G. Huber (1984) : Microorganisms as a source of pharmacologically active agents, in *Developments in Industrial Microbiology,* Vol. 25, chap. 23, pp. 293-304, Society for Industrial Microbiology, Arlington, VI.

H. Umezawa (1982) : Low-molecular-weight enzyme inhibitors of microbial origin, in *Annual Review of Microbiology,* Vol. 36, pp. 75-99, Annual Reviews Inc., Palo Alto, CA.

H. Umezawa (1984) : Studies on low-molecular-weight immunomodifiers produced by microorganisms : results of ten years' effort, *Reviews of Infectious Diseases, 6* (3), 412-420.

CHAPTER 3

Marcel Florkin : *Theodor Schwann et les débuts de la médecine scientifique* (Palais de la Découverte, 4 February 1956, pp. 7, 8, 9, 17)

CHAPTER 4

René Dubos : *The Professor, the Institute and DNA* (The Rockefeller University Press, 1976, p 155)
Hommage à Jacques Monod, les origines de la biologie moléculaire (Etudes vivantes, 1980, pp 106, 127, 128)

CHAPTER 5

Les Prix Nobel 1978 (Almquist and Wiksell International, p. 180)

ARTICLE 3

(1) Burnett, F.M. (1959) *The Clonal Selection Theory of Aquired Immunity.* London, Cambridge university Press.
(2) Littlefield, J.W. (1964) *Science, 145,* 709.
(3) Köhler, G., and Milstein, C. (1975) *Nature,* 256, 495.
(4) Gerhard, W. (1977) *Topics in Infectious Diseases 3,* 15.
(5) Koprowski, H., and Wiktor, T. (1980) in *Monoclonal Antibodies,* Plenum Press, Eds. Kenneth, R.H., McKearn, T.J., and Bechtol, K.B.
(6) *Monoclonal Antibodies,* Plenum Press, Eds. Kenneth, R.H., and Bechtol, K.B. (1980).
(7) Moseley, C.M. (1984) *Biologist, 31,* 274.

ARTICLE 6

(1) Homandberg, G.A., J.A. Mattis, and M. Laskowski : Synthesis of peptide bonds by proteinases. Addition of organic cosolvents shifts peptide bond equilibria toward synthesis. *Biochemistry 17,* 5220-5227 (1978).
(2) Morihara, K.T. Oka, and H. Tsuzuki : Semisynthesis of human insulin by trypsin-catalysed replacement of Ala-B30 by Thr in porcine insulin. *Nature 280,* 412-413 (1979).
(3) Gattner, H.-G., W. Danho, and V.K. Naithani : Enzyme-catalysed semisynthesis with insulin derivatives. In Brandenburg, D., A., and Wollmer (eds.) : *Proceedings 2nd International Insulin Symposium,* Aachen. Walter de Gruyter, Berlin and New York, 117-123 (1980).
(4) Markussen, J. : Process for preparing insulin esters. United Kingdom Patent Application GB 2 069 502A (1980).
(5) Schmitt, E.W., and H.-G. Gattner : Verbesserte Darstellung von Desalanyl B30 – insulin. *Hoppe-Seyler's Zeitschrift Physiologische Chemie 359,* 799-802 (1978).
(6) Markussen, J., U. Damgaard, M. Pingel, L. Snel, A.R. Sørensen, and E. Sørensen : Human insulin (Novo) : chemistry and characteristics. *Diabetes Care,* Vol. 6, 4-8 (1983).
(7) Nicol, D.S.H.W., and L.F. Smith : Amino acid sequence of human insulin. *Nature 187,* 483-485 (1960).
(8) Markussen, J., U. Damgaard, K.H. Jørgensen, E. Sørensen, and L. Thim : Paper presented at the 16th Annual Meeting of the Scandinavian Society for the Study of Diabetes (1981).
(9) Schlichtkrull, J., J. Brange, Aa. Christiansen, O. Hallund, L.G. Heding, K.H. Jørgensen, S. Munkgaard Rasmussen. E. Sørensen, and Aa. Vølund : Monocomponent insulin and its clinical implications. *Horm. Metab. Res. (Suppl.) 5,* 134-143 (1974).
(10) Schernthaner, G., M. Borkenstein, M. Fink, W.R. Mayr, J. Menzel, and E. Schober : Immunogenicity of human insulin (Novo) or pork monocomponent insulin in HLA-DR-typed insulin-dependent diabetic individuals. *Diabetes Care,* Vol. 6, Suppl. 1, 43-48 (1983).

ARTICLE 7

(1) D.A. Jackson, R.H. Symons, and P. Berg, *Proc. Natl. Acad. Sci. U.S.A., 69,* 2904 (1972).
(2) S.N. Cohen, A. C. Y. Chang, H.W. Boyer, and R.B. Helling, *ibid, 70,* 3240 (1973).
(3) U.S. patent number 4,237,224.
(4) P. Berg, D. Baltimore, S. Brenner, R.O. Roblin III, and M.F. Singer, *Science, 188,* 991 (1975).
(5) R. Crea, A. Kraszewski, H. Tadaaki, and K. Itakura, *Proc. Nat. Acad. Sci. (Wash.), 75,* 5765 (1978).
(6) D.V. Goeddel, D. G. Kleid, F. Bolivar, H.L. Heyneker, D.G. Yansura, R. Crea, T. Hirose, A. Kraszewski, K. Itakura, and A.D. Riggs, *ibid, 76,* 106 (1979).
(7) D.C. Williams, R.M. Van Frank, W.L. Muth, and J.P. Burnett, *Science, 215,* 687 (1982).
(8) R.E. Chance, J.A. Hoffmann, E.P. Kroeff, M.G. Johnson, E.W. Schirmer, W. W. Bromer, J.M. Ross, and R. Wetzel. In : *Peptides, Synthesis, Structure and Function.* Proceedings of the Seventh American Peptide Symposium. Eds. D.H. Rich, and E. Gross, pp. 721-728. Pierce Chemical Company, Rockford, IL, 1981.
(9) B.H. Frank, J.M. Pettee, R.E. Zimmerman, and P.J. Burck, *ibid,* 729-738, 1981.
(10) J.-C. Du, R.-Q. Jiang, and C.-L. Tsou, *Sci. Sin., 14,* 229 (1965).
(11) P.G. Katsoyannis, *Science, 154,* 1509 (1966).
(12) P.G. Katsoyannis, and A. Tometsko, *Proc. Nat. Acad. Sci. (Wash.), 55,* 1554 (1966).
(13) J. Meienhofer, E. Schnabel, H. Bremer, O. Brinkhoff, R. Zabel, W. Sroka, H. Klostermeyer, D. Brandenburg, T. Okuda, and H. Zahn, *Z. Naturforsch., 18b,* 1120 (1963).
(14) W. Kemmler, J.D. Peterson, and D.F. Steiner, *J. Biol. Chem., 246,* 6786 (1971).
(15) S.A. Chawdhury, E.J. Dodson, G.G. Dodson, C.D. Reynolds, S. Tolley, and A. Cleasby. In *Hormone Drugs.* Proceedings of the FDA-USP Workshop on Drug and Reference Standards for Insulins, Somatropins, and Thyroidaxis Hormones, pp. 106-115. U.S. Pharmacopeial Convention, Bethesda, MD, 1982.
(16) S.A. Chawdhury, E.J. Dodson, G.G. Dodson, C.D. Reynolds, S.P. Tolley, T.L. Blundell, A. Cleasby, J. E. Pitts, I.J. Tickle, and S.P. Wood, *Diabetalogia, 25,* 460 (1983).
(17) R.E. Chance, E.P. Kroeff, J.A. Hoffmann, and B.H. Frank, *Diabetes Care, 4,* 147 (1981).
(18) R.S. Baker, J.M. Ross, J.R. Schmidtke, and W.C. Smith, *Lancet, II,* 1139 (1981).
(19) J.W. Ross, R.S. Baker, C.S. Hooker, I.S. Johnson, J.R. Schmidtke, and W.C. Smith. In *Hormone Drugs.* Proceedings of the FDA-USP Workshop on Drug and Reference Standards for Insulins, Somatropins, and Thyroidaxis Hormones, pp. 127-138. U.S. Pharmacopeial Convention, Bethesda, MD, 1982.
(20) I.S. Johnson, *Diabetes Care, 5,* (Suppl. 2), 4 (November-December, 1982).

ARTICLE 9

H.G.W. Leuenberger and K. Kieslich : Biotransformationen. In *Handbuch der Biotechnologie* (Eds: P. Prave, U. Faust, W. Sittig and D.A. Sukatsch), 2. Aufl., Oldenbourg Verlag, München, 1984, pp. 453-482.
H.G.W. Leuenberger : Methodology. In *Biotechnology* (Eds: H.J. Rehm and G. Reed). Vol. 6a : *Biotransformations,* Verlag Chemie, Weinheim, 1984, pp. 5-29.

CHAPTER 14

Biofutur July-August 1983, p. 27.

DATES

1630 : Proof of the mechanism of photosynthesis by Van Helmont.
1663 : First description of "cells" by Hooke.
1797 : Jenner inoculates a child with a viral vaccine to protect him from smallpox.
1830 (circa) : Discovery of proteins by Muller.
1833 : Isolation of the first enzyme by Payen and Persoz.
1839 : Schwann and Schleiden's cellular theory.
1855 : Discovery of the *Escherichia coli* bacterium by Escherich.
1858 : Virchow publishes his book on cellular pathology.
1860 : Johannsen invents the word "gene."
1866-1869 : Mendel's laws.
1869 : Discovery of "nucleine" by Miescher.
1876 : Publication of Pasteur's "Study on Beer."

1878 : • The term microbe invented by Sédillot.
• First centrifuge developed by Laval.
1883 : First vaccination against rabies.
1888 : "Chromosomes" so named by Waldeyer.
1897 : Bucher founds biochemistry.
1900 : Discovery by Landsteiner of blood groups.
1902 : Appearance of the word "immunology."
1903 : Sutton studies chromosomes.
1905 : Bateson names the science of heredity "genetics."
1911 : Rous discovers the first cancer-causing virus.
1915 : Discovery of phages.
1920 : • Morgan proposes the chromosome theory of heredity.
• Discovery of the growth hormone by Evans and Long.
1920-1940 : Development of ultracentrifugation by Svedberg.
1921 : Insulin isolated by Banting and Best.
1927 : Muller explores the phenomenon of genetic mutation by x rays.
1929 : Discovery of penicillin by Fleming.
1938 : • Cortisone isolated by Kendall and Reichstein.
• Culture of undifferentiated tissues by Gautheret.
1940 : Formation of the Phage Group.
1941 : • Beadle and Tatum establish the concept that will become: one gene, one enzyme.
• First use of penicillin.
1944 : • Identification of DNA as the vehicle of genetic heritage by Avery, MacLeod, and McCarty.
• Discovery of streptomycin by Waksman.
1949 : Pauling advances the idea that all hereditary illness could be molecular illness.
1950 : • Chargaff establishes the basic components of DNA.
• Pauling and Corey establish the three-dimensional structure of the alpha helix.
1950s : Discovery of jumping genes by McClintock.
1952 : The Hershey-Chase experiment.
1953 : • Discovery of the "double helix" by Watson and Crick.
• Creation at the University of Tokyo of the Institute of Applied Microbiology.
1954 : Technique for the cultivation of isolated cells developed by Muir, Hildebrandt, and Riker.
1956 : • Kornberg and Lehmann discover DNA polymerase.
• Hoagland and Zamenick discover transfer RNA.
• Kornberg and Ochoa discover the mechanisms of the biological synthesis of RNA and DNA.
• Fermentation process perfected in Japan by Kinoshita.
1957 : Discovery of interferon in Great Britain by Isaacs and Lindenmann.
1958 : Discovery of an "inhibition factor" (interferon) by Nagaro in Japan.
1960 : • Discovery of messenger RNA by Brenner and Jacob, Gros, and Gilbert.
• Technique of fusing protoplasts perfected by Cocking.
1960-1961 : • Marmur, Doty, and Schildkraut demonstrate that the two strands of DNA can be separated, then rearranged or hybridized.
• Monod and Jacob determine the mechanism for the genetic regulation of the synthesis of proteins.
• Nirenberg and Ochoa determine the chemical nature of the genetic code.
1966 : Discovery and immobilization of enzymes by Chibata at Tanabe.
1967 : The first automatic protein sequencer perfected by Edman.
1968 : • Discovery of brain hormones by Guillemin and Schally.
• Existence of the repressor determined by Gilbert.
1969 : First synthesis of an enzyme (ribonuclease of bovine pancreas) simultaneously at Rockefeller University and at Merck.
1970 : • Discovery of restriction enzymes by Arber, Smith, and Nathans.
• Temin, Mizutani, and Baltimore discover reserve transcriptase.
• For the first time, an entire gene is synthesized by the Khorona team at the University of Wisconsin.
• Cantell's first tests with interferon.
1970s : Electrophoresis on gel and HPLC.
1971 : • Foundation of the first NBF, Cetus.
• Inauguration in Japan of the Mitsubishi Kasei Institute of Life.
1972 : • Jackson, Seymours, and Berg determine DNA recombination techniques.
• Initial research on embryo transfers.
• First applications of immobilized enzymes.
• Kohn, Hoyer, and Chison, by hybridization, compare the DNA of humans with those of gorillas and chimpanzees; they discover that there is but a 1% difference between the bases of human DNA and those of the other primates and obtain the first estimate of the percentage of the sequence of DNA modified during the evolution of the primates.
1973 : First successful experiment in cloning genes by Cohen and Chang, Helling and Boyer.
1974 : Localization of the Ti plasmid by Schell and Van Montagu.
1975 : • Asimolar conference on the possible dangers of genetic engineering.
• Discovery of the hybridoma and monoclonal antibodies by Milstein and Kohler.
• Coulson and Sanger's method of sequencing genes.
• Discovery of the first "brain morphines," the enkephalins, by Hughes and Kosterlitz.
1976 : • Genentech founded by Swanson and Boyer.
• Discovery of endorphins by Guillemin.
• HFCS (High Fructose Corn Syrup) developed.
1977 : • Itakura, Riggs, and Boyer announce the expression of somatostatin by a bacterium as the result of genetic manipulation.
• Baxter isolates the gene coding for the human growth hormone.
• The Maxam-Gilbert method of sequencing genes.
• Discovery of mosaic genes.
1978 : • Gilbert and the Joselin Foundation for Diabetes announce the production of rat proinsulin by genetic engineering.
• Genentech and Eli Lilly announce the production of human insulin by genetic engineering.
1979 : Goodman and Baxter (University of California at San Francisco) and Goeddel and Seeburg (Genentech) announce the production of the human growth hormone by biosynthesis.
1980 : • Cloning and expression of alpha interferon by Weissmann at Biogen.
• Patent granted Stanford University on genetic recombination techniques.
• The Chakrabarty patent.
1981 : • Discovery of GRF by Guillemin.
• Hood and Itakura develop machine for synthesizing genes.
1982 : • Giant mice obtained by Brinster and Palmer as the result of genetic manipulation.
• Commercialization of the first vaccine to be obtained by genetic engineering, a veterinary vaccine.
• Development by Chedid and Sela of the first synthesized vaccine.
1983 : • Cloning and expression of Interleukin 2 by four research teams simultaneously.
• In July, Eli Lilly launches Humulin, the first product of genetic engineering to go on sale.
1984 : • Caron and Preer's research on paramecia invalidates the concept of the universality of the genetic code.
• Constitution of the first international data bank on hybridomas.
• Second Cohen-Boyer patent granted Stanford University, this time covering the "products" of genetic engineering.
1985 : Pääbo at the University of Uppsala clones the DNA of an Egyptian mummy.
1986 : A.J. Zaug and T. Cech (University of Colorado) discover that an RNA may produce an enzymic action, thus causing the very definition of enzymes to be re-examined.

BIBLIOGRAPHY

A

Books, pamphlets, reports:
Ackerknecht E.: *Rudolph Virchow, Doctor, Statesman, Anthoropologist* (Madison, 1953).
Arber W.: *Von Fluss der Gene* (Hans Erni-Stiftung, 1983).
Asimov. I.: *The Genetic Code* (Clarkson N. Potter, 1962).
Asimov I.: *The Wellsprings of Life* (Abelard-Schuman, 1960).
Atlan H.: *Entre le cristal et la fumée* (Seuil, 1979).

Articles:
Allain-Regnault M.: "Nobel: La vieille dame aux maïs" *(Science et Avenir, 1984).*
Alper J.: "Vaccine research gets new boost" *(High Technology, April 1983).*
Alper J.: "Bioengineers are off to the mines" *(High Technology, April 1984).*
Angier N.: "Helping children reach new heights" *(Discover, March 1982).*
Arber W.: "Promotion and limitation of genetic exchange" *(Science, 27 July 1979).*
Arnon R. and Sela M.: "Les antigènes et vaccins synthétiques" *(La Recherche, March 1983).*

B

Books, pamphlets, reports:
Barfoed H.C.: *Production of Enzyme by Fermentation* (John Wiley and Sons, 1981).
Baskin Y.: *The Gene Doctors* (William Morrow and Co., 1984).
Biomedical Technology Industrial Analysis (Kidder Peabody and Co, 2 April 1984).
Bodanis D.: *The Body Book* (Little Brown and Company, 1984).
Buican D.: *Sur-être ?* (Serge Fleury-L'Harmattan, 1983).

Articles:
Bach J.F.: "Les pionniers de l'immunogénétique" *(La Recherche, December 1980).*
Balkwill F.: "Interferons: from common cold to cancer" *(New Scientist, 14 March 1985).*
Barlet A.: "Le verrouillage des marchés internationaux" *(Sciences et Techniques, February 1984).*
Basta N.: "Biopolymers challenge petrochemicals" *(High Technology, February 1984).*
Bauman D. and Bruce Currie W.: "Partitioning of nutrients during pregnancy and lactation" *(J. Dairy Sci. 1980).*
Bayen M.: "Capital-risque: prudence aux Etats-Unis" *(Sciences et Techniques, May 1984).*
Begley S.: "Greening the gene" *(Newsweek, 12 November 1984).*
Bérézine I. et Kasanskaya N.: "La photographie enzymatique" *(La Recherche, January 1984).*
"Biotech comes of age" *(Business Week, 23 January 1983).*
Blanc M.: "Gregor Mendel: la légende du génie méconnu" *(La Recherche, January 1984).*
Blanc M.: "Clonage des mammifères" *(La Recherche, April 1981).*
Brierley C.: "Microbiological mining" *(Scientific American, August 1982).*
Brumdy P. and Hancock J.: "The galactopoetic role of growth hormone in dairy cattle" *(Journal of Science and Technology, 1955).*
Bylinsky G.: "Biotech breakthroughs in detecting disease" *(Fortune, 9 July 1984).*

C

Books, pamphlets, reports:
Cavalieri L.: *The Double-Edged Helix* (Columbia University Press, 1981).
Commercial Biotechnology (OTA, 1984).
"Conference on life sciences and mankind" (The Japan Foundation, 19-22 March 1984).
"Connaissance de la bière" (Société de nutrition et de diététique de langue française, 1975).
Crick F.: *Life Itself* (Simon and Schuster, 1981).

Articles:
"Cells bug the future of biotechnology" *(The Economist, 26 January 1985).*
Charrier A.: "L'amélioration génétique des cafés" *(La Recherche, September 1982).*
Chase M.: "After slow start, gene machines approach a period of fast growth and steady profits" *(The Wall Street Journal, 13 December 1983).*
Chase M.: "Altering heredity" *(The Wall Street Journal, 26 January 1984).*
Chilton M.D.: "L'introducton de gènes étrangers dans les plantes" *(Pour la Science, August 1983).*
Christen Y.: "ADN: La grande pagaille" *(Science Digest, February 1982).*
Collier J. and Kaplan D.: "Les immunotoxines" *(Pour la Science, September 1984).*
Curtin M.E.: "Microbial mining and metal recovery" *(Biotechnology, May 1983).*
Curtin M.E.: "Harvesting profitable products from plant tissue culture" *(Biotechnology, October 1983).*

D

Books, pamphlets, reports:
Davis J.: *Endorphins* (Doubleday and Company, 1984).
Debru C.: *L'esprit des protéines* (Hermann, 1983).
Dubos R.: *Les célébrations de la vie* (Stock, 1982).

Articles:
Dalgliesh C.: "La biochimie de la bière" *(La Recherche, April 1980).*
Darnell J.: "La maturation des ARN" *(Pour la Science, December 1983).*
Demain A.: "A new era of exploitation of microbial metabolites" *(Biochem. Soc. Symp., G.-B.).*
Dennett J.: "Microbe miners" *(AMM Magazine, 2 July 1984).*
Dommergues Y., Dreyfus B., Hoang Gia Diem, and Duroux E.: "Fixation de l'azote et agriculture tropicale" *(La Recherche, January 1985).*
Dottori F.: "Se rinasce la mummia" *(Panorama, 19 May 1985).*
Dulbecco R.: "La biologie du cancer" *(La Recherche, May 1979).*

E

Books, pamphlets, reports:
Edelhart M.: *Interferon, the New Hope for Cancer* (Ballantine Books, 1982).

Article:
Erni F.X.: "Knacknuss für Immunologen" *(Glückspost, 6 December 1984).*

F

Books, pamphlets, reports:
Faibis L.: *Biotechnologies et bio-industries en France* (Biofutur, département conseil, September 1984).
Florkin M.: *Theodor Schwann et les débuts de la médecine scientifique* (Les Conférences du Palais de la Découverte, 4 February 1956).
Fond O.: *La Biotechnologie au Japon* (Office Franco-japonais d'études économiques, October 1981).
Frédéricq L.: *Theodor Schwann, sa vie et ses travaux* (Université de Liège, 1884).

Articles:
Feder J. and Tolbert W.: "La culture en masse des cellules de mammifères" *(Pour la Science, March 1953).*
Fukui S. et Tanaka A.: "Enzymatic reactions in organic solvents" *(Endeavour, New Sciences, Vol. 9, 1, 1985, Pergamon Press).*
Fukui S. and Tanaka A.: "Application of biocalysts immobilized by prepolymer methods" (Springer 1984).

G

Books, pamphlets, reports:
Galdston I. (editor): *The Impact of the Antibiotics on Medicine and Society* (International University Press, 1958).
Grall J. and Lévy B.R.: *La guerre des semences* (Fayard, 1985).
Guyot L.: *La biologie végétale* (P.U.F., 1978).

Articles:
Garel P.: "La surprise des petits gènes" *(La Recherche, February 1981).*
Garrett De Young H.: "Mystery out of enzymes" *(High Technology, April 1984).*
Gisler R.: "Das Immunsystem" *(Der Informierte Arzt, 1983).*
Gordon E.: "Plastiques sans pétrole" *(Le Monde, 22-23 January 1984).*
Gordon E.: "Financiers pas aventuriers" and "Chez les alchimistes de Silicon Valley" *(Le Monde-Aujourd'hui, 10-11 June 1984).*
Gregory G.: "Big is beautiful in biotechnology" *(New Scientist, 6 December 1984).*

H

Books, pamphlets, reports:
Harsanyi Z. and Hutton R.: *Genetic Prophecy: Beyond the Double Helix* (Rawson, Wade Publishers, 1981).
Hérédité et manipulations génétiques (Bibliothèque Pour la Science, Diffusion Belin, 1984).
Hilts P.: *Scientific Temperaments* (Simon and Schuster, 1982).
Hoagland M.: *The Roots of Life* (Houghton Mifflin, 1977).
Hommage à Jacques Monod, les origines de la biologie moléculaire (Etudes Vivantes, 1980).

BIBLIOGRAPHY

Articles:
Hall P.: "The business of biotechnology" (*Financial World*, 21 March 1984).
Hall S.S.: "Biochips" (*United Airlines Magazine*, December 1983).
Hasegawa M.: "Industry gets off the ground toward full-blown commercial production" (*Industrial Review of Japan*).
Hirth L.: "Les viroïdes" (*La Recherche*, February 1981).
Hochhauser S.: "Bringing biotechnology to market" (*High Technology*, February 1983).
Hovarth S., Hunkapiller T., Tempst P., and Hood L. "A microchemical facility for the analysis and synthesis of genes and proteins" (*Nature*, 1984).
Hughey A.: "Burgeoning field" (*The Wall Street Journal*, 10 May 1983).
Hunter T.: "Les protéines des oncogènes" (*Pour la Science*, October 1984).
Hutton J.: "The effect of growth hormone on the yield and composition of cows' milk" (*J. Endocrin.* 1957).

I

Articles:
Illmensee K., Arber W., Peacock W.J., and Starlinger P.: "Genetic manipulation: impact on man and society" (ICSU Press, Cambridge, 1984).
"Interferon" (*Time*, 31 March 1981).

J

Books, pamphlets, reports:
Jaworski E. and Tiemeier D.: "Genetic engineering" (in *Encyclopedia of Chemical Technology*, John Wiley and Sons, 1984).

Articles:
Johnson I.: "Human insulin from recombinant DNA technology" (*Science*, 11 February 1983).
Jones J.S.: "Mummified human DNA cloned" (*Nature*, 18 April 1985).

K

Books, pamphlets, reports:
Komiya R. and Yamamoto Y.: *Japan: The Officer in Charge of Economic Affairs* (History of Political Economy, Duke University Press, 1981).
Kosuge T., Meredith C. et Hollaender A. (editors): *Genetic Engineering of Plants* (Plenum Press, 1983).

Articles:
Kellogg R.: "Les enzymes artificielles" (*La Recherche*, June 1984).
Kelly F.: "La longue marche des gènes sauteurs" (La Recherche, December 1983).
Kelly F.: "Les souris géantes ont-elles un avenir?" (*La Recherche*, April 1983).
Klibanov A.: "Enzymes: nature's chemical machines" (*Technology Review*, November-December 1983).
Koenig R.: "Scientists ability to design new proteins" (*The Wall Street Journal*, 10 October 1984).
Kourilsky P.: "Le génie génétique" (*La Recherche*, April 1980).

L

Books, pamphlets, reports:
La mélasse de betteraves (special number of *La Sucrerie Française*, June 1984).
La Recherche en biologie moléculaire (Le Seuil, 1975).
La Révolution biologique (special number of *Pour la Science*, November 1981).
La révolution biologique (special number of *Science et Vie*).
Leach G.: *Les biocrates* (Le Seuil, 1970).
Le boom des biotechnologies (special number of *Science et Avenir* 50).
Les enjeux de la biologie (*Magazine Littéraire*, April 1985).
L'évolution (Bibliothèque Pour la Science, Diffusion Belin, 1980).
Lhéritier P.: *La grande aventure de la génétique* (Flammarion, 1984).
L'industrie pharmaceutique européenne et le développement de la biotechnologie (Rapport de la Fédération Européenne des Associations de l'Industrie Pharmaceutique, February 1984).
Les nouveaux moyens de la médecine (Bibliothèque Pour la Science, Diffusion Belin, 1983).
Luria S., Gould S., and Singer S.: *A View of Life* (Benjamin/Cummings Publishing,, 1981).

Articles:
Labich K.: "Monsanto's brave new world" (*Fortune*, 30 April 1984).
"La génétique et l'hérédité" (special number of *La Recherche*, May 1984).
Land H., Parada L.F., and Weinberg R.: "Cellular oncogenes and multistep carcinogenesis" (*Science*, 18 November 1983).
Lavigne M.: "Recombinant DNA" (*Columbia Magazine*, 1983).
Leder P.: "L'origine génétique de la diversité des anticorps" (*Pour la Science*, May 1983).
Lenoir J., Lamberet G., and Schmidt J.L.: "L'élaboration d'un fromage: l'exemple du camembert" (*Pour la Science*, July 1983).
Lerner R.: "Les vaccins de synthèse" (*Pour la Science*, April 1983).
Leventer M.: "Plantes: les jardiniers en blouse blanche" (*Le Point*, 25 February 1985).
Lipinski M. and Hertzenberg L.: "Les hybridomes et leurs applications" (*La Recherche*, September 1981).

M

Books, pamphlets, reports:
Mendel A.: *Les manipulations génétiques* (Le Seuil, 1980).

Articles:
Machlin L.: "Effect of growth hormone on milk production and feed utilization in dairy cows" (*Journal of Dairy Science*, 1973).
Marbach W.: "The bust in biotechnology" (*Newsweek*, 26 July 1984).
Marcus S.: "Bugs that make chemicals" (*The New York Times*, 2 March 1984).
Martin C.: "High-tech winemaking" (*Discover*, March 1983).
Martin C.: "La culture des plantes en éprouvette" (*La Recherche*, November 1984).
McKie R.: "Trying to clone pete marsh" (*The Observer*, 21 April 1985).

Melchers G., Sacristan M., and Holder A.: "Somatic hybrid plants of potato and tomato regenerated from fused protoplasts" (*Carlsberg Res. Commun.*, Vol. 43, pp. 203, 218, 1978).
Merceau-Puijalon O.: "Des bactéries productrices de protéines animales" (*La Recherche*, November 1978).
Montagnier L., Brunet J.B., and Klatzmann D.: "Le Sida et son virus" (*La Recherche*, June 1985).

N

Articles:
Nagano Y.: "Studies on virus-inhibiting factor or interferon" (*The Kitasato Archives of Experimental Medicine*, June 1981).
Nicolau C. and Parat A.: "Les liposomes, agents thérapeutiques de demain" (*La Recherche*, June 1981).
Nitsch J.P.: "Les plantes sans mères" (*La Recherche*, June 1971).

O

Books, pamphlets, reports:
Ochoa S.: "The pursuit of a hobby" (*Ann. Rev. Biochem*, 1980).
Omenn G.S. and Hollaender A. (editors): *Genetic Control of Environmental Pollutants* (Plenum Press, 1984).

P

Books, pamphlets, reports:
Pfeiffer J. *La cellule* (Time-Life, 1972).
Poilâne L.: *Guide de l'amateur de pain* (Laffont, 1981).
Portugal F. and Cohen J.: *A Century of DNA* (The MIT Press, 1979).
Prentis G.: *Biotechnology* (George Braziller, 1984).

Articles:
Papaioannou V.E.: "Le destin des cellules chez l'embryon" (*La Recherche*, November 1978).
Parks D.R. and Hertzenberger L.A.: "Fluorescence activated cell sorting" (Academic Press, 1984).
Parks D.R., Hardy R.R., and Hertzenberger L.A: "Three color immunofluorescence analysis of mouse B-lymphocytes subpopulations" (*Cyclometry*, Alan R. Liss, 1984).
Parks D.R., Hardy R.R., and Hertzenberger L.A.: "Dual immunofluorescence – new frontiers in cell analysis and sorting" (*Immunology Today*, May 1983).
Peel C., Fronk T., Bauman D., and Gorewit R.: "Lactational response to exogenous growth-hormone" (*American Inst. of Nutrition*, 1982).
Peel C., Bauman D., Gorewit R. et Sniffen C.: "Effect of exogenous growth-hormone on lactational performance in high yielding dairy cows" (*The Journal of Nutrition*, 1981).
Perpich J.: "Genetic engineering and related biotechnologies" (*Technology in Society*, Pergamon Press, 1983).
Pestka S.: "Interferon für klinische Tests" (*Spektrum der Wissenschaft*, October 1983).
Plagnol H.: " Bio: les cinq plus beaux marchés" (*L'Usine Nouvelle*, 28 March 1985).
Porter S.: "Structures et origines des anticorps" (*La Recherche*, May 1971).

Poulsen C., Porath D., Sacristan M.D., and Melchers G.: "Peptide mapping of the ribulose biphosphate carboxylase small subunit from the somatic hybrid of tomato and potato" *(Carlsberg Res. Commun.* Vol. 45, pp. 249-267, 1980).
Ptashne M., Johnson A., and Pabo C.: "Un commutateur génétique d'un virus bactérien" *(Pour la Science,* January 1983).

R

Books, pamphlets, reports:
Rehm H.J. and Reed G. (editors): *Biotechnology* (Verlag Chemie, 1983. 8 volumes).
Rosnay J. de: *Biotechnologie et bio-industrie* (Le Seuil/Documentation Française, 1979).
Ruffié J.: *Traité du vivant* (Fayard, 1982).

Articles:
Renard J.P. and Heyman Y.: "Les banques d'embryons" *(La Recherche,* February 1982).
Robertson M.: "Split genes: the meaning of nonsense DNA" *(New Scientist,* 10 July 1983).
Roissart H.B. de: "Que sont les bactéries lactiques?" *(La Technique laitière,* September 1983).
Rosenberg E., Gottlieb A., and Rosenberg M.: "Inhibition of bacterial adherence to hydrocarbon and epithelial cells by Emulsan" *(Infection and Immunity,* March 1983).
Rosnay J. de: "Les biotransistors: la micro-électronique du XXIe siècle" *(La Recherche,* July-August 1981).

S

Books, pamphlets, reports:
Salomon M.: *L'avenir de la vie* (Seghers, 1981).
Sasson A.: *Les biotechnologies, défis et promesses* (Unesco, 1983).
Sayre A.: *Rosalind Franklin and DNA* (W.W. Norton, 1975).
Schrödinger E.: *Ma conception du monde* (Mercure de France, 1982).
Shigeru N., Swain D., and Eri Y.: *Science and Society in Modern Japan* (The MIT Press, 1974).
Surbled G.: *Biotechnologies: le défi industriel* (Crédit Commercial de France, June 1983).
Sylvester E. and Klotz L.: *The Gene Age* (Charles Scribner's, 1983).

Articles:
Sauclières G.: "Traitement des maladies génétiques" *(La Recherche,* January 1985).
Sauclières G.: "Le génie génétique au secours des hémophiles" *(La Recherche,* February 1985).
Serre J.L.: "La genèse de l'œuvre de Mendel" *(La Recherche,* September 1984).
Smillie R.M., Melchers G., and von Wettstein D.: "Chilling resistance of somatic hybrids of tomato and potato" *(Carlsberg Res. Commun.* Vol. 44, pp. 127-132, 1979).
Somerville C.R. and Somerville S.C.: "Les photosynthèses des plantes" *(La Recherche,* April 1984).
Stehelin D.: "Les oncogènes cellulaires" *(Théorie et pratique thérapeutiques,* May-June 1984).

Stehelin D.: Interview *(Recherche et Santé,* December 1984).

T

Books, pamphlets, reports:
Temin H.: "L'origine des rétrovirus" *(La Recherche,* February 1984).
The Carlsberg Laboratory, 1876/1976, editors: H. Holter and K. Max Möller (The Carlsberg Foundation, 1976).
The Growth of a Global Enterprise (Alfa-Laval, 1983).
Théodoridès G.: *Histoire de la biologie* (PUF, 1984).
The Omni Interviews, edited by P. Weintraub (Ticknor and Fields, 1984).
Thomas L.: *The Lives of a Cell* (Penguin Books, 1982).
Trommsdorff E.: *Dr Otto Röhm, Chemiker und Unternehmer* (Econ Verlag, 1976).

Articles:
"The next revolution in medicine is almost here" *(Business Week,* 8 October 1984).
"Thirty years of DNA" *(Nature,* 21 April 1983).
Téoule R.: "Les gènes artificiels" *(La Recherche,* March 1982).
Tucker J.B.: "Biochips: can molecules compute?" *(High Technology,* February 1984).
Tucker J.B.: "Military rDNA research stirs debate" *(High Technology,* July 1984).

U

Books, pamphlets, reports:
"Über die Verwendung von Enzymen in Waschmitteln" (Burnus Geselschaft).

Article:
Ulmer L.: "Protein engineering" *(Science,* 11 February 1983).

V

Articles:
Vidal G.: "Les premières cellules eucaryotes" *(Pour la Science,* April 1984).

W

Books, pamphlets, reports:
Wade N.: *The Ultimate Experiment* (Walker, 1979).
Wahl J.: *Le pétrole vert français* (Flammarion, 1983).
Waksman S.: *Ma vie avec les microbes* (Albin-Michel, 1964).
Watson J.: *La double hélice* (Laffont, 1968).
Watson J.: *Biologie moléculaire du gène* (Interéditions, 1978).
Watson J. and Tooze J.: *The DNA Story* (W.H. Freeman, 1981).

Articles:
Waldholz M.: "New drug could reshape medicine" *(The Wall Street Journal,* 16 December 1983).
Watson J.: Interview *(Omni,* May 1984).
Weaver R.F.: "Beyond supermouse" *(National Geographic,* December 1984).
Weinberg R.: "Une base moléculaire du cancer" *(Pour la Science,* January 1984).

Y

Books, pamphlets, reports:
Yoxem E.: *The Gene Business* (Harper and Row, 1984).
Yanchinski S.: *Biotechnology* (Penguin Books, 1985).

Articles:
Yaniv M.: "Chimie et manipulation du gène" *(La Recherche,* December 1980).
Yanchinski S.: "Protein engineering makes its mark on enzymes" *(New Scientist,* 6 December 1984).

Z

Books, pamphlets, reports:
Zimmerman B.: *Biofuture* (Plenum Press, 1984).

Articles:
Zosim Z., Gutnik D., and Rosenberg E.: "Uranium binding by Emulsan and Emulsanosols *(Biotechnology and Bio-engineering,* Vol. XXV, John Wiley and Sons, 1983).
Zotov A.: "Les Hépatites" *(La Recherche,* June 1983).

We also call the reader's attention to all issues of the magazines *Biofutur* and *Biotechnology.*

NAME INDEX

Words in italics correspond to companies or organizations, numbers in italics to picture captions

NAME INDEX

ACKNOWLEDGMENTS

Scientific articles:

Richard Axel (Director, United States Cancer Research Institute, and Professor of Biochemistry and Biopathology, Columbia Univesity).

Hans C. Barfoed (Director of Scientific and Technical Information, Novo).

Jean Bébin (Assistant Director, Lyonnaise des Eaux-Degrémont Research Center).

Teruhiko Beppu (Director of Research, Fermentation and Microbiology Department, Faculty of Agriculture, University of Tokyo).

Pierre-Étienne Bost (Director of Research Programs and Projects, Rhône-Poulenc Santé) and **Jean Florent** (Assistant Head of Biochemical Research, Rhône-Poulenc Santé).

Jacques Chardon (in charge of developing industrial fermentation, Roussel-Uclaf) and **Dominique Mison** (Doctor-Engineer, INSA, Production Division Roussel-Uclaf).

Jean Dausset (Professor of Experimental Medicine, Collège de France, member of the Académie des Sciences, Nobel Prize in Medicine 1980).

Olivier Fond (Doctor of Agronony).

Ernest G. Jaworski (Director of Biological Sciences, Research and Development Department, Monsanto).

Irving S. Johnson (Vice President, Eli Lilly Research Laboratories, member of the American Association for the Advancement of Science, the Academy of Sciences of New York).

Bernard Langley (responsible for University-Industry Relations, ICI).

H.G. Leuenberger (Assistant Director of Research, Hoffmann-La Roche).

Kenneth Mitchell (Director of Research, Du Pont de Nemours).

Roméo Roncucci (Director of Research and Development, Sanofi), **P. Gros** (Director of Immunology/Oncology Research, Sanofi), and **F.K. Jansen** (Chief, Immunotoxin Project, Sanofi).

Toshinao Tsunoda (Vice President and General Manager, Ajinomoto).

Shigetoshi Wakaki (Vice President and General Manager, Kyowa Hakko).

The autors would like to extend their special thanks to **France Normand,** Mission Chief, CNRS, for her help throughout this book, as well as to the following companies:

Abbott Laboratories: C.P. Weber
Air Liquide (Eurozyme): D. Gambet
Ajinomoto: A. Nakanishi, M. Sugimori, Y. Tamura
Alfa-Laval: A. Armand, D. Letourneau
Allied Breweries: R. Fricker
AMT (Advanced Mineral Technologies): C.L. Brierley
Applied Biosystems: S. Eletr
Arco Plant Cell Research Institute: J.E. Fox
Asahi Chemical: Y. Ichikawa, M. Shibukawa, Y. Yano, A. Ichiyama
Barberet-Ducloux: M. Rudelle
BASF: V. Floret
Baxter Travenol: E. Shiman
Bayer: M. d'Huart, I. Casse
Biogen: J.E. Davies, T. Dash

Biogenex: K. Kalra
Biosystems Associates: E. Shnéour
BioTechnica International: D. Glass, R. Hardy
BRGM: M. Bouchi-Lamontagne
Calgène: D. Facciotti, R. Heer
Carlsberg Laboratory: D. von Wettstein
CEA: J. Normand, Roche
Celltech: S. Walder
Centre Culturel Suédois: G. Sjöstedt
Cetus: K.A. Russel
Ciba-Geigy: R. Porchet
Comité Interprofessionnel des Vins de Champagne: M. Valade
Compagnie Générale des Eaux: M. Jouany
Daicel Chemical Industries: A. Kamibayashi, H. Ohnishi
Du Pont de Nemours: R. Hamilton, D. Andriadis, C. Gabail
Ecogen: J.E. Davies
Ecole Nationale d'Industrie Laitière: J. Lablée
Elf-Aquitaine: J.P. Brochier, J.P. Turbil, F.P. Bernard
Eli Lilly: M.P. Grein
Eurozyme: D. Gambet
Genencor: O.G. Midler
Genetic Systems: K. Carr
Genex: K. Ulmer
Gentronix: J. McAlear
Gen-Probe: G. Koestler, H. Birndorf
Getty: W.R. Taylor
Gist-Brocades: H.J. Kooreman
Hayashibara: S. Kamiya, M. Raees
Hoechst: H.H. Schöne, F. Feick, J. Fricke
Hoffmann-La Roche (et Roche Institute): E. Gwinner, J. Doorley, F. Gissinger
Hybritech: J. Martinis
IBA (Industrial Biotechnology Association): H. Price
ICI: C. Reece, P.S. Rodgers, M.H. Demay
Idianova: G. Nebot
IMREG: D. Whitney
INRA: M. Roumengou, J. Nioré
INSERM: M. Depardieu
Institut Curie: L. Bonnot
Integrated Genetics: F.D. Hudson
Intelligenetics: P. W. Armstrong
IPRI: L. Huntley
JITA (Japan Industrial Technology Association): M. Suzuki
John Brown: C.S. Mullins
Kyodo News Service: K. Tamura
Kyowa Hakko Kogyo: H. Samejima, Y. Mori
Lafarge Coppée: P. Juston
Meloy Laboratories: W. Terry, J.G. Perpich, W. Drohan
Merck: W.H. Helfand, C. High
Mitsubishi Chemical Industries: W. Yamaya, A. Tanaka, T. Takagi, T. Kunori
Mitsubishi Kasei Institute: G. Nomoto, K. Nakamura
Mitsui and Co: Y. Takeda, K. Ishihara
Mitsui Petrochemical Industries: H. Fujii, J. Hiratsuka, Suzuki
Mitsui Toatsu Chemicals: H. Sugama, Shoji
Molecular Genetics: J.L. Holland
Monsanto: M.G. Carnes, F.W. Morgan, K.K. Rogers, G.A. Young, F. Blanty
Mycogen Corp.: G.G. Soares
Nippon Soda: M. Hiraoka
Nippon Zeon: M. Saburai, T. Kawai
Novo: A. Cognard, B. Bena, P. Kennedy, M. Ringsted

Orsan: R. Magnan, P. Malléjac, C. Renaud
Pernod-Ricard: J. Bricout, D. Berthu, Y. Ménoret
Pharmacia: C. Calvot
Pierre Fabre: Couzinier
Pointet-Girard: E. Boschetti
Polybac: J.N. Zikopoulos
PPG Industries: L.P. Galanter
Rhône-Poulenc: G. Strain, F. Guynot, B. Gagnard-Masserand
RIBI Immunochem Research: K. Dyszynski
Rockefeller University: F. Bardossi
Röhm: H. Uhlig
Roquefort (Laboratoires de la Société des Caves et des Ets Louis Rigal): L. Assénat
Roussel-Uclaf: C. Euvrard, J.P. Raynault
Sandoz: H. Winkler, S. Guttmann, A. Einsele
Sanofi: R. Sautier
Schering Plough: H. Ostrowski
Scripps Clinic: A. Olson
SGN: C. Camilleri, M. Henry, E. Haye
Shell: J.C. Clément, L. Hoffacker
Signal UOP Goup: D. Byrne
SmithKline Beckman: T. Kulak, B.P. Rorres
Speichim: M. Revuz
Stanford News and Publication: B. Beyers
Sumitomo Chemical: R. Ishiwata, M. Tanabe
Sungene: D. Mc Connell
Suntory: T. Noguchi, M. Morita, S. Nagata, K. Naiki
Takeda Chemical: E. Ohmura, Y. Sugino
Toyo Jozo: J. Abe, T. Watanabe
Transgene: E. Eisenmann, J.F. Lecocq
UCB: A. Loffet
UNCEIA (Union Nationale des Coopératives d'Elevage et d'Insémination Artificielle): M. Thibier
Unilever Research: D.J. Frost
Universal Food Corporation: A. Mundstock
University of California, Berkeley: W. Ravven
University of California, Davis: W. Rains, C. Meredith
Whitehead Institute: J. Pratt
Yamasa Shoyu: A. Kuninaka

And personally to:

Kiyoshi Aoki (Sophia University)
Ananda Chakrabarty (University of Illinois, College of Medicine)
Dr. Ron Coleman (Government Chemist, British Department of Trade and Industry)
Michel Dagonneau (French Embassy to Japan)
Jacqueline de Durand-Forest (Chargée de Recherches, CNRS)
Dr. Jack Edelman (Technical Director, RHM)
Edgar Engleman (Stanford University Medical Center)
Mr. Gérard Fairtlough (Chief Executive, Celltech)
Gérard Gelff (University of Compiègne)
Roger Gilmour (Chief Executive, Agricultural Genetics Company)
Gene Gregory (Sophia University)
Hélène Haon (SGN)
Dr. Gilmour Harris (The Mathilda and Terence Kennedy Institute of Rheumatology)
Mr. Wensley Haydon-Baillie (Chairman, Porton International)
Sir William Henderson (Secretary, London Zoo)
Masamishi Kohiyama (University of Paris VII)
Danielle Longin (Chamber of Commerce)
John Maddox (*Nature*)

Dr. Robert Margetts (Research Director, ICI Agricultural Division)
Alain Molinier
Yasuichi Nagano
Dr. Charles Reece (Director of Research, ICI Plc)
Hiuga Saito (University of Tokyo)
Geoffrey Surbled (Crédit Commercial de France)
Iwao Tabushi (Kyoto University)
Dr. Alasdair Thomson (Imperial Cancer Research Fund Laboratories)
Itaru Watanabe (Kitasato University)
Robert Weinberg (MIT)

We thank the following people for reading all or part of this book:

W. Arber (University of Basel, Nobel Prize in Medicine, 1978)
A. Armand (Alfa-Laval)
L. Assénat (Laboratoires de la Société des Caves et des Établissements Louis Rigal, Roquefort)
J. Bébin (Lyonnaise des Eaux-Degrémont)
D. Bellet (Institut Gustave Roussy, Villejuif)
J. Berger (Institut Mérieux)
M. Bouchi-Lamontagne (BRGM)
J.R. Brochier (Elf-Aquitaine)
S. Brun (Faculty of Pharmacy, University of Montpellier)
C. Calvot (Pharmacia)
F. Caron (CNRS)
L. Chédid (Institut Pasteur)
J.C. Clément (Shell)
J. Dausset (Collège de France, Nobel Prize in Medicine, 1980)
E. Eisenmann (Transgène)
S. Eletr (Applied Biosystems)
M. Fenech-Hamelin (Institut Mérieux)
J. Florent (Rhône-Poulenc Santé)
R. Gicquiaud (Novo-Enzymes)
P. Juston (Lafarge-Coppée)
J. Keilling (member of the Académie d'Agriculture)
J.M. Lehn (University of Strasbourg)
Y. Mazières (Delbard)
Y. Ménoret (Pernod-Ricard)
O. Midler (Genencor)
P. Monsan (University of Toulouse, BioEurope)
L. Poilâne
A. Sedent (Rhône-Poulenc)
D. Stéhélin (Institut Pasteur, Lille)
J. Tektoff (Institut Mérieux)
M. Verry (Sanofi)

Photo credits:
Ajinomoto: 120-121 no. 3, 4, 5, 136 no. 2, 141 no. 2, 146, 202-203 no. 5, 210 no. 1, 2
Alfa-Laval: 173 no. 1, 175 no. 1, 2, 4
Applied Biosystems: 185 no. 5, 6
Barberet et Blanc: 126 no. 1
Bayer: 33 no. 1, 132 no. 3, 6 (Max Planck Institut): 155 no. 2
Biogen: 83 no. 5, 193 no. 1
Biogenex: 85 no. 1, 113 no. 2
Calgene: 132 no. 4 (Laboratory of Dr. Ann Crossway), 132 no. 5 (Bio-Tec Images)
Caltech: 184
CEA: 173 no. 3
Celltech (Infopress): 101 no. 4, 182, 195 no. 2
Centre de génétique moléculaire de Gif/Yvette: 49 no. 2

Cetus: 73 no. 3, 79, 172 no. 2, 185 no. 1, 2, 190, 193 no. 4
Ciba Geigy: 126 no. 4, 211 no. 1
CIDIP (Centre d'Information et de Documentation Israël-Proche-Orient): 18 no. 1
CNRI (Centre National de Recherches Iconographiques): 34, 41 no. 1 (Beulacque)
CNRS: 127 no. 2
Cold Spring Harbor Laboratory Library Archives: 45 no. 2, 3, 47 no. 1 (A.C. Barrington Brown), 3, 5
Cornell University (Department of Microbiology, B. Seaman Eaglesham and W.C. Ghiorse): 155 no. 1
Cosmos: 39 no. 1, 2 (E. Gravé/Phototake), 41 no. 2 (SPL Dr Brian Eyden), 44 (J. Hinsch/SPL), 45 no. 4 (SPL, Anderson/Simon), 52 (T.R. Broken/Phototake), 59 no. 3 (H. Sochurek/Wookin Camp), 95 no. 5 (SPL), 6 (Impact Photos, D. Reed), 114 no. 1 (A. Meininger/Contact), 3, 4 (Dimitrius/Contact), 169 no. 6 (A. Tsiaras), 193 no. 3 (D. Kirkland/Contact)
D. Bellet (Institut Gustave-Roussy): 70 no. 1, 2
Delbard: 127 no. 1
Du Pont de Nemours: 73 no. 4, 164, 208
Elf Aquitaine: 169 no. 2, 3
Eli Lilly: 101 no. 2, 3, 105
General Electric: 166, 169 no. 1
Genetic Systems: 71 no. 2, 5, 113 no. 4, 172 no. 1, 173 no. 5
Gentronix: 168
Gist-Brocades: 30 no. 2, 3
Hakushika/Ajinomoto: 13 no. 4
H. Harada: 39 no. 3, 4, 126 no. 3, 127 no. 4, 5, 6, 128, no. 1, 2, 3, 132 no. 1
Hayashibara: 199
Hoechst: 94 no. 3, 148, 160 no. 1, 161 no. 2a, 2b, 2c, 2d
Hoffmann-La Roche (and Roche Institute): 41 no. 3, 47 no. 6, 57 no. 2, 71 no. 1, 83 no. 1, 2, 4, 139 no. 4, 5, 6, 142-143 no. 1, 2, 3, 5, 172 no. 5, 196, 198 no. 1
Hologramme (P-Y Dhinaut): 16
Hondashi Ajiwai Dokuhon/Ajinomoto: 211 no. 3
Hybritech: 71 no. 3, 4, 73 no. 2, 5, 113 no. 3, 193 no. 2
ICI: 124 (Joannis Schneider Conseil), 150, 169 no. 4, 5
IFP: 160 no. 3
INRA: 151 no..2, 4, 158-159 no. 1 (O. Sébart), 160 no. 2 (O. Sébart)
INSERM: 28 (J-P Mornon), 39 no. 5 (C. Carré), 73 no. 1 (N. Blanc)
Institut Mérieux: 95 no. 2, 4, 7, 97 no. 2, 115 no. 2, 3, 120 no. 7, 198 no. 3
Institut Pasteur: 47 no. 4, 131 no. 1
Integrated Genetics: 59 no. 1, 2
J-L Charmet: 13 no. 5, 14 (collection J. de Durand-Forest), 17 no. 1, 36, 94 no. 1, 2, 112, 115 no. 1, 161 no. 3, 189 (collection J. de Durand-Forest)
J-P Mornon, I. Morize, E. Surcouf, M.C. Vaney: 97 no. 1, 102 no. 2, 167 no. 1, 2, 3, 4
Kyowa Hakko: 13 no. 1, 3, 18 no. 2, 21, 23, 116, 120-121 no. 1, 6, 138 no. 1, 161 no. 1
Laboratoires de la société des caves et des établissements Louis Rigal: 17 no. 2
Laboratoires Carlsberg: 128 no. 4, 5
Loup: 121, 211 no. 4
Magnum: 10 (R. Kalvar), 32 (E. Hartmann), 49 no. 3 (E. Hartmann), 92 (Riboud), 118 (Griffiths), 152 (Abbas), 156 (S. Salgado),

185 no. 3 (E. Hartmann), 186 (Moratia), 193 no. 5 (E. Hartmann)
Medical Research Council (MRC): 188 no. 3
Meloy Laboratories: 55 no. 1, 2, 3, 58 no. 1, 2, 3, 4
Merck: 57 no. 3, 120 no. 2, 173 no. 4
MIT: 57 no. 1, 113 no. 1, 188 no. 1, 2, 218 no. 2
Mitsui Petrochemical: 129 no. 1, 2, 3, 4, 5, 6
Mitsui Toatsu: 138 no. 3
Moët et Chandon: 13 no. 2, 15 no. 3
Monsanto: 98
Mycogen (Dr. G.G. Soarès): 131 no. 2, 3
Nippon Zeon: 185 no. 4
Novo Industri A/S: 30 no. 1, 33 no. 3, 105, 198 no. 2
Orsan: 202-203 no. 1, 2, 3, 4
Pernod-Ricard: 139 no. 1, 2, 3
Pharmacia: 170, 216 (Kjell Svensson), 219 no. 1
RHM (Headline Creators): 136
Rhône-Mérieux: 95 no. 1, 3, 200 no. 2
Rhône-Poulenc: 19 no. 1, 2, 141 no. 3, 4, 5, 142-143 no. 4, 200 no. 1
Ribi ImmunoChem: 97 no. 3, 4
Roussel-Uclaf: 102 no. 3
Salk Institute: 101 no. 1
Sandoz: 25, 86-87 no. 1, 2, 3, 4, 5, 6
Sanofi: 76, 101 no. 5
Schering Plough Corporation: 42 (P. Harrington), 56 (P. Harrington), 68 (P. Harrington), 80, 83 no. 3 (P. Harrington), 172 no. 3, 219 no. 2
Science Museum: 47 no. 2
Scientific American and Pour la Science: 155 no. 3
Searle: 49 no. 1, 172 no. 4
SGN: 159 no. 2, 3
Signal UOP Group (Ken Whitmore): 19 no. 4, 141 no. 1, 172 no. 6, 173 no. 2
SmithKline Beckman: 85 no. 2, 110, 198 no. 4
Sumitomo Chemical: 211 no. 2
Sungene: 132 no. 2, 195 no. 3
Suntory: 19 no. 3, 102 no. 1, 210 no. 3
The Rockefeller University Archives: 45 no. 1
Toyo Jozo: 33 no. 2
Transgène: 195 no. 1
UNCEIA: 151 no. 1, 3
Unilever: 126 no. 2, 127 no. 3
United Breweries Ltd: 15 no. 1, 2

Translations:
Melvin Wallace (Text of Elizabeth Anbébi).
Philippa Crutchley-Wallis (scientific articles)

Art direction:
Tilman Eichhorn

Computer designs:
Visuel Images

Composition:
SCG Paris: Jacques Audebert, Michel Grignon